FLORA OF UPPER LIDDAR VALLEYS OF KASHMIR HIMALAYA

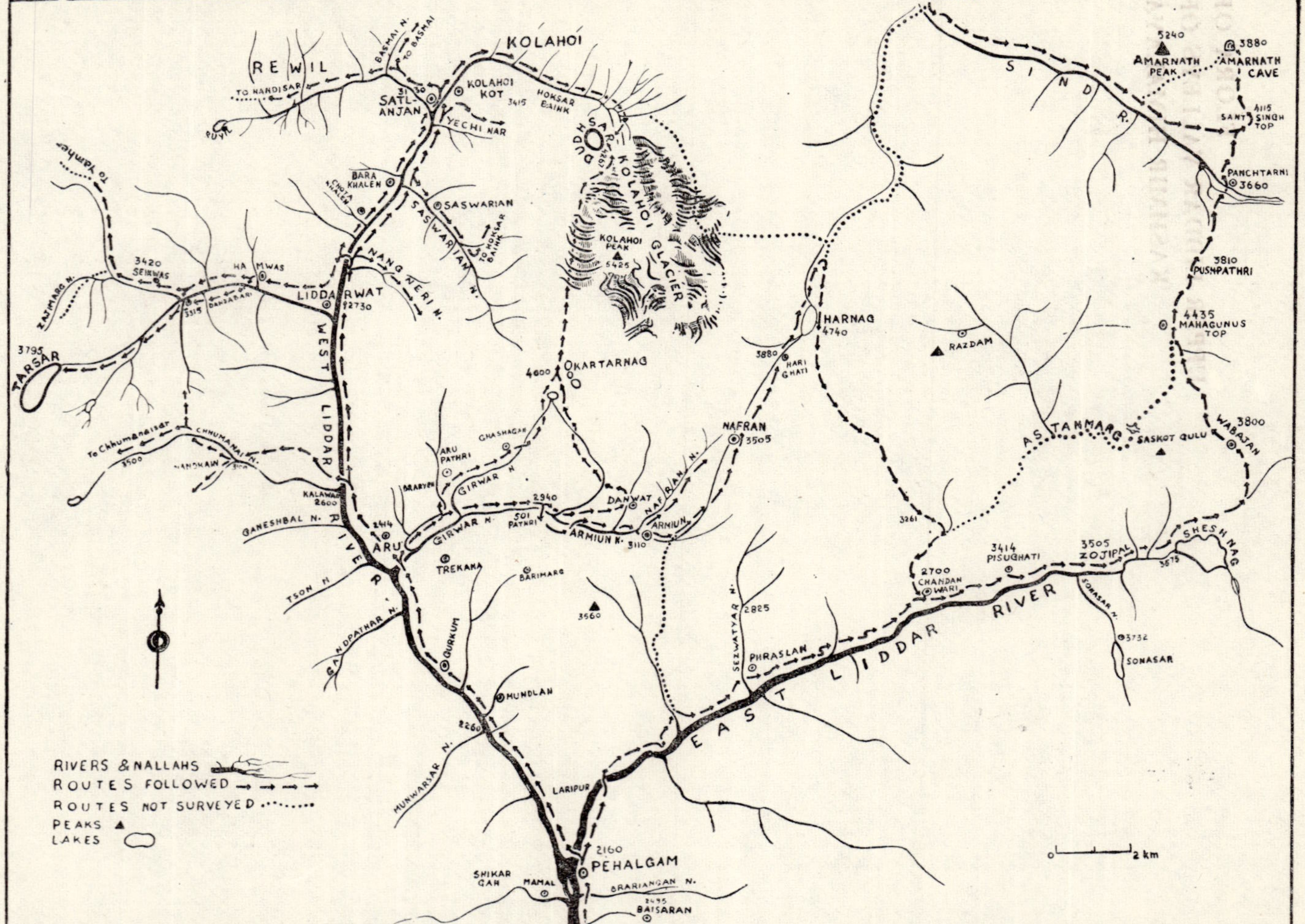

Map : Upper Liddar Valleys of Kashmir Himalaya

Journal of Economic and Taxonomic Botany Additional Series IV

FLORA OF UPPER LIDDAR VALLEYS OF KASHMIR HIMALAYA

Vol. I

BRIJ MOHAN SHARMA
and
P. S. JAMWAL
Floristic Studies Division
Regional Research Laboratory
Jammu - 180 001

1988

SCIENTIFIC PUBLISHERS, JODHPUR-342 001

Published by
SCIENTIFIC PUBLISHERS
Maan Bhawan, Ratanada Road,
P. O. Box 91
JODHPUR - 342 001

ISSN : 2050-9768 **Journal of Economic and Taxonomic Botany**
ISBN : 81-85046-63-8

Printed at
Alok Press, Jodhpur

The senior author dedicates his portion of

work to a friend whom he met after

32 long years of patient wait

Message from Professor R.R. Stewart

I have read the first four fascicles of the Kolahoi Valley Flora with a good deal of interest because I think that it is one of the most beautiful areas in the whole of the Himalayas. In my years of plant hunting in the Western Himalayas my first summer spent in collecting in Liddar Valley was in 1919. My wife and I also spent the summers of 1925 and 1927 camping in a grove of blue pine trees at Pehalgam which is a wonderful centre as a base for exploring the valley. My last summer in the valley was in 1945.

In going through the fascicles which have so far been prepared I was pleased to see that the flora is well done in many ways and well illustrated. The flora is provided with keys and there is a good description of each species. The authors are to be congratulated for providing a useful piece of work for those interested in the alpine flora of Kashmir.

May, 1981
Ann Arbor, 48109
The Herbarium
University of Michigan
North University Building

Ralph R. Stewart

Ralph R. Stewart
Principal Emeritus of
Gordon College, Rawalpindi

FOREWORD

Floristic botany in India is in a renaissance. There was a great flowering with Hooker's Flora of British India, including its sequelae, the regional floras that ended before the Second World War. During the last 20 years there has been another flowering, now of more local floras (such as this one), coming out of an astonishing number of smaller research and educational institutes throughout India. The many floras, appearing at an increasing pace, are clear evidence of this renaissance.

Important as the publications are, I consider that the vitality of the research and the expanding experience of the workers is really the most important. Floristic research tends to die when embalmed in books and only comes to life when workers are actively collecting and studying plants. The books on the shelves are tools and are important, but the life of floristic and systematic botany is in its workers. It is they, not books that give life to the unending synthesis that is systematic botany.

Congratulations on bringing your work to this stage !

Dan H. Nicolson Ph.D.
Vice-President,
International Association of Plant Taxonomists;
Curator, Department of Botany,
Smithsonian Institution,
Washington, D.C., U.S.A.

April, 1986

PREFACE

In the wake of recent awareness about the importance of plants and plant cover in preserving the ecosystem, the botanically rich vale of Kashmir has attracted the attention of quite a few workers who have adequately contributed their share through investigations dealing with various floristic and ecological aspects of the valley. Among the published manuals and Floras, however, there is none perhaps where all the more important aids to plant identification, viz, illustrations of plants, their descriptions and the analytical keys at the familial, generic and specific levels etc. are to be found encompassed within the covers of any single work. In attempting to achieve this systematically for different areas of Kashmir Himalaya, our first preference for the Upper Liddar Valleys has been inspired by the eagerness of many people, both at home and abroad, to know about the plants abounding this picturesque, predominantly alpine tract. Whereas Kolahoi glacier and the route from Liddarwat to Sind valley through Seikwas and Yamher has been a source of fascination for a multitude of foreign trekkers, the holy cave of Amarnath attracts several thousand pilgrims annually from all the different corners of the country itself. Incidentally this huge tourist influx is synchronal with the time when majority of our temperate and alpine species happen to be in full bloom. Bewildered by the radiant hues, varied odours and peculiar life forms exhibited by several of these Himalayan elements; the interested passer by, in his enthusiasm to know their correct identity, is not unjustly inclined to complain of the non existence of a work which could ensure for him a ready comprehension of this veritable array.

Above consideration apart, there has of late been an increasingly painful realisation that the famous health resort of Pehalgam—the majestic gateway to the Upper Liddar Valleys—and its immediate environs are under increasing pressures of rapidly expanding tourist infrastructural facilities and, in the absence of a timely and appropriate step, it can be too long before it is possible to study and document the existing plant cover. That no apology is needed to extend this step to the entire area under these valleys of great scenic charm is evident from the vanishing natural habitats of many a species consequent to heavy and indiscriminate extraction of plants for their medicinal and other uses, coupled with deforestation and subsequent land erosion resulting from the unabated spoils of the graziers and herdsmen.

Charged by the foregoing and related exigencies we explored the area in periodical forays extending over the years 1976-1980. Whereas the materials collected during these field trips form the primary basis of the present work, account has been taken of the earlier collections housed in the herbaria of Forest Research Institute, Dehradun (DD) and Regional Research Laboratory, Jammu (RL. Herb.). Our own collections have been deposited in the latter named herbarium. With a view to make the study more comprehensive, taxa not seen by us but authentically recorded from the area by previous collectors have as well been accordingly incorporated.

Keeping with arrangement of natural orders in major Indian herbaria and important Floras published in the country, the system followed in the present work is essentially that of BENTHAM and HOOKER. However, plausibility of conformity with certain latest publications has induced us to split certain families in accordance with the scheme of HUTCHINSON without violating the superstructure of the said system. The sequence of genera and species is of necessity our own.

Nomenclature has been brought upto-date through adopting the currently accepted names. To avoid overburdening of the text with synonymy only more important synonyms have been given. Of course, until demanded otherwise by the situation, reference has invariably been made to the name used in HOOKER's *Flora of British India*. Where personal verification of the nomenclature of species was not within our means, we have applied names on the authority of generally reliable workers.

Descriptive account of personally examined species is in each case followed by a brief note on its occurrence, flowering and wherever possible fruiting period, the exsiccata examined together with reference to the altitude from where the species was collected, and finally the information on medicinal and other uses, if any, as gathered from literature or from the word of mouth.

In our endeavour to illustrate the work, we have made use of the experience of one of us (B.M.S.) in preparing the drawings for his *Flora of Jammu and Plants of Neighbourhood*. One species for each genus might have proved to be the ideal practice for making the sketches but the factor of form appeal has in some instances urged us to deviate from strict adherence to this procedure. Since preparation of figures for the complete work would have necessarily entailed further deference of publication, we are obliged to seek recourse to the part wise submission of the account.

Barring six species of Gymnospermae (3 families), all the rest 295 circumscribed in this little volume belong to Polypetalae alone, comprised of 124 genera belonging to 32 families and illustrated with 94 line drawings. With this, we look forward to helpful suggestions and constructive criticism from the users.

Jammu
31st March, 1986

Authors

ACKNOWLEDGEMENTS

This book is the outcome of a zestful suggestion from our Director Dr. C.K. Atal who, in his teaching days at the Punjab University, left no opportunity unavailed to visit the Upper Liddar Valleys, in his admiration for their flora which captivated his mind right since he undertook his first trip to the area as a student and a mountaineer. We gladly take the opportunity of recording our deep indebtedness to him for his keen interest in the progress of the work and every possible facility he provided for its successful culmination. To our acting Director, Dr. C.L. Chopra, we are grateful for enthusing us to work with redoubled zeal in the absence of Dr. C.K. Atal, who, at present, is on deputation as Chief Advisor at UNIDO. To Mr. B.M. Kapur, Chairman, Technical Cell of this laboratory, we tender our sincerest thanks for his hearty co-operation during editing and various other stages of implementation of the publication of this volume.

We are also indebted to Prof. R.R. Stewart, formerly of Ann Arbor Michigan, for his message which he so kindly sent to us after going through the manuscript when it was in its fascicles stage. To Dr. Dan H. Nicolson we owe our special thanks for obliging us with the foreword to the book. Prof. S.K. Shah of Geology Department of Jammu University helped us with the geological account of the area. For this, we express our sincere gratitude to him.

We are also under obligation to the authorities of Forest Research Institute and Botanical Survey of India (Northern Circle), Dehradun for allowing us the use of their herbaria and library facilities. Also, we are under an obligation to the Royal Botanic Gardens, Kew for the help received with regard to identification of certain plant species.

Grateful acknowledgement is made to Mr. K.L. Gupta, Plant Operator with us, for his invaluable field assistance, and to Mr. Anil Khanna for the pains he took in typing out the manuscript.

Authors

Contents

Alphabetical list of families dealt with in this volume

INTRODUCTION

The interest in the knowledge and use of the plants of Kashmir has been as old as the recorded history. However, it was not untill W. Moorcroft, accredited as the first European to have set foot in the valley in 1822, made his collections that the materials started getting amassed for the preparation of the floristic accounts of Kashmir. The first serious work to have been undertaken in the latter direction was that of John Forbes Royle who, in the years 1833-39, published a good number of Kashmir plants together with their coloured illustrations. Nevertheless, it was for the first time in Hooker's *Flora of British India* (1872-97) that information gleaned from the collections of all the botanists and plant hunters, who had explored the region till that date, was brought together to give anything approaching a complete flora of Kashmir. In this, of necessity, our plants are scattered through all its seven volumes. Also, besides its unweildiness in the field, it has by now become much out-of-date, though still holding its sway as the principal work for determination of majority of our plant species. The closing years of the century were marked by the appearance of Duthie's reports (1893-94) on his botanical trips to Kashmir.

The first half of the present century witnessed the publication of two important books, B.O. Coventry's *Wild Flowers of Kashmir* (1923-30) and E. Blatter's *Beautiful Flowers of Kashmir* (1927-28). Both these illustrated works are, however, known better for their share in the promotion of popular knowledge rather than their taxonomic merit. Other notable additions to the botanical literature of the period include those by Meebold (1909), Stewart (1916-17, 1945), Lambert (1933) and the information available in contributions by Bamber (1916), Mukerji (1940) and Pennell (1943).

In the years that have followed, consolidation of the results of his almost sixty years' effort into an annotated check list by Prof. Stewart (1972) and the publication of handbooks on the flora of various parts by Singh and Kachroo (1976), Kachroo et al. (1977) Dhar and Kachroo (1983) have been the major steps towards the advancement of the knowledge of Kashmir botany. To these, of course, has to be added the more recent, though general work, *Flowers of the Himalaya* (1984) by Polunin and Stainton, describing, among others, several Kashmir plants and gorgeously embellished with colour photographs and line illustrations. Besides, during the period, accounts of the flora and vegetation have appeared in various journals and periodicals as well. Notable among the contributors are Stewart (1951, 61, 67, 70), Ludlow (1951), Rao (1960), Dutt et al. (1964), Javeid (1965, 68), Kaul and Zutshi (1966, 67), Ahmad (1968), Kaul (1971, 75, 83) and Gupta and Kachroo (1985). Revisions and monographs by Raven (1962, van Soest (1961, 63), Grierson (1964), Kazmi (1970-71), and Koul and Bakshi (1984), as also the reports of new plant records by quite a few botanists and references to distribution of certain Kashmir plants in the *Flora of Pakistan* edited by Nasir and Ali (1970–) are of further significance in the enhancement of our knowledge during the present era.

Students, teachers, professional botanists and botanically minded laymen, desirous of trekking through the scenic Upper Liddar Valleys, as also those who ever did avail such opportunity, have often expressed their genuinely felt concern about the absence of a specific book which could prove as a useful companion for field identifications and location of flowering plants occupying one's particular interest. Upper Liddar Valleys left alone, in fact there has been no Flora exclusive for any of the side valleys of Kashmir, barring of course the brief enumeration lists attempted by Wali and Tiku (1964), Singh (1970) and Singh and Wafai (1973). Precisely it is for remedying this lacuna, in whatever measure possible, that we have endeavoured to prepare the account that follows.

THE AREA

Discussing spots worthy of commanding interest in Kashmir, Sir Francis Younghusband (1909) made a special reference to the area embraced by the present study; wrote he, '*There is a large choice...to points of interest...First in the immediate vicinity...ruins of Martand which occupy the grandest site of any of the world...; to Liddar Valley, Pehalgam, the Kolahoi glacier and the caves of Amarnath...*' By Upper Liddar Valleys is to be understood that tangled, though accessible, tract of montane land which has for its gateway the famous resort of Pehalgam situated 96 km to the east of Srinagar, the summer capital of Jammu and Kashmir state. Extending from 34° to 34° 13′ 24″ N and 75° 8′ 25″ to 75° 30′ E, the area derives its name from the east and west Liddar torrents which, after their confluence at Pehalgam, form the Liddar proper trending in a southwesterly direction for roughly

36 km to drain off finally in river Jehlum at a point just few kilometers to the north of Anantnag town. For a wider coverage, however, the terrain stretching 18 km from the head of the east Liddar valley to the north lying Amarnath shrine (alt. 3880 m), and thenceafter 15 km towards northwest along Sind (Amravati stream) to Baltal (alt. 2760 m) at the base of Zoji la, has also been investigated. None of the points in the area under investigation lies below 2100 m altitude, the highest elevation explored during treks being achieved about Harnag lake situated at an altitude of 4740 m. Arrow marks shown in the map indicate the course followed during collection trips.

The entire tract thus encompassed is readily approachable from Pehalgam, which is ideally located for a base at 2160 m altitude. From here the mountains become increasingly lofty, the maximum altitude (5425 m) in the area being attained by the precipitous Kolahoi massif standing as the crow flies, about 22 km to the north of Pehalgam. Closing the west Liddar valley and extending southeast from the foot of Mt. Kolahoi is the colossal glacier of this name, with west Liddar issuing from its snout (Pl. 1.a) at 3415 m altitude. Strewn with a confused mass of morainal boulders Kolahoi is a small, more or less flattened, rather open valley with sparingly vegetated mountains (Pl. 1.b) rising steeply on its either side. East of the valley is a 4890 m high imposing peak, forming the eastern terminus of a ridge and line of summits ending in west at Lesser Thajwas. To the northwest is located Basmai peak composed mostly of bare rock of poor quality. In lower reaches of Kolahoi, the stream rushes through gorges and cascades in a narrow defile which, after about 14 km, opens for a while into the small Liddarwat camping ground surrounded by thickly forested slopes and pinnacled ridges springing out in nearly all directions. At this point, the Liddar adds to its waters a swift torrent, which, trickling from Tarsar lake (Pl. 2 a; alt. 3795 m) at 16 km to west of Liddarwat, carries with it the one flowing through Seikwas meadowland, to be overarched by the snow bridges at Hamwas (Pl. 2.b). Seikwas rises into the Yamher pass in its north, the structure of the hills around being such as to enclose a deep gigantic oval basin roughly 6 km in length.

Twelve kilometers downstream, the volume of Liddar further swells up at Aru as a result of the Girwar nallah draining into its left bank after rolling down out of Kartarnag over the Danwat mountains. Aru (alt. 2414 m), the last upland village in this valley, and situated in a small rolling meadow overtowered in the northeast by snow capped Kartarnag ridge, makes the thickly wooded Armium glen and alpine areas about Nafran and Harnag easily available. The combined flow of west Liddar is joined by several minor water courses from the conifer clad slopes abutting on its 11 km long course before its final union with east Liddar stream in the small, triangulated, more or less flattened Pehalgam valley.

The narrow defile ascending obliquely in the north easterly direction from Pehalgam affords the most practicable approach to the dolomite cave of Amarnath and the small deep and rocky valley of that name, bounded by steep mountains and traversed by a torrent from a large glacier at its upper end. The ascent along the east Liddar stream is along a gentle gradient flanked by well-wooded hill sides for about 16 km to Chandanwari (alt. 2700 m), a small shut-up valley at the base of the steep rock cliff of Pisu Ghati which, actually, is the face of a spur truncated by some ancient glacier At Chandanwari, the Liddar is joined by a tributary torrent from Astanmarg, an elevated debris filled miniature vale lying at the foot of the south eastern buttresses of Mt. Kolahoi and enclosed by sheer cliffs, screes and precipitous slopes.

From Pisu top stretches the 8 km long glacial terrace of Zojipal elevated above the tree line and leading to holy Sheshnag lake (Pl. 3; alt. 3575 m)—the apparent source of East Liddar—nestled in a cirque formed by a precipitous ridge whose strata are much twisted and distorted to its very summits. The stream, rising from actually the snowy southern slopes of Panchtarni, drains into the north-eastern corner of Sheshnag lake which itself is linked with the small glacier fed Zamtinag lake, the latter lending its peculiar whiteness to the Liddar waters. Above Sheshnag, an ancient moraine lying across the valley slices the latter in its middle. The ascent from Sheshnag leads to the 4435 m high Mahagunus pass which constitutes the water shed between the East Liddar and Sind valleys. About 8 km down hill lies the flat, circular, desolate valley of Panchtarni (alt. 3660 m) surrounded by snowy ranges and traversed by cris-crossing streamlets taking their origin from the Koun Nag and the glaciers between the Liddar and Sind valleys. Broad morainal terraces spread on all the sides of Panchtarni. The Panchtarni streamlets affect their union about the foot of Santsingh top and, receiving a tributary from the adjacent Amarnath valley, form an awe inspiring torrent which drives through the narrow defile leading to Baltal in the Sind valley.

Owing to the facilities of travelling afforded by the construction of a jeepable road from Pehalgam to Aru in the West Liddar valley, and the one upto Chandanwari, the access to the more difficult mountainous areas further afield has been in some measure rendered easier.

GENERAL GEOLOGY AND GEOMORPHOLOGY

The area constitutes the south westerly slope of the Great Himalayan range which runs in a northwesterly direction along the general orographic trend of the Himalaya. The Great Himalayan range bifurcates into the smaller southerly branch, the Pir Panjal range and the main range, enclosing in between the oval strike valley of Kashmir. Beyond this bifurcation, the south westerly slope is drained by the Liddar and the Sindh rivers. The Pehalgam—Kolahoi belt falls along the Liddar valley system.

The rocks that constitute this slope are essentially Palaeo-Mesozoic sediments and volcanics of Tethys facies which are severly folded into two major anticlines and an intervening syncline. The major fold axis is NW-SE and the two anticlines, viz., Kolahoi-Basmai anticline to the north and the Kotsu anticline to the south expose lower Palaeozoic (Cambrian to Silurian) rocks in their cores. These rocks are essentially argillaceous and comprise shales, sandy shales and slates. They are found along Basmai belt in Sindh valley and Hapatnar–Margan pass belt in the Liddar valley. The intervening area from Aishmuqam through Pehalgam to Kolahoi is synclinal and comprises Upper Palaeozoic and Triassic rocks. The upper Palaeozoic rocks contain Carboniferous limestone (Syringothyris Limestone) succeeded by Carboniferous quartzite and shale (Fenestella shale), agglomerate and lavas (Panjal Traps) and Permian and Triassic Limestone. On the whole, bulk of rocks in this region comprises limestone, especially the belt between Pehalgam and Kolahoi is mainly limestone associated only occasionally with volcanic rocks of andesitic composition. The two kinds of rocks can be easily located from a distance by their colouration and bedding pattern. The Panjal volcanics are universally green in colour (hence locally called '*Nilawat*', meaning green stone) while the Triassic limestone is dark grey and bedded and tightly folded. The geomorphological pattern of the two is also quite different, the latter eroding easily and producing valleys and the former standing out in high relief and constituting the ridges. All the major springs in the area are located in limestone terrain.

The Liddar river flows across the strike of the rocks and cuts through all the formations. In its higher reaches it follows what appears to be a glacial valley which was fed by the tributaries. This is indicated by the U-shaped design at the upper level which is cut by the river.

The entire geomorphological pattern of the area has been affected to a considerable extent by the glacial activity of late Pleistocene which has modified the topography. The evidences of a widespread glaciation are noticeable in the presence of huge morainic deposits. In fact the town of Pehalgam itself rests on morainic material and sevaral lateral moraines along the stream valleys especially near their junction can be seen in Pehalgam, Aru and Liddarwat region.

One of the important aspects of glacial topography is the presence of large amphitheatres and small cirques on the mountain slopes along the river valleys. While most of them are thickly forested and thereby concealed, some of them can be easily spotted. They are generally covered by soil and clay and give rise to large savannahs and grasslands (smaller ones called '*pathri*' locally and the large ones '*marg*'). Most of the grasslands and *margs* dot the mountain slopes throughout the area surrounding the Liddar valley.

CLIMATE

The principal features of the climate of the investigated tract are short but pleasant summers, severe and prolonged winters, heavy winter precipitation and lastly the moderate summer rains resulting from the advent in July of the already nearly exhausted southwest monsoon. The area experiences frequent spring showers

Table : CLIMATOLOGICAL DATA OF PEHALGAM (averages for 19 79-83).

Month	Mean of		Relative humidity (percentage)		Total Rainfall mm
	Daily max. temp. (°C)	Daily min. temp. (°C)	0.830 hrs.	17.30 hrs.	
1	2	3	4	5	6
January	4.52	—6.6	83.2	74.6	104.78
February	5.18	—5.68	90.4	84.0	171.6
March	8.38	—2.28	88.4	74.2	307.94
April	17.12	2.98	77.4	58.0	112.82
May	20.06	5.68	74.8	56.2	137.88
June	23.98	8.18	74.6	53.0	82.94
July	24 82	11.76	84.8	64.8	120.06
August	25.68	12.1	88.4	65.0	86.76
September	23.8	7.96	84.6	55.4	68.94
October	18.64	3.17	81.6	55.8	45.66
November	12.06	—1.52	82.4	66.4	65.24
December	5.06	−4.66	89.2	77.2	55.98
Mean	18.05	2.59	38.31	65.43	Total 1361.14

Source : Tourist Meterological Observatory, Pehalgam (Kashmir). Latitude 34° 2′ N. Longitude 75° 23′ E. Elevation 2160 m above m.s.l.

as well. The winter precipitation being received largely in the form of snow, much of the area remains snowbound upto the Ist week of April, but in the major portion of its alpine part the snow persists till only the first week of June, the areas of perpetual snow, however, remaining rcstricted to Kolahoi glacier and the small pockets about the mountain peaks.

The chief climatic factors influencing the montane vegetation being rainfall, relative humidity and temperature, a fair idea of the climate obtaining in the Liddar valleys, on the whole, can be gathered from the below tabulated data pertaining to these elements as recorded for the period 1979-83 at Pehalgam, the only station in the area to possess a Meteorological Observatory.

From the above it follows that dispite its severe winters and accompanying snowfall, the area is predominantly temperate climatically.

COLLECTORS

In the 165 years since the time of Moorcroft, literally hundreds of plant hunters including botanists, professional collectors and amateurs have made collections of plants, large or small, from various parts of Kashmir. Dozens at least, out of these, have made Liddar valleys their sojourn at one time or another for their collecting pursuits. Hugh Falconer

(1838), C.B. Clarke (1876) and J.R. Drummond are the more famous 19th century names associated with the exploration work in these valleys. Col. A.E. Ward and his daughter, Miss Grace, also did some collecting in this part towards the close of the century, but during this period the most indefatigable of the botanists to have substantially collected from the area under reference was John Firminger Duthie (1893) who, for the purpose, also utilized the services of his equally ardent and skilled collector Munshi Inayat Khan (1901).

In our own times, the name that out shines the others, on account of collecting zeal particularly for these valleys, is that of Prof. R.R. Stewart, then of Gordon College, Rawalpindi (Pakistan). During the several summers (1919, 25, 27-28, 45) he collected in our area, he was able to amass a stupendous wealth of Liddar valley specimens, many of these having been cited in his *Annotated Catalogue of the Vascular Plants of Pakistan and Kashmir.* His main collection is at Rawalpindi, with large sets at New York, Kew, Michigan and Edinburgh. In many of his collecting trips he was assisted by his wife, Isabelhae Darrow Stewart. The others to have collected in company with him, at one occasion or other, were Mr. Canon Stokoe (Chaplain of Srinagar, c. 1920-40), Prof. F.G. Dickason (1927-28) who was then teaching at Gordon College, and students including Eugene Nasir, now professor at the same college.

Stewart and his coadjutors apart, in the galaxy of the visitors who collected in the Liddar valleys during the present century, the names that deserve special mention are A. Meebold (1905), Col. Houston (1939), M.V. Laurie (1941), S.L. Nayar (1945), L.D. Kapoor (1945, 46, 53), K.S. Ahluwalia (1951), K.M. Vaid (1954), T.A. Rao (1956), H.D. Thapliyal (1958), van Soest and Harbhajan Singh (1959) and more recently the students of the Botany Department of Kashmir University.

To be fair to others who picked up a solitary or only a few specimens of their interest, paucity of information permits the names of Dawarka Nath, L. Deo (1954), V. Puri (1956), Irshad, Kabir, J.B. Fuller, J. R. Harlan and Oleg Polunin only to be recorded.

VEGETATION

From the foregoing it is evident that in its vertical aspect the tract of exploration is sufficiently elevated above sea level to support the growth of both temperate as well as alpine vegetation. Barring the portion overlying *c.* 3150 m, the mountain slopes in the entire area are predominantly under a coniferous cover composed chiefly of *Abies pindrow*, with *Pinus wallichiana* being the next commonest conifer to enter the forest composition. In consonance with the condition obtaining on the northern slopes of Pir Panjal range, there is no *Cedrus deodara* of any significance in the Liddar valleys, a feature sought to be explained by Sher Singh (1929) on climatic grounds but ascribed by Puri (1948) to the absence of much of the flood plain deposits and the past glaciation.

During the Pleistocene Ice Age four main glaciations were experienced in the north western Himalaya and likewise the whole Kashmir valley also witnessed four main oscillations of climate in the glacial cycle (De Terra & Peterson, 1939.) Palaeobotanical evidence strongly favours the opinion that Pir Panjal barrier, which now effectively obstructs the free flow of prevailing monsoon winds was low enough during early Pleistocene to admit of the growth of tropical elements like *Woodfordia*, *Mallotus*, *Pittosporum*, *Myrsine* etc. in the valley. Pleistocene glaciation together with the uplift of the Pir Panjal range by about 1800 m, and the formation of Karewa lake deposits, is believed to have resulted in the extermination of the tropical flora (Puri, 1943, 47) and the generation of the substrata and climate so congenial to the present day vegetational complex, dominated at higher altitudes by a luxuriant growth of conifers. That the elevations above the present Karewas might well have supported oaks as climax species, with laurels probably as undergrowth, has been advocated on the basis of the discovery of fossilised remains of *Quercus* and *Litsaea lanuginosa* (*cf.*

Inayat Ullah & Tikoo, 1964) from the Panjal slopes. The gradual wiping off of the oaks from Kashmir has, however, according to Vishnu - Mittre (1963), been a consequence of the combined action of man and changed climate.

Despite blocking up of a free entry of southwest monsoon by uplifting of Pir Panjal to its present height (*c.* 4,500 m), the valley is not entirely free from summer precipitation owing to orientation of ridges and the rain bearing winds gaining access into the region through the Jehlum gorge near Baramulla (Puri, 1960). Thus it is not surprising that the topography of Liddar valleys, as that of other similar recesses in the north facing Panjal slopes, should materially modify the rainfall to create in the area a climate under which mesophytic flora represented by broad leaved deciduous temperate trees like *Sorbus*, *Aesculus*, *Acer*, *Juglans* etc. and the shrubs like *Viburnum*, *Cotoneaster*, *Syringa*, *Parrotia* etc. are able to exist.

The slopes skirting the Pehalgam town are clothed chiefl with *Pinus wallichiana* (blue pine) Pl. 4, leaving the inhabited valley floor dotted largely with cultivated trees like *Platanus orientalis*, *Juglans regia*, *Morus* sps. and the roadside plantations of *Aesculus indica*, *Ailanthus altissima*, *Robinia pseudoacacia*, *Populus nigra* and the willows of various kinds. A few small trees and shrubs of *Crataegus songarica* encountered along the commencement of the road to Aru, however, appear to be the remnants of the vegetational cover cleared from the valley bottom for the purpose of habitation. *Viburnum grandiflorum*, a common shrub of adjacent slopes, is also distributed here and there in the form of isolated individuals. In wetter and shady situations along Liddar banks, it is frequently found associated with *Sorbaria tomentosa*. Other shrubs of contiguous slopes contributing to the vegetation of Pehalgam include *Rubus* sps., *Spiraea canescens* and *Lonicera quinquelocularis*. These, however, are met with chiefly among the fences bordering the cultivated fields which besides are impregnated with the woody climbing species exemplified by *Clematis grata* and *Rosa brunonii*. On the islets in the western Liddar and also along its banks *Myricaria germanica* scrub forms nearly pure patches which only occasionally are interrupted by the bushes of *Berberis orthobotrys*. The rock outcrops along the left bank of eastern Liddar here are, however, under a rich growth of *Indigofera heterantha* and the low lying ground by the torrent, being unusually moist, enables herbs like *Delphinium roylei*, *Dipsacus mitis*, *Mentha longifolia* etc. to grow to gigantic dimensions. In certain stony and gravelly situations along the stream bed, localised patches of *Paeonia emodi* figure out conspicuously on account of the large showy flowers of this species, while scattered individuals of *Phytolacca acinosa* (Pl. 5.a) and small groups of *Sambucus wightiana* plants (Pl. 5.b) arrest attention owing to the shrubby habit of them both coupled with prominence of the white umbellate inflorescences of the latter. With the melting of snow in early spring, the first to bloom in the vacant spaces on the surrounding lower slopes of Pehalgam are *Corydalis diphylla*

Viola sylvestris, *Colchicum luteum*, *Tulipa stellata* and *Gagea* sps. while in the *Moister niches* a little higher up in the defiles *Callianthemum pimpinelloides*, *Berginia ciliata* and *Primula denticulata* are among the first to put forth their flowers during this season. At this time *Tussilago farfara* is yet another conspicuous herb which, owing to its gregarious occurrence, lends the yellow splash of its flowers to the eroded curvatures of the roads to Aru and Chandanwari.

On the hill sides, especially such as those facing the north, seral blue pine forms belts of fine forests which at the lower elevations alone, more so along the approach to Aru, may be sprinkled with occasional individuals of *Cedrus deodara*. Thin stands of another conifer, *Picea smithiana*, tend to break the continuity of blue pine about the stream banks untill along the lower slopes near Aru, Gha Shangan and Chandanwari they manifest themselves into a conspicuous component of the coniferous cover by virtue of their swollen numbers. In the area, it is also not unusual to encounter *Taxus wallichiana* figuring sporadically at the lower levels of the blue pine forest.

The shrubby component of the blue pine forest is comprised of the usual deciduous elements listed above besides a few other notable additions including *Cotoneaster affinis*, *Rosa macrophylla*, thinly represented *Indigofera hebepetala* and *Viburnum cotinifolium*, gregariously occurring *Rabdosia rugosa* and, on the steeper slopes, prolific thickets of *Parrotiopsis jacquemontiana* in the understrata of both the eastern as well as the western defiles. While all others restrict themselves to altitudes below 2400 m, *Rabdosia* has been found to colonize the rocky blanks at slightly higher elevations also, and *Rosa macrophylla* alongwith extensive patches of *Viburnum grandiflorum* and *Sambucus wightiana* forms the undergrowth of the *Abies pindrow* forests well above 2750 m and then extends further higher to become a conspicuous element of the subalpine-alpine zone.

The blue pine along with its usual measure of deciduous tree associates may grow upto even 2800 m, but above an elevation of about 2350 m it rapidly begins getting replaced by *Abies pindrow* (silver fir), the latter advancing from its lower limit to become a dominating feature of the landscape (Pl. 6.a) in the interior of both the valleys The tree line, which in the area under discussion recedes sometimes to elevations as low as 2975 m, may be exclusively under silver fir but in several places, e.g., Dudhsar and Seikwas in the western and Pisu Ghati in the eastern portion of the tract, it is the patches of *Betula utilis* (white birch), which occupy the uppermost limit of the tree growth (Pl. 6 b). At one point on the east facing slope opposite to Dudhsar, however, a solitary specimen of *Sorbus cashmiriana* was located to be growing higher even than a patch of birch trees.

The composition of the shrub layer of the low level silver fir is much the same as that discussed for the blue pine, but about the middle level at nearly 2750 m it undergoes a discernible alteration through extensive strips of *Salix denticulata* gaining hold on steeper ground and *Skimmia laureola* forming dense expanses on the gentler slopes along the nallahs. Besides, a little higher up, the appearance of isolated individuals of *Syringa emodi* and, on the rocky ground, small patches of low growing *Cotoneaster microphylla* and *Lonicera myrtillus* accentuate this change still further. The others to achieve prominence at this level include *Berberis jaeschkeana* in the eastern and *Ribes himalayense* and the lowly species of *Rubus*, e.g., *R. irritans*, *R. pungens* and the herbaceous *R. saxatilis* in both the valleys.

Although in certain places with better soil and moisture conditions the coniferous cover of the slopes is dotted with the deciduous trees, yet it is remarkably uniform on account of the conifers occurring mostly in pure stands. As against this, the broad leaved forests supported by the valley bottoms in the entire temperate zone comprise a diversified assemblage of a number of arboreal elements. In Kashmir, majority of these ascend upto alts. 2800-3100 m but *Euonymus hamiltonianus*, so common in the eastern Liddar valley is conspicuous by its near absence in the Western defile where *Aesculus indica*, *Sorbus lanata*, *Ulmus wallichiana* and *Corylus jacquemontii* too preponderate largely upto 2450 m altitude. *Prunus cornuta*, marked by its relatively low frequency in Pehalgam, becomes increasingly common with rising altitude untill near Liddarwat and Chandanwari it becomes one of the major components of the deciduous cover of the valley floors where *Populus ciliata* also adds to its numbers considerably along the stream banks. Hitherto occurring in intermixture with these species, *Acer caesium* tends to organize itself into fine clumps in the part of the valley slightly beyond Liddarwat and in spots at almost similar elevations near Chandanwari and below the Chatponsal ridge enroute Amarnath cave. In the alpine zone *Betula utilis* continues to restrict itself to the slopes and the ridges. Occasionally, however, portions of the valley floor also, e.g., the one between Bara Khalen and Satlanjan, may be under exclusive occupation of this species (Pl. 7).

Flowering of a spectacular variety of annuals and perennials from July onwards to the end of September enables the herbaceous ground cover of the blue pine and low level silver fir to achieve a striking prominence during this period. Notable among the species that manifest themselves most frequently on the slopes are *Androsace rotundifolia*, *Bistorta amplexicaule*, *Campanula colorata*, *Cucubalus baccifer*, *Cynoglossum* sps., *Fragaria nubicola*, *Galium asperuloides*, *Geranium wallichianum*, *Geum urbanum*, *Lamium album*, *Lotus corniculatus*, *Myriactis wallichii*, *Nepeta linearis*, *Oxalis corniculata*, *Stellaria monosperma*, *Strobilanthes attenuatus*, *Taraxacum officinale*, *Tricholepis elongata*, *Valeriana hardwickii*, *Veronica melissifolia*, *Viola pilosa*, *Trillium govanianum* and a host of others, among which *Ranunculus laetus* and the fine looking but inedible grass *Stipa sibirica* occupy vast expanses of meadows and forest clearings, *Saussurea albescens* occupies dry sunnier blanks, *Ainsliaea aptera*, *Polygonatum verticillatum* and *Valeriana jaeschkei* colonize humus laden rocks supporting *Sedum ewersii* tufts from their crevices, *Seseli libanotis* abounds on rocky preipices. *Aconogonum alpinum*, *Persicaria polystachya*, *Thalictrum minus* var. *majus* and shrubby *Aralia cachemirica* flourish in moist shelters and elements like *Artemisia absynthium*, *A. incisa*, *A. laciniata*, *Astragalus chlorostachys*, *Cannabis indica*, *Cirsium falconeri*, *C. wallichii*, *Clinopodium umbrosum*, *Corydalis ramosa*, *C. stewartii*, *Lactuca brunoniana*, *Nepeta elata*, *Plantago lanceolata*, *Siegesbeckia orientalis*, *Urtica dioica* and *Verbascum thapsus* tend to become relegated more commonly to the forest borders along the valley bottoms. Whereas the borders of the major streams traversing the latter support, among others, herbs examplified by *Carduus edelbergi*, *Cardamine macrophylla*, *Impatiens brachycentra*, *I. glandulifera*, *Ligusticum thomsoni* var. *evolutior*, *Myosoton aquaticum*, *Senecio ligularia* etc., on some of the islets in stream beds *Heracleum candicans* displays itself as an almost pure crop and, on patches of nitrophilous ground under cattle pens on the banks, *Geranium nepalense* and *Rumex nepalensis* grow gregariously to near exclusion of the other species of the valley floor. By the edges of the lesser water courses draining the slopes *Epilobium cylindricum*, *E. royleanum* and *Impatiens edgeworthii* are at their best, while along irrigation channels in the villages these species share common ground with *Achillea millefolium*, *Agrimonia eupatoria*, *Filipendula vestita*, *Mentha longifolia*, *Plantago major*, *Prunella vulgaris* and *Sium latijugum*.

About village sites *Arctium lappa*, *Datura stramonium* and *Malva verticillata* are common in weedy places, *Myosotis palustris*, *Persicaria hydropiper* and creeping *Potentilla reptans* thrive on damp ground along the drains and *Chenopodium album*, *Cichorium intybus*, *Crepis sancta* sps. *bifida*, *Lactuca dissecta*, *Lamium amplexicaule*, *Lapsana communis*, *Silene connoidea*, *Spergularia rubra*, *Thlaspi arvensis*, *Torilis japonica*, *Turritis glabra* and *Veronica biloba*, besides the grasses *Avena fatua*, *Chrysopogon echinulatus*, *Dactylis glomerata*, *Digitaria cruciata*, *Glyceria tonglensis*, *Poa annua*, *Pennisetum flaccidum*, *Phacelurus speciosus*, *Setaria viridis* etc. represent the commonplace weeds of the grain fields.

Acorous calamus, *Alisma plantago-aquatica*, *Nasturtium officinale*, species of *Juncus*, various sedges, *Veronica beccabunga* and the marsh orchid *Dactylorhiza hataigera* are among the common denizens of the wet and swampy localities in the zone under discussion. Climbing herbs are scantily represented the more striking ones being exemplified by *Bilderdykia dumetorum*, *Codonopsis rotundifolia*, *Convolvulus arevensis*, *Cynanchum auriculatum* and parasitic *Cuscuta reflexa*. Of the few parasites, the minute *Arceuthobium minutissimum* deserves special reference on account of its abundance and the damage it causes to the valuable timber yielding blue pine.

Common ferns and fern allies which enrich the composition of the herbaceous cover of the forest floor in the temperate zone include *Adiantum venustum*, *Asplenium adiantum-nigrum*, *A. trichomanes*, species of *Dryopteris* including remarkably beautiful 'shuttle cock fern' *D. sinofibrillosa*, on damper sites *Athyrium acrostichoides*, by drains and other water courses *Equisetum arvense* and in the rock crevices *Asplenium pseudofontanum*, *A. septentrionale*, *Cystopteris fragilis* and *Pleopeltis calthrata*. On dry sunny slopes abounds *Pteridium aquilinum*. Abundantly occurring *Selaginella sanguinolenta* grows gregariously in patches on the rocks and cliffs in this zone.

In the forest undergrowth at Liddarwat in the western defile and Chandanwari in the eastern one, the intermixing of the temperate herbs with occasional individuals of *Aquilegia fragrans* and *Megacarpaea polyandra*, the commonly occurring *Cerastium dahuricum*, *Hackelia uncinata*, *Thlaspi cochleariforme*, *Trillium govanianum* and clumps of *Aconitum laevae*, *Actaea spicata*, *Thalictrum cultratum* and the 'yellow violet' *Viola biflora* (Pl. 8.a) portend the transition into the immediately overlying alpine zone. In these localities, the other plant indicators that foreshadow the approaching high altitude habitats include clumps of *Iris hookeriana* (Pl. 8.b) left alone by the cattle in the grazing grounds, matted perennials like *Alchimella ypsilotoma*, *Anemone obtusiloba* and *Sibbaldia cuneata* with a fair sprinkling of the brillantly blue flowered *Gentiana marginata* in wet and open grassy blanks, and gregarious growth of the white variety of *Caltha palustris* (Pl. 9.a) interspersed with *Juncus membranaceous*, *Lindelofia spectabilis* and *Prunella vulgaris* along the drains and streams. Also, it is precisely at these two places that at *c.* 2900 m alt. that the silver fir starts rapidly giving way to the already intruding white birch.

The striking contrast in the botany above the zone of transition is ushered in by the meadowlands becoming chequered with the brilliancy of the colour splashes of the flowers of many different alpine herbs and the trees all but the firs and the birches left behind. There is a marked decline in the representation of the shrubs too, the noteworthy to persist from the lower level being species of *Cotoneaster*, *Lonicera*, *Viburnum* and *Salix* and the characteristic ones to amplify the eontrast being the impressive patches of *Rhododendron campanulatum* (Pl. 9.b), *R. hypenanthum* and the now increasingly gaining *Juniperus squamata* whose impenetrable thickets speckle acres together (Pl. 10.a) of the alpine landscape on largely the east facing slopes in these valleys.

The alpine fir-birch forest, representing the culmination of tree zone, is well represented near Satlanjan (Pl. 10.b) and along the Dandabari and Pisu Ghati gradients. Although *Abies pindrow* and *Betula utilis* are its principal components, it is not unusual for *Pinus wallichiana* individuals too to figure in this forest occasionally. The more common shrubby associates here are much the same as referred to in the preceding paragraph, but with the *Rhododendron campanulatum* component of lower levels (*c.* 3100 m alt.) of this forest getting replaced in its higher reaches by extensive patches of gregariously occurring *R. hypenanthum* (Pl. 11.a), until it is this latter alone, in certain places of course with some *R. lepidotum*, that coexists with birches at their highest limit. Although in the ground cover of the slopes under these alpine forests, predominance of elements like *Aconitum laevae*, *Aconogonum alpinum*, *Actaea spicata*, *Aquilegia moorcroftiana*, *Aster thomsonii*, *Leonurus cardiaca*, *Persicaria polystachya*, *Podophyllum hexandrum*, *Saussurea candolleana*, *S. lappa*, *Valeriana hardwickii* and the low growing *Anemone obtusiloba*, *Fragaria nubicola*, *Jaeschkea canaliculata*, *Gentianella pedunculata*, *Polygonatum geminiflorum*, *Sibbaldia cuneata*, *Trillium govanianum*, *Valeriana dioica*, *Vicatia conifolia* and species of *Corydalis* and *Potentilla* afford adequate comprehension of the herbaceous growth of this part of the alpine zone, the valley floors and the borders of the glacial water courses running at this level are ablaze with brightly coloured flowers of still further variety of herbs which transform the whole tract into a veritable alpine garden during July to September. The case of Kolahoi valley may serve here as perhaps one of the most illustrative examples. Whereas the morainal rubble littering the flat valley bottom is during this period prevailingly under yellow flowers of *Potentilla argyrophylla*, *Senecio chrysanthemoides* and *Euphorbia pilosa*, the islands formed by the stream splitting into many channels after debouching from the glacier snout are under the striking deep pink of the flowers of *Epilobium latifolium* manifesting itself in exclusive masses This pink is relieved only here and there by the yellow flowered *Pedicularis punctata*, the rich concentration of *Corydalis govaniana* and *C. thyrsiflora* bedecking the stream banks and *Picrorhiza kurrooa* from rocky slopes finding vacant gravelly patches on the islands suitable for its growth.

Other attractive herbs most in evidence in the valley are *Anemone polyanthes*, *A. tetrasepala* and *Swertia petiolata* clumped in a profusion of *Polemonium coerulum* immediately at the base of the Dudhsar slope, *Androsace mucronifolia* (Pl. 11.b) in matted patches among boulders, *Meconopsis aculeata* (Pl. 12.a) in shade of rocks, meadowland elements like *Inula royleana* (Pl. 12.b) with its immense heads of yellow flowers featuring on adjoining lowermost slopes and *Salvia hians* alongwith *Achillea millefolium*, *Lindelofia longifolia* and species of ferns affecting associations (Pl. 12.c) with a variety of other herbs like *Geum elatum* etc. beside the lesser water courses running down the bouldered slopes.

Where the main stream and the Rewil nallah cut through the rocky basin to affect steeper water falls, the moist slopes and the steeps edging the torrents are under exclusive expanses of *Bergenia stracheyi*, whose fleshy leaves turn this habitat into a flaming red in autumn due to their changed colour during this season. The swampy flats in portions of valley are in places matted with *Stellaria alsine*, *S. cherleriae* and *Veronica* sps., with an occasional near neat array of *Primula rosea* by the sluggishly rippling drains across these flats. In lower reaches up to Bara Khalen, where birches take down to valley floor, *Astragalus himalayanus* is common along paths, *Campanula cashmeriana* and *Thymus serpyllum* by steeper stony sides of the stream and *Androsace primuloides*, *A. sempervivoides* and *Rhodiola crassipes* together with stately *Epilobium angustifolium* on gravelly banks, islets and on the *Sedum ewersii* supporting moist rocks. Among rock pieces interspersed with *Selinum filicifolium* and *S. vaginatum* in open situations, *Codonopsis ovata* is yet another common element to attract notice owing to the captivating sky blue of its flowers.

In its higher reaches at *c.* 3350 m the alpine forest is succeeded by the alpine scrub with *Juniperus squamata*, *Rhododendron hypenanthum*, *Cotoneaster microphyllus* and dwarf species of *Salix* including *S. flabellaris* along alpine lakes being its major components. Among the scrubs whereas *Rosa macrophylla* var. *minor* is seen growing through the *Juniper* thickets, the heath-like growth of *Cassiope fastigeata*—so common on exposed rock faces—takes to breaking the continuity of *Rhododendron* patches. Apart from usual *Corydalis govaniana*, *Meconopsis aculeata* etc. growing beside, the taller herbs that manage to emerge from between the scrubs include *Aconogonum rumicifolium*, *Actaea spicata*, *Aquilegia moorcroftiana*, *Dipsacus mitis*, *Inula royleana*, *Rhodiola heterodonta*, *Saussurea candolleana*, *Thalictrum cultratum* (Pl. 13.a) *Rheum webbianum* and *Heracleum candicans*, the latter two (Pl. 13.b & 14) also finding the nearby inaccessible rock interspaces their favourable recourse.

In the alpine meadows, which essentially are degraded from alpine forests and extend above and into the alpine scrub zone, despite the ground being covered everywhere with a sward of herbaceous plants, the quality of cover is such as to be reminiscent of human interference and uninterrupted grazing pursuits extended over years together. As a consequence, the herbs that are predominantly evident are either poisonous or of inedible or unpalatable kinds. Thus, in places like Hamwas it is the formidable thistle *Cirsium falconeri* (Pl. 15.a) which holds its sway over vast areas of alpine pasture; the slopes leading to Pisu top are, in parts, prevailingly under loose mats of *Scutellaria prostrata;* the drier gradients on Baltal route near Sangam are overrun by sizeable patches of spinescent yellow-flowered *Morina coulteriana*, gregariously occurring *Anemone rupicola* and patchy growth of the prostrate rosettes of *Saussurea atkinsonii*, with parts of the terrain here and along the Zojipal-Sheshnag route either exclusively under the golden-yellow flowered clumps of *Adonis chrysocyathus* or associations of the latter with *Iris hookeriana*. Likewise, whereas many places along the ascent from Hamwas are by and large monopolized by the blue-flowered Boraginaceae, especially *Myosotis* sps., still other portions of ground are overgrown with a luxuriance of *Anaphalis nepalensis*, *Cerastium cerastoides* and *Euprasia* sps. At *c.* 4200 m about Dudhsar, while *Aster falconeri* groups are evident for the colour appeal of their large-sized heads, the interspaces between the *Junipers* on the opposite slopes are overwhelmingly scattered with ashy-green rosettes of *Jurinea himalaica* at lower levels, with *Arnebia benthamii* being only but one of the striking herbs of the slopes and screes higher up.

In addition to the alpine pasture herbs already referred to in the discussion of

Plate 1 a Kolahoi glacier with west Liddar issuing from its snout.
b Panoramic view of a portion of the upper reaches of Kolahoi valley.

Plate 2 a Tarsar lake (alt. 3795 m)
b Crossing a snow bridge at Hamwas.

Plate 3 Sheshnag lake (alt. 3575 m).

Plate 4 Almost pure crop of *Pinus wallichiana* A.B. Jackson on one of the slopes about Pehalgam. In the foreground scrubby growth of *Myricaria germanica* Desv. is evident.

Plate 5 a *Phytolacca acinosa* Roxb.
b *Sambucus wightiana* Wall. occurring gregariously.

Plate 6 a Pure stand of *Abies pindrow* Royle on one of the flanks of western Liddar valley. In the foreground (left) are a couple of blue pines.

b *Betula utilis* D. Don in a highly disturbed state at upper-most limit of tree growth near Seikwas.

Plate 7 Valley floor near Satlanjan under exclusive occupation of *Betula* forest.

Plate 8 a The 'Yellow violet' *Viola biflora* Linn.
b *Iris hookeriana* Foster left alone by grazing animals in the subalpine—alpine pastures.

Plate 9 a Association of *Caltha palustris* Linn. *Juncus membranaceous* Royle, *Lindelofia longiflora* Baill. and *Prunella vulgaris* Linn. along an alpine stream.

b Close up of a patch of *Rhododendron campanulatum* D. Don.

Plate 10 a Procumbent thickets of *Juniperus squamata* Buch.-Ham. occupying vast areas of high alpine landscape.

b Alpine *Abies-Betula* forest near Satlanjan

Plate 11 a Patches of gregariously occurring *Rhododendron hypenanthum* Balf. with *Betula utilis* D. Don on an alpine slope.

b A patch of *Androsace mucronifolia* Watt among boulders in Kolahoi valley.

Plate 12 a *Meconopsis aculeata* Royle

b *Inula royleana* C.B. Clarke

c Association of *Salvia hians* Royle, *Achillea millefolium* Linn., *Lindelofia longiflora* Baill. and a fern species by an alpine water course.

Plate 13 a *Thalictrum cultratum* Wall. growing among junipers in alpine zone.
b *Rheum webbianum* Royle emerging from among the alpine scrub.

Plate 14 *Heracleum candicans* Wall. among alpine rock interspaces.

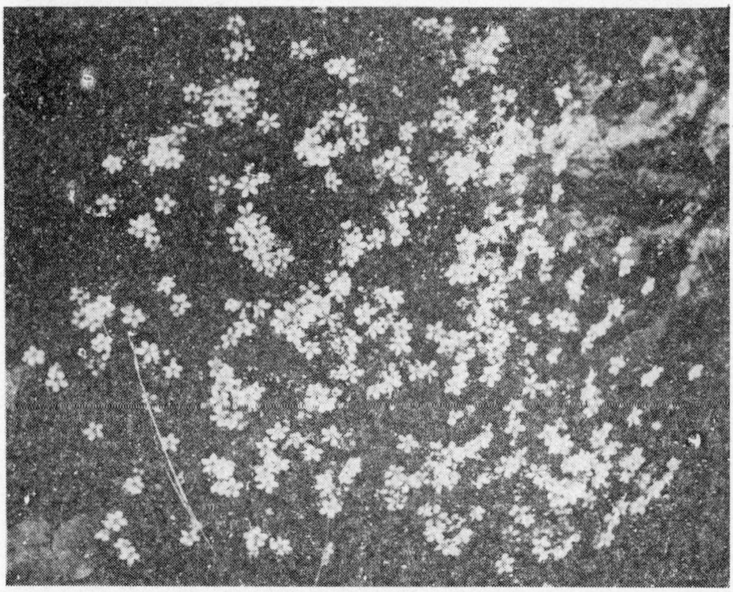

Plate 15 a An alpine pasture exclusively under *Cirsium falconeri* Petrak.
b *Minuartia lineata* Bornm. on a rocky cliff.

Kolahoi valley vegetation, the other common elements with which the alpine meadow is liberally stippled include *Aconitum violaceum*, *Bupleurum longicaule*, *Campanula aristata*, *Corydalis cashmeriana*, *Cremanthodium decaisnei*, *Erigeron* sps, *Fritillaria roylei*, *Geranium pratense*, *Leontopodium alpinum*, *Pedicularis oederi*, *Phlomis bracteosa*, *Primula involucrata*, *P. macrophylla*, *Rhodiola* sp. (*R. quadrifida ?) Sedum oreades*, *Saxifraga sibirica*, *Senecio jacquemontianus*, *Thalictrum alpinum*, *Trollius acaulis* and a host of others, of which *Aconogonum rumicifolium* and *Chenopodium foliosum* occupy nitrophilous habitats, *Nepeta connata* and *N. eriostachya* thrive on well drained but moist slopes, *Thalictrum platycarpum* among Junipers on drier steeps, *Minuartia lineata* (Pl. 15.b) and *Potentilla* of different species abound on cliffs, *Gentiana cachemirica* restricts to rock crevices, *Ranunculus munroanus* preferably to shaded nooks and *Caltha palustris* alongwith *Saxifraga moorcroftiana* and *Bistorta vivipara* affect marshy spots. In other such wet spots, its abundance and deep purple spikes render *Dactylorhiza hatagirea* worthy of notice.

Whereas *Agropyron canaliculatum*, *Alopecurus arundinaceus*, *Eremopoa persica*, *Hierochloe laxa*, *Koeleria gracilis*, *Oryzopsis lateralis*, *O. munroi*, *Pennisetum flaccidum*, *Phleum alpinum* and *Poa alpina* are the chief grasses of the alpine zone, the more common ferns here are, among others, represented by *Athyrium wallichianum*, *Dryopteris barbigera*, *Polystichum lachense*, *P. prescottianum* and the handsome *Osmunda claytoniana* growing gregariously on the grassy hillsides.

The floristic composition of the alpine vegetation of either of the flanks of the Upper Liddar Valleys is much the same but for the moisture conditions and aspect governing the presence, absence or rarity of certain components in one or the other of the flanks. Thus, *Euphorbia wallichii*, conspicuous by its absence in the western Liddar valley, is a common herb of hillsides near Pisughati, and similar is the case with *Gentianella moorcroftiana* and *Delphinium cashmerianum* which adorn the Sangam slopes with a bright splash of their bluish flowers. Another to belong to this category is perhaps *Bistorta affinis* which gregariously strikes the rocky surround of Sheshnag lake but is apparently wanting in the west section of these valleys. In the desolate valley of Amarnath, where the unstable habitat is characterized by screes and shifting bits of rock, the most extraordinary of the fumitories *Corydalis crassifolia* is a common place herb which has only but occasionally been collected from Kolahoi. Other plants to be taken note of in Amarnath valley are species of *Sedum* and *Androsace*, sporadically occurring *Oxytropis cachemiriana*, and on the sides of rocks a dense growth of moss-like cushions of *Saxifraga pulvinaria* to remind one of the alpine stony desert conditions.

ARTIFICIAL KEY TO THE FAMILIES

1. Ovules not enclosed in an ovary but naked and fertilized by direct contact with pollen tubes (GYMNOSPERMAE) :
   2. Female flowers in cones; seeds non-arillate :
      3. Trees; cones woody, made up of spirally arranged scales; ovules inverted — I. **Pinaceae**
      3. Shrubs; cones more or less succulent when ripe, made up of opposite scales which at length become coalescent into a berry-like, globose fruit; ovules erect — II. **Cupressaceae**
   2. Female flowers solitary; seed surrounded by a fleshy, cup-shaped aril — III. **Taxaceae**
1. Ovules enclosed in an ovary and fertilized by the pollen tube entering through the stigma (ANGIOSPERMAE) :
   4. Leaves usually net veined; flowers mostly 5-merous or 4-merous; embryo within 2 cotyledons (DICOTYLEDONES) :
      5. Plants without chlorophyll :
         6. Twining herbs — LII. **Cuscutaceae**
         6. Erect herbs :
            7. Flowers 1-sexual, minute — LXV. **Balanophoraceae**
            7. Flowers 2-sexual, evident :
               8. Flowers zygomorphic; stamens 4 — LV. **Orobanchaceae**
               8. Flowers actinomorphic; stamens 8-10 — XLIV. **Monotropaceae**

5. Plants with chlorophyll :
   9. Parasitic on branches of trees — LXIV. **Loranthaceae**
   9. Not parasitic :
      10. Gynoecium of 2 or more separate or nearly quite separate carpels with separate styles and stigmas :
         11. Petals present :
            12. Petals free
               13. Stipules present, mostly adnate to the petiole, sometimes minute — XXVI. **Rosaceae**
               13. Stipules absent, rarely if present then membranous :
                  14. Stamens numerous, more than 12 :
                     15. Shrubs — XXVI. **Rosaceae** *(Spiraea)*
                     15. Herbs, sometime woody at base :
                        16. Disk present: stamens centrifugal; seeds arillate — VI. **Paeoniaceae**
                        16. Disk absent; stamens centrifugal seeds non-arillate :
                           17. Carpels with more than 1 ovule; fruit follicular — IV. **Helleboraceae**
                           17. Carpels with only 1 ovule; fruit a bunch of achenes — V. **Ranunculaceae**
                  14. Stamens 12 or fewer :
                     18. Carpels 2 — XXVII. **Saxifragaceae** *(Saxifraga cernua)*
                     18. Carpels more than 2 :
                        19. Leaves succulent; carpels 4-5, with more than 1 ovule; fruit follicular — XXX. **Crassulaceae**
                        19. Leaves not succulent; carpels many, with 1 ovule; fruit a bunch of achenes — V. **Ranunculaceae**

12. Petals united at the base into a short tube — XXX. **Crassulaceae** (*Sedum oreades*)

11. Petals absent, though sepals sometimes petaloid :
   20. Carpels in fruit forming a single whorl of 5-8 fleshy berries — LXII. **Phytolaccaceae**
   20. Carpels in fruit achenial or follicular, not as described above :
      21. Carpels with more than 1 ovule; fruit follicular — IV. **Helleboraceae**
      21. Carpels with 1 ovule; fruit achenial :
         22. Stamens many, hypogynous — V. **Ranunculaceae**
         22. Stamens not more than 5; perigynous — XXVI. **Rosaceae** (*Alchemilla*)

10. Gynoecium of 1 carpel or of 2 or more united carpels with free or united styles, or if the carpels free below then the styles or stigmas more or less united :
   23. Ovules attached to the outer wall or walls of the ovary or to the partly intrusive placentas, if only 1 carpel then attached to the adaxial suture :
      24. Petals absent; flowers 1-sexual in catkins — LXXIII. **Salicaceae**
      24. Petals present; flowers 2-sexual :
         25. Ovary inferior — XXIX. **Grossulariaceae**
         25. Ovary superior, rarely (*Parnassia*) adnate only at its base to the calyx tube :
            26. Flowers zygomorphic :
               27. Stamens 10, 9 united in one bundle and tenth free or partly adnate to the rest; fruit a legume — XXV. **Fabaceae**
               27. Stamens 6 or less; fruit a capsule :
                  28. Sepals 2; petals 4; stamens 2, each 3-partite (often referred to as 6, diadelphous) — X. **Fumariaceae**
                  28. Sepals, petals and stamens 5 each — XII. **Violaceae**
            26. Flowers actinomorphic :
               29. Petals united into a corolla — XLVIII. **Gentianaceae**
               29. Petals free :
                  30. Leaves simple, palmately lobed to the middle or base, peltate — VIII. **Podophyllaceae**

30. Leaves simple or variously divided but not as described above :

31. Fertile stamens more than 6 :

32. Sepals 2; juice milk-like — IX. **Papaveraceae**

32. Sepals more than 2; juice not milk-like :

33. Leaves ternately divided, carpel 1, baccate in fruit — IV. **Helleboraceae** *(Actaea)*

33. Leaves not ternately divided, ovary of 2 united carpels divided by a false septum; fruit a broadly winged silicula — XI. **Brassicaceae** *(Megacarpaea)*

31. Fertile stamens 6 or less :

34. Sepals united into a 5-lobed calyx; stamens 5, alternate with an equal number of staminodes — XXVIII. **Parnassiaceae**

34. Sepals free; stamens 4 or 6; staminodes 0 :

35. Stamens 4; inner two petals slipper-shaped, with flat proximal and tubular distal end; ovary 1-celled — VII. **Berberidaceae** *(Epimedium)*

35. Stamens 6, tetradynamous, only very rarely 4 by abortion; petals all similar, flat; ovary mostly divided into 2 cells by a false septum — XI. **Brassicaceae**

23. Ovules attached to the central axis or to or near the base of the ovary, sometimes only 1 ovule, sometimes the ovule or ovules pendulous from the top of a basal placenta in a 1-locular ovary :

36. Petals present :

37. Ovary inferior or very rarely semi-inferior :

38. Stamens 10 or more :

39. Herbs; leaves exstipulate; stamens 10 — XXVII. **Saxifragaceae**

39. Shrubs or trees; leaves stipulate; stamens many — XXVI. **Rosaceae**

38. Stamens 8 or less :

40. Petals free :

41. Flowers in racemes or axillary and solitary; petals 2 or 4; stamens 4 or 8 — XXXIII. **Onagraceae**

41. Flowers in umbels; petals and stamens 5 each :

42. Ovary 2-celled, styles 2; fruit schizocarpic, of 2 indehiscent mericarps XXXIV. **Apiaceae**
42. Ovary 5-celled; styles 5; fruit a 5-angled drupe XXXV. **Araliaceae**

40. Petals united into a tube or cup :
43. Anther connate into a tube sheathing the style XL. **Asteraceae**
43. Anthers free :
44. Leaves stipulate XXXVI. **Caprifoliaceae** *(Sambucus)*
44. Leaves exstipulate :
45. Shrubs or small trees XXXVI. **Caprifoliaceae**
45. Herbs :
46. Leaves in whorls of 4 or more; ovary 2-celled with 1 erect ovule in each cell XXXVII. **Rubiaceae**
46. Leaves (cauline) alternate or opposite, rarely if whorled then ovary 1-celled with 1 pendulous ovule :
47. Stamens 5; ovary multi-ovulate XLI. **Campanulaceae**
47. Stamens 4 or less; ovary uni-ovulate :
48. Stamens 3; ovary with 1 fertile and 2 empty cells XXXVIII. **Valerianaceae**
48. Stamens 4 or 2; ovary 1-celled, without additional empty cells XXXIX. **Dipsacaceae**

37. Ovary superior, rarely only partly immersed in the disk or calyx :
49. Stamens 12 or more :
50. Stamens all united into 1 bundle XVII. **Malvaceae**
50. Stamens free or united into more than 1 separate bundles :
51. Stamens united into 3 separate bundles XVI. **Hypericaceae**
51. Stamens free :
52. Trees or shrubs; leaves alternate; epicalyx absent; fruit a drupe or follicular XXVI. **Rosaceae**
52. Herbs; lower leaves opposite or whorled; epicalyx present; fruit a capsule XXXII. **Lythraceae**
49. Stamens 10 or less :
53. Petals free or very slightly connate at the base alone or connate in pairs but not into a complete tube :
54. Flowers zygomorphic :
55. Trees; leaves digitately compound XXIV. **Hippocastanaceae**

55. Herbs; leaves simple :
56. Posterior sepal neither spurred nor saccate; stamens 8 XIII. **Polygalaceae**
56. Posterior sepal produced into a spur or at least saccate; stamens 5 XX. **Balsaminaceae**

54. Flowers actinomorphic :
57. Armed; perfect stamens as many as the petals and opposite to them VII. **Berberidaceae** (*Berberis*)
57. Unarmed; perfect stamens as many as the petals and alternate with them or twice as many, rarely fewer :
58. Leaves alternate or radical or radical and alternate :
59. Juice acidic; leaves 3-foliolate XIX. **Oxalidaceae**
59. Juice not acidic; leaves simple :
60. Leaves mostly radical; anthers opening by apical pores XLIII. **Pyrolaceae**
60. Leaves all cauline or radical and cauline; anthers opening by slits lengthwise to the base :
61. Leaves strongly aromatic when crushed; stamens as many as the petals; fruit a fleshy drupe XXI. **Rutaceae**
61. Leaves not aromatic; stamens twice as many as the petals; fruit capsular :
62. Shrubs; stamens monadelphous for a part of their length below XV. **Tamaricaceae**
62. Herbs; stamens free XXVII. **Saxifragaceae**
58. Leaves opposite, never all radical :
63. Trees :
64. Leaves palmately lobed; fruit of 2 conjoined 1-seeded samaras XIII. **Aceraceae**
64. Leaves not lobed; fruit a capsule XXII. **Celastraceae**
63. Herbs :
65. Leaves pinnate or palmately (rarely ternately) lobed or divided; ovary completely 5-celled XVIII. **Geraniaceae**
65. Leaves undivided and entire; ovary 1-celled or incompletely septate XIV. **Caryophyllaceae**

53. Petals all more or less equally united into a lobed tube or cup :
66. Stamens 6 or more XLII. **Ericaceae**
66. Stamens 5 or less :

67. Ovary vertically 4-lobed or, partite; style inserted between the lobes; fruit separating into 4 or fewer 1-seeded nutlets at maturity :
   68. Leaves alternate; flowers actinomorphic; fertile stamens 5 — L. **Boraginaceae**
   68. Leaves opposite; flowers mostly zygomorphic; fertile stamens 4 or 2 — LVIII. **Lamiaceae**
67. Ovary not vertically 4-lobed or, partite; style terminal; fruit not of nutlets :
   69. Flowers distinctly zygomorphic :
      70. Ovules numerous or many in each cell of the ovary — LIV. **Scrophulariaceae**
      70. Ovules 1-2 in each cell of the ovary :
         71. Corolla limb 2-lipped; stamens 2; ovule 1 in each cell of the ovary — LVII. **Selaginaceae**
         71. Corolla limb spreading; stamens 4; ovules 2 in each cell of the ovary — LVI. **Acanthaceae**
   69. Flowers actinomorphic or nearly so :
      72. Corolla with a corona; carpels free; fruit follicular — XLVII. **Asclepiadaceae**
      72. Corolla without a corona; carpels united; fruit capsular or baccate :
         73. Stamans fewer than the corolla lobes :
            74. Herbs — LIV. **Scrophulariaceae**
            74. Shrubs — XLVI. **Oleaceae**
         73. Stamens the same number as the corolla lobes :
            75. Stamens opposite to the corolla lobes — XLV. **Primulaceae**
            75. Stamens alternating with the corolla lobes :
               76. Leaves all radical; corolla labes and stamens 4 each — LIX. **Plantaginaceae**
               76. Leaves cauline or both radical and cauline on the same plant; corolla lobes and stamens 5 each :
                  77. Ovary 3-locular; stigmas 3 — XLIX. **Polemoniaceae**
                  77. Ovary 2-locular or spuriously 4-locular; stigmas 2 or 1 :
                     78. Ovules 2 in each cell of the ovary — LI. **Convolvulaceae**
                     78. Ovules numerous in each cell of the ovary :

79. Corolla infundibuliform or campanulate LIII. **Solanaceae**

79. Corolla rotate :

 80. Anthers connivent; fruit a berry LIII. **Solanaceae** (*Solanum*)

 80. Anthers not connivent; fruit a capsule LIV. **Scrophulariaceae** (*Verbascum*)

36. Petals absent :

 81. Flowers 2-sexual :

  82. Trees or large shrubs :

   83. Stamens 5-6; ovary superior; fruit a samara LXVII. **Ulmaceae**

   83. Stamens usually 15; ovary semi-inferior; fruit a woody capsule XXXI. **Hamameliadaceae**

  82. Herbs :

   84. Leaves opposite XIV. **Caryophyllaceae**

   84. Leaves alternate or radical :

    85. Leaves stipulate :

     86. Ovary included in the calyx tube; style basal or ventral XXVI. **Rosaceae** (*Alchemilla*)

     86. Ovary free; styles 2-4, terminal LXIII. **Polygonaceae**

    85. Leaves exstipulate :

     87. Flowers solitary; ovary 2-celled, adnate to the calyx in lower part; leaves imbricated XXVII. **Saxifragaceae** (*Saxifraga microphylla*)

     87. Flowers clustered; ovary 1-celled, free; leaves not imbricated LXI. **Chenopodiaceae**

 81. Flowers 1-sexual, rarely (*Parietaria*) polygamous :

  88. Leaves palmatipartite LXVIII. **Cannabaceae**

  88. Leaves simple or pinnate :

   89. Leaves pinnate LXX. **Juglandaceae**

   89. Leaves simple :

    90. Plants with stinging hairs; leaves opposite LXIX. **Urticaceae** (*Urtica*)

    90. Plants without stinging hairs; leaves alternate or radical :

     91. Leaves stipulate :

      92. Herbs; flowers not in catkins LXIII. **Polygonaceae** (*Rumex acetosa*)

      92. Shrubs (sometimes small) or trees; at least the male flowers in catkins :

93. Usually dioecious; perianth O — LXXIII. **Salicaceae**
93. Monoecious; perianth present in male or female flowers :
    94. Male flowers 3 to each bract, without perianth; female perianth O — LXXI. **Betulaceae**
    94. Male flowers 1 to each bract, without perianth; female perianth adnate to the ovary — LXXII. **Corylaceae**

91. Leaves exstipulate :
    95. Sap milky flowers in a cyathium margined with glands; male flowers each represented by a jointed stamen, the female solitary and represented by a 3-celled ovary — LXVI. **Euphorbiaceae**
    95. Sap not milky; flowers not in a cyathium; ovary 1-celled :
        96. Female perianth tubular; fruit an achene — LXIX. **Urticaceae** *(Parietaria)*
        96. Perianth free; fruit utricular — LX. **Amaranthaceae**

4. Leaves usually parallel veined; flowers typically 3-merous or the perianth members in multiples of 3; embryo with 1 cotyledon (MONOCOTYLEDONES) :
    97. Plants consisting of a discoid, free-floating thallus, not rooted in mud — LXXXIII. **Lemnaceae**
    97. Aquatic or terrestrial plants but not as described above :
        98. Perianth O or of scales or bristles :
            99. Inflorescence a spadix, lateral on a leaf-like stem or terminal and more or less enclosed by a conspicuous spathe — LXXXII. **Araceae**
            99. Inflorescence not a spadix, flowers in variously arranged spikelets and in the axils of glumaceous bracts :
                100. Stems usually triquetrous and nearly always solid; leaf-sheaths usually closed; flowers in the axil of a single bract — LXXXV. **Cyperaceae**
                100. Stems usually terete with hollow internodes; leaf-sheaths usually open; flowers enclosed by a bract above and below — LXXXVI. **Poaceae**
        98. Perianth present; not of scales or bristles :
            101. Gynoecium of free carpels arranged in a single whorl — LXXXIV. **Alismataceae**
            101. Gynoecium syncarpous :
                102. Ovary superior :
                    103. Perianth dry, membranous or scarious — LXXXI. **Juncaceae**

104. Flowers in a scapose umbel subtended by more or less membranous, spathaceous bracts LXXVI. **Amaryllidaceae**

104. Flowers not in umbel; if umbellate then bracts not spathaceous :

 105. Leaves disposed in a solitary, 3-membered, terminal whorl LXXX. **Trilliaceae**

 105. Leaves (scaly in *Asparagus*) radical, alternate, opposite or in whorls but never disposed as above :

  106. Leaves reduced to minute scales bearing axillary tufts of cladodes LXXIX. **Liliaceae** *(Asparagus)*

  106. Leaves normal, not reduced to scales :

   107. Herbs; leaves with parallel nerves and veins; flowers 2-sexual LXXIX. **Liliaceae**

   107. Undershrubs; leaves with reticulate venation; flowers 1-sexual LXXVIII. **Smilacaceae**

102. Ovary inferior :

 108. Twining herbaceous vines; flowers 1-sexual LXXVII. **Dioscoreaceae**

 108. Erect herbs; flowers 2-sexual :

  109. Flowers zygomorphic; stamens 2 or 1, confluent with the style into a column; ovary 1-locular LXXIV. **Orchidaceac**

  109. Flowers actinomorphic; stamens 3; not as described above; ovary 3-locular LXXV. **Iridaceae**

DESCRIPTIVE TEXT

I. PINACEAE

1. Cones pendulous, falling off entire :
   2. Leaves 10-20 cm long, in bundles of 5, triquetrous — **1. Pinus**
   2. Leaves 2.5-5 cm long, single, 4-sided — **2. Picea**
1. Cones erect, breaking up on ripening to leave behind a persistent woody axis:
   3. Leaves flat, more or less distichous — **3. Abies**
   3. Leaves triquetrous, fascicled on dwarf shoots, single and spirally arranged on long shoots — **4. Cedrus**

1. PINUS Linn.

P. wallichiana A.B. Jackson, Kew Bull. 1938 : 85, 1938; Raizada, Indian For. 84(8) : 503, 1958. *P. excelsa* Wall. ex. D. Don in Lamb., Descr. Gen. Pinus, ed. 2 : 5, t. 3, 1824 (non Lamk.); Hk. f., FBI. 5 : 651, 1888. *P. griffithii* Mc Clelland in Griff. Notul. 4 : 17, 1854; For. Fl. Sgr. 230.

Tall, evergreen tree; bark smooth on young trees, greyish-brown, corky and shallowly fissured into irregular plates on old individuals; branches whorled. Leaves slender, bluish-, or greyish-green, minutely serrulate, drooping on old shoots; basal sheath deciduous. Male catkins ovoid, 0.8-1 cm long. Cones solitary or 2-3 together, erect when young, ripening pendulous, cylindrical, 15-30 cm long; scales spathulate, 5-6 cm long, with slightly thickened apex. Seeds *c.* 8 mm long, with obliquely truncate, about thrice as long wing.

Occ : A major component of the coniferous forest cover of the area; more abundant up to *c.* 3000 m and on certain slopes forming almost pure crops.

Fls. April-June. Cones ripening in autumn of next year. Aru (2400 m) *B.M. Sharma* 14241; Aru (2450 m) *P.S. Jamwal* 16385

Next to *Cedrus deodara* this is the most important coniferous timber tree of Kashmir.

2. **PICEA** A. Dietr.

P. smithiana (Wall.) Boiss., Fl. Or. 5 : 700, 1884; Parker, For. Fl. ed. 2 : 544, 1924. *Pinus smithiana* Wall., Pl. As. Rar. 3 : 24, t. 246, 1832. *Picea morinda* Link, Linnaea 15 : 522, 1841; Hk. f., l.c. 653.

Tall, evergreen, pyramidal tree; bark brownish-grey, shallowly furrowed into small plates; branches whorled, horizontal, with pendulous, tassel-like branchlets. Leaves dark green, stiff, acicular, forward directed and scattered all round the branchlets, pungent. Male catkins solitary, erect, 2.5 cm long, sessile in the axils of upper leaves. Cones cylindric, dark brown when ripe, 10-15 cm long, solitary and pendulous from the ends of the branches; scales rounded at the tip, broadly obovate from a cuneate, slightly thickened base. Seeds 5 mm long, with a broad, obliquely spathulate, *c.* 1.2 cm long wing.

Occ : Associated with *Pinus wallichiana* but more frequently with *Abies pindrow* up to *c.* 3000 m; its long, pendulous, dark green branchlets impart it a very distinctive appearance.

Fls. April. Cones ripening in Oct-Nov.

Aru (2350 m) *B.M. Sharma* 14363

The wood is used for planking and packing cases.

3. **ABIES** Miller

A. pindrow Royle, Ill. Bot. Himal. t. 86, 1836, p. 350, 1839; Spach, Hist. Phan. 11 : 423, 1842; Raizada, l.c. 468. *A. webbiana* Lindl. var. *pindrow* Brandis, For. Fl. 528, 1876; Hk. f., l.c. 655.

Lofty, evergreen tree with dense, conical crown and whorled branches; bark dark-grey and deeply vertically furrowed on old stems. Leaves linear, spiral but more or less bifariously disposed, 2.5-6 cm long, above dark-green and shining, beneath pale-green with a silvery grey band running along either side of the raised midrib; tip notched. Male catkins clustered, 1.1-1.8 cm long. Cones solitary or distantly paired, erect, at maturity dark-purple, cylindric, 10-18 × 5-7 cm blunt; scales thin, obovate with rounded edge. Seeds *c.* 1 cm long, with longer, rounded broad wing.

Occ : Forms dominant forests at 2400-3100 m, occurring pure or, at places, mixed with *Picea smithiana* and other conifers.

Fls. April. Cones ripening in Sept.-Oct.

Liddarwat (2800 m) *B.M. Sharma* 14242

Uses similar to those of the previous species.

4. **CEDRUS** Trew.

C. deodara (Roxb.) G. Don in Loud., Hort. Brit. 388, 1830; Parker, l.c. 547. *C. libani* Barrel. var. *deodara* Hk.f., Himal. Jour. 1 : 257, 1854 & FBI. 5 : 653, 1888.

Large, evergreen tree; bark dark, on old stems rough, thick and vertically fissured; branches not whorled, the leading shoot and branchlets pendent. Leaves acicular, 2.5-5 cm long, dark green with age. Male and female flowers on the same or on separate trees. Male catkins erect, solitary at the ends of branchlets, at maturity cylindric and *c.* 4 cm long. Mature cones dark brown, ovoid or ovoid oblong, blunt, 10-12×7-10 cm, solitary at the ends of branchlets; scales densely imbricate, thickened towards the base. Seeds broadly winged; wing triangular, longer than the seed.

Occ : Sporadic; associated with *Pinus wallichiana* in and around Pehalgam and coexisting also with *Picea smithiana* elsewhere up to *c.* 2700 m.

Fls. Sept.-Oct. Cones ripening in Oct.-Nov. of the following year.

Slopes near Mundlan (2350 m) *B.M. Sharma* 16687; Pehalgam (2200 m) *P.S. Jamwal* 16397.

Most important of the coniferous timber trees of Kashmir, its wood being extensively used for building.

II. Cupressaceae

1. **JUNIPERUS** Linn.

1. **J. squamata** Buch.-Ham. ex Lamb., Descr. Gen. Pinus 2 : 17, 1824; Rehder, Man. Cult. Tr. & Shrubs ed. 2 : 63, 1940. *J. recurva* D. Don, Prodr. Fl. Nep. 55, 1825 var. *squamata* Parlat. in DC., Prodr. 16 : 482, 1868; Hk.f., FBI. 5 : 647.

Bushy, evergreen, aromatic shrubs with long, decumbent stems giving off numerous, erect, short branches. Leaves in whorls of threes, decurrent, overlapping and pointing forwards, linear or lanceolate, somewhat curved, 3-5 mm long, pungent, above concave often glaucous and usually with a faint midrib, beneath green, convex and usually furrowed. Male catkins and berries terminal or terminating short lateral branches. Fruit ovoid, 0.8-1 cm long, blackish-purple and shining when ripe; seed solitary.

Occ. : Very common on open, steep hillsides at alts. 3100-3700 m; often occurring in large patches and forming almost impenetrable thickets.

Fls. June-July. Berries ripen Aug.-Oct. of the following year.

Saswarian (3250 m) *B.M. Sharma* 14175; Hamwas (3250 m) *P.S. Jamwal* 16362.

Used chiefly as fuel.

III. Taxaceae

1. **TAXUS** Linn.

1. **T. wallichiana** Zucc. in Abhand. Bayer. Acad. Classe Math. Phys. 3 : 805, t. 5, 1843; Kitam., Fl. Afgh. 20, 1960. *T. baccata* Linn. ssp. *wallichiana* (Zucc.) Pilger in Engl. Das

Pflanzenreich 4-5, Ht. 18 : 112, 1903. *T. baccata* Hk f., l.c. 648.

Dioecious, small or medium, evergreen tree with horizontal branches; bark thin, smooth, reddish-grey. Leaves flattened, distichous, linear, 2-4 cm long, dark-green and shining above, pale or rusty beneath, acute, narrowed into a short, decurrent petiole; margins recurved. Male flowers in solitary, axillary, subglobose catkins each consisting of an axis terminating into a cluster of stamens; female flowers solitary, axillary, each of a single erect ovule surrounded at base by a cupular disc at the apex of a scaly peduncle; disc enlarging in fruit into a red, fleshy cup surrounding the olive-green seed.

Occ : Isolated individuals are found in *Pinus wallichiana* stands and also in association with *Abies pindrow* upto *c*. 2900 m.

Fls. April-May.; *Frt*. ripens Oct.-Nov. Liddarwat (2800 m) *B.M Sharma* 14319; near Aru (2350 m) *P.S. Jamwal* 16383.

IV. Helleboraceae

1. Flowers ac inomorphic :
   2. Each petal drawn into a usually backwardly directed or bent spur — **3. Aquilegia**
   2. Petals, if present, not spurred :
      3. Ovary solitary; fruit a berry — **4. Actaea**
      3. Ovaries several, free, distinct :
         4. Leaves undivided; petals absent — **1. Caltha**
         4. Leaves palmatipartite; petals small — **2. Trollius**
1. Flowers zygomorphic :
   5. Upper sepal produced at the base into a spur — **5. Delphinium**
   5. Upper sepal vaulted into a helmet or hood, not produced at the base — **6. Aconitum**

1. CALTHA Linn.

C. palustris Linn., Sp. Pl. 558, 1753 var. **alba** (Jacq. ex Camb.) Hk.f. et Th. in Hk.f., FBI. 1 : 21, 1872. *C. alba* Jacq. ex Camb. in Jacq., Voy. Bot. Inde. 6 : 4, 1844. **Fig. 1**

Glabrous, succulent, perennial herb from a stout rootstock bearing thick, fibrous roots; stems fistular, erect or ascending, 10-60 cm tall. Leaves reniform to almost deltoid, 3-15 cm wide, glossy, crenate or dentate, the radical and lower cauline petiolate, upper small and nearly sessile. Flowers apetalous, white, 2-4 cm across; sepals 5 or 6. Follicles flat, beaked with persistent styles.

Occ : Abundant and gregarious in marshy habitats; also in shallow, slow running water throughout in the valley from Pehalgam to Kolahoi-n-kot.

Fls. April-July.; *Frts*. July.-Aug.

Liddarwat (2750 m) *B.M. Sharma* 19358; Liddarwat (10,000′) *K.M. Vaid* DD. no. 117538; Sindthan (Liddarwat) *K.S. Ahluwalia* RL. Herb. nos. 250 & 251; Aru (2400 m) *P.S. Jamwal* 15645.

2. TROLLIUS Linn.

1. **T. acaulis** Lindl., Bot. Reg. 1842, Misc. 56, 1842; Hk.f. et Th., l.c. 22. **Fig. 2**

Glabrous, up to 30 cm high herbaceous perennial, clothed at the base with brown, fibrous remnants of old leaf-bases. Leaves orbicular-reniform, 3-8 cm wide, palmati-5-partite into incised and toothed, cuneate segments, radical and the lower cauline long-petioled, the upper sessile or nearly so. Flowers solitary, 3.5-5.5 cm across, with calycine and corolline whorls both golden yellow; sepals 5-8, broadly oval, spreading; petals oblong, 12-16, shorter than the numerous.

Occ. : In damp alpine pasture lands; also on sides of stony nallahs. Frequent.

Fls. June-July.

Nangneri nallah (2850 m) *B.M. Sharma* 14112; Zaji marg, *Falconer* 1/1873; Kolahoi, Dudsar (11000′) *R.L. Badhwar* 3729; Nangneri nallah (2850 m) *P. S. Jamwal* 15646.

3. AQUILEGIA Linn.

1. Spurs strongly hooked or incurved; flowers bicolorous or concolorous, not violet :
   2. Spur short-projecting above the calyx lobes; sepals purple : **1. A. pubiflora**
   2. Spur long-projecting above the calyx lobes; sepals not purple :
      3. Sepals blunt, deep sky-blue; petals white **2. A. moorcroftiana**
      3. Sepals and petals both white; the former acute or drawn into long sharp point **3. A fra rans**
1. Spurs straight or only bent, never hooked or inourved; flowers deep violet **4. A. nivalis**

1. **A. pubiflora** Wall. ex Royle, Ill. Bot. Himal. 55, 1834; Kitam., Fl Afgh. 121, 1960. *A. vulgaris* ssp. *pubiflora* Hk.f. et Th., l.c. 24.

Thinly hairy, 24-60 cm high, perennial herb with leafy, usually branched, erect flowering stems. Leaves glaucous, 2-ternate, progressively short-stalked upwards, with deeply 3-lobed, coarsely crenate leaflets from a cuneate or almost truncate or cordate base. Flowers nodding, pale to deep-purple, 2.5-3 cm across, solitary or in a loose panicle. Sepals caducous, elliptic-ovate, 2.5-2.8 × 0.9-1.2 cm; petal limb longer than the stamens, 1.5 × 1.1 cm; spur conical, with its hooked base shortly projecting between the calyx lobes. Follicles 6, covered with mixed glandular and simple hairs, tipped with long, subulate, recurved styles.

Occ. : Usually on humus rich soil; common among thickets and on roadside slopes in Aru-Pehalgam Chandanwari track.

Fls. May-June

Aru (2300 m) *B.M. Sharma* 14335; Mundlan-Aru route (2300 m) *P.S. Jamwal* 15660.

2. **A. moorcroftiana** Wall. ex Royle, Ill. Bot. Himal. 55, t. 21, 1834; Stewart, Annot. Cat. Vasc. Pl. W. Pak. & Kash. 262, 1972. *A. vulgaris* ssp. *viscosa* Hk.f. et Th., l.c. 24 (non Gouan). **Fig. 3**

Perennial, 50-80 cm tall, with general habit of the preceding species; root horizontal or inclined; petiole, undersurface of leaflets and at least the upper part of stem hairy. Flowers *c.* 6 cm across, nodding; sepals deep sky-blue, 3.6 × 1.5-1.7 cm, ovate, blunt; petals white with 1.7 × 1.7 cm limb with its strongly hooked, knob-tipped spur long projecting between the calyx lobes. Follicles 6-8, hairy, marked with strong transverse veins.

Occ. : Abundant and almost gregarious in shaded places along the slope below Dudhsar; also on humus laden rocks along the torrent near Nangneri and Saswarian nallahs.

Fls. July.-Aug.; *Frts.* Aug -Sept.

Nangneri (2900 m) *B.M. Sharma* 14113; Amarnath, *K.S. Ahluwalia* RL. Herb. no. 331; Below Dudhsar (3600 m) *P.S. Jamwal* 15674.

3. A. **fragrans** Benth. in Maund Bot. 4 : t. 181, f. 21, 1840; Mukerjee, Bull. bot. Surv. India, 2 (3-4) : 296, 1960. *A. vulgaris* ssp. *pyrenaica* Hk.f. et Th., l.c. 24.

Glaucous perennial, much resembling the preceding but with white, fragrant flowers, acute or sharply long-pointed sepals and the stamens protruding beyond the petal limb; spurs more slender.

Occ. : On forested slopes abuting on the right bank of Tarsar torrent near Liddarwat; not common.

Fls. July.

Liddarwat (3000 m) *B. M. Sharma* 14267

4. A. **nivalis** Falc. ex Baker, Gdnrs. Chr. 2(10) : 76, 1878; Mukerjee, l.c. 296. *A. vulgaris* ssp. *jucunda* Hk.f. et Th., l.c. 24.

Perennial herb, hardly attaining 30 cm height; flowering stems 1-2-leaved, solitary or few from the root. Basal leaves long-petioled, 2-ternate; leaflets deeply 3-lobed; lobes coarsely crenate. Flowers intense violet *c.* 4.5 cm across, nodding; sepals ovate, 2.5 × 1.5 cm, blunt; petal limb about equalling or slightly longer than stamens; spur conic-obtuse, shortly projecting above the calyx lobes. Follicles mostly 5-8, hairy, strongly transverse veined.

Occ. : In shaded nooks on the slope below Dudhsar and in similar habitats about Satlanjan and Nangneri.

Fls. July.-Aug.

Nangneri (2800 m) *B. M. Sharma* 14126; Dudhsar (4000 m) *B.M. Sharma* 14643. Kainmul, Lidder Vy. (10-11000′) *Duthie* s.n. DD; near Satlanjan (3080 m) *P.S. Jamwal* 15661.

4. ACTAEA Linn.

1. A. **spicata** Linn., Sp. Pl. 504, 1753; Hk.f. et Th., l.c. 29. **Fig. 4**

Erect, 60-90 cm high, bushy perennial from a stout, tuberous rootstock; stems usually branched. Leaves alternate, 2-pinnately compound, 20-40 cm long, smaller and sessile upwards; leaflets ovate-lanceolate, 1.5-9 × 1.7 cm, sharply dentate and often lobed, pubescent beneath. Flowers white, 6 mm across, in short, dense, terminal racemes; sepals and petals 4 each, similar but the former longer and caducous. Berry ovoid, 0.8-1.2 cm long, black when ripe.

Occ. : Common and often associated with *Aconitum laeve* in the limestone tract beyond Satlanjan; also on humus laden ground under firs at Nangneri.

Fls. June-July.; *Frts.* Aug.-Sept.

Kolahoi (3350 m) *B.M. Sharma* 14044; Nangneri slope (2800 m) *B.M. Sharma* 14133.

5. DELPHINIUM Linn.

1. Flowers in elongate racemes; sepals below 1.5 cm in length; spur cylindric ... 1. D. roylei
2. Flowers in lax corymbs; sepals 2.5 cm and above in length; spur conical, inflated 2. D. cashmirianum

1. **D. roylei** Munz, J. Arnold Arbor., 48 : 292, 1967; Stewart, l.c 276; Nair, Fl. Bash. Himal. 12, 1977. *D. incanum* Royle, Ill. Bot. Himal. 55, 1835; Hk.f. et Th., l c. 25. **Fig. 5**

Perennial, 0.5-1.4 m tall, hoary herb. Radical leaves withered at flowering time, like the lower cauline long-stalked, orbicular and 3-5-partite into pinnately segmented lobes, 5-7 cm in diam., upper cauline nearly sessile. Flowers bright blue, about 3 cm long including their nearly straight spur. Follicles 3, pubescent.

Occ. : Common near streams and on humus laden scrubby slopes about Aru and similar habitats along Pehalgam-Chandanwari route.

Fls. July.-Aug.; *Frts.* Aug.

Aru (2350 m) *B.M. Sharma* 14092 and 19361; Liddar Vy. below Tanin (8-9000′) *Duthie* DD. no. 13440; Pehalgam (7000′) *K.M. Vaid* 7140.

2. **D. cashmirianum** Royle, Ill. Bot. Himal. 55, 1834; Hk.f. et Th., l.c , 26.

Erect, 15-40 cm tall, perennial herb with softly strigose to near glabrous, sparingly leafy stem. Radical and lower cauline leaves long-stalked, 5-10 cm across, palmately divided, the former orbicular and 5-7 lobed, the latter reniform and 3-5 lobed; lobes cuneate, 3-segmented and towards apex acutely cut. Flowers bluish purple, 3-4.5 cm long including their broad, obtuse, decurved spur; sepals hairy. Follicles 3-7, hairy.

Occ. : Very common on open alpine slopes throughout from Chandanwari to Sangam and Santsingh Pari.

Fls. Aug.-Sept.

Pisu top (3400 m) *B.M. Sharma* 14625; Chandanwari to Wabjan, *L.D. Kapoor* RL. Herb. nos. 386 & 387; Wabjan (3800 m) *P.S. Jamwal* 15649.

6. ACONITUM Linn.

1. Helmet produced dorsally into a high, cylindrical or conical obtuse peak ... 1. A. laeve
1. Helmet rounded, navicular or slightly depressed, not produced into a peak :
   2. Root elongate, cylindric, ultimately breaking up into separate or anastomosing strands; carpels 3 ... 2. A. moschatum
   2. Root tuberous, paired; carpels 5 :
      3. Leaves on middle portion of stem sessile, undivided, carpels crispo-pubescent with adpressed hairs ... 3. A. heterophyllum
      3. All leaves except the uppermost distinctly to long-stalked, deeply dissected; carpels densely tomentose ... 4. A. violaceum

1. **A. laeve** Royle, Ill. Bot. Himal. 45 et 56, 1834; Stapf, Ann. R. bot. Gdn Calcutta, 10(2) : 136, pl. 92, 1905. *A. lycoctonum* Hk.f. et Th., Fl. Ind. 55, 1855 et in Hk.f., FBI. 1 : 28, 1872. **Fig. 6**

Perennial, up to 1.8 m tall, stout herb, branched and viscidly hairy upwards at least; root cylindric, ultimately separating into anastomosing strands. Basal and lower cauline leaves glabrous, orbicular-cordate or reniform, 15-36 cm across, subpedately 5-9-parted into the outer trapezoid and inner oblong or obovate-cuneate, lobed, sharply inciso-dentate segments. Upper cauline leaves similar but gradually smaller with shorter petioles, the uppermost pubescent and nearly sessile. Flowers dull purple, 1.5-2.5 cm long, in usually leafy panicles; helmet 1-1.6 cm long, tapering suddenly from an abruptly widened base pointed anteriorily, recurved to almost straight. Follicles 3, 1-1.6 cm long, glabrous; seeds dark brown, 3 mm long, transversely rugose, shining.

Occ. : Very common in forest shade throughout the left bank of the torrent from Nangneri to Satlanjan, also in association with *Actaea spicata* in open ground beyond Satlanjan; frequent on slopes near Zojipal and in Sangam valley.

Fls. July.-Sept.; *Frts.* Aug.-Sept.

Nangneri slope (2850 m) *B.M. Sharma* 14043; Zojipal (3550 m) *B. M. Sharma* 14626; Satlanjan (2950 m) *P.S. Jamwal* 15654; Sangam Valley (3500 m) *P.S. Jamwal* 15652.

The root is suspected to be poisonous but it has not been chemically evaluated so far.

2. **A. moschatum** Stapf in Ann. R. bot. Gdn Calcutta, 10(2): 139, pl. 94, 1905. *A. ferox* Wall. ssp. *moschatum* P. Bruhl in Ann. R. bot. Gdn Calcutta, 5(2) : 109, pl. 111, figs. A & 20, 26, 1896.

Erect, unbranched, perennial herb, glabrous below, upwards covered with soft, spreading hairs. Leaves hairy, orbicular-cordate or subreniform, cauline similar to basal but with much shorter stalks, up to 9 cm across, 5-parted into outer round-trapezoid and inner broadly obovate-cuneate, lobed, inciso-crenate segments; crenatures spiculate. Flowers lurid purple, in a lax, narrow, more or less viscid-hairy, up to 40 cm long raceme; helmet navicular, 1.4-1 6 cm high, longer than the lateral sepals.

Occ. : A denizen of rock crevices; rare.

Fls. Sept.

Tulian Sar (12000′) *W.J. Lambert* DD nos. 44703 and 44704; cited by Stewart (l.c. 258) From Sonsal nallah Liddar Valley, Astanmarg and Sheshnag.

3. **A. heterophyllum** Wall. ex Royle, Ill. Bot. Himal. 56, t. 13, 1834; Hk.f. et Th. in Hk.f., l.c. 29. **Fig. 7**

Erect, mostly simple, usually 15-75 cm tall, leafy herb from paired, tuberous root; mother tuber shrivelled; stem glabrous, upwards crispo-pubescent. Leaves heteromorphic, lower long stalked, orbicular-, or broad ovate-cordate, usually 5-lobed, 3-8 cm across with blunt or shortly pointed, rounded teeth, upper ovate or ovate-lanceolate from a cordate, sessile base, up to 13×9 cm, crenate or inciso-crenate, the uppermost small, lanceolate amplexicaul, sharply toothed. Flowers 2.5-3 cm long, dull green-blue or greyish green-purple with deeper purple veins, in many-flowered racemes; helmet rounded at top, pointed in front. Follicles 1.5-2 cm long, downy; seeds 3-4 mm long, blackish-brown.

Occ. : Common in the forest and on open slopes about Saswarian, Kolahoi and Zojipal, Wabjan and near Sangam.

Fls. July.–Sept.

Dudhsar (3600 m) *B. M. Sharma* 14042; Liddarwat (9-10,000′) *Duthie* 13536; near Mahagunus (13,000′) *K.M. Vaid* DD no. 117526; Saswarian nallah (3300 m) *P.S. Jamwal* 15659.

The roots contain atisine- an amorphous, non toxic alkaloid-and are eaten as a tonic to remove general debility. They are also reported to be used as febrifuge and in cases of diarrhoea, dysentery and cough.

4. A. violaceum Jacq. ex Stapf in Ann. R. bot. Gdn Calcutta, 10(2) : 144, pl. 97 1905 incl. var. **robustum** Stapf, l.c. 146; *A. napellus* Hk.f. et Th., l.c. 28 p.p. (non Linn.)

Slender or robust, 12-50 cm high, erect or ascending, leafy herb with narrowed hypogeous base from a tuberous, paired root; mother tuber shrivelled. Stem glabrous, fulvo-tomentose upwards. Leaves orbicular-cordate or reniform, 3-9 cm across, 5-palmati partite into laciniated lobes; laciniae acute or subobtuse. Flowers violet or deep-, to sky-blue, few or in many flowered, dense, 6-20 cm long, stiff raceme; helmet navicular or rounded at the top, 1.5-2 cm high. Follicles truncate, erect, 1-1.2 cm long, hairy.

Occ. : Mostly in damp places running water on open alpine slopes.

Fls. Aug.-Sept.; *Frts.* Oct.

Amarnath (3900 m) *B.M. Sharma* 14659; Dudhsar (3600 m) *B.M. Sharma* 14662; Mahagunus Pass (12-14,700′) on way to Amarnath, *K.M. Vaid* DD no. 117525; Panchtarni (12000′) *K.M. Vaid* DD. no. 117524; Thuliyan Sar (12000′) *Lambert* DD no. 44705; Panchtarni, *K.S. Ahluwalia* R.L. Herb. nos. 457, 459 & 5010; Dudhsar (3600 m). *P.S. Jamwal* 19275; near Amarnath, *S.L. Nayar* RL, Herb. no. 450.

V. RANUNCULACEAE

1. Leaves opposite; climbing shrubs ... **1. Clematis**
1. Leaves alternate or radical; herbs :
    2. Perianth of one whorl of green or petaloid segments :
        3. Flowers moderately large, subtended by an involucre of bracts; perianth segments longer than stamens ... **2. Anemone**
        3. Flowers small, without an involucre; perianth segments shorter than stamens ... **3. Thalictrum**
    2. Perianth of two dissimilar whorls, the inner comprising petals :
        4. Sepals petaloid :
            5. Petals with nectaries at base ... **4. Callianthemum**
            5. Petals without nectaries ... **5. Adonis**
        4. Sepals herbaceous :
            6. Achenes with an empty cell on either side; beak 2-3 times as long as achenes ... **6. Ceratocephalus**
            6. Achenes without any empty cells; beak less than twice as long as the achenes ... **7. Ranunculus**

1. CLEMATIS Linn.

1. Flowers less than 3 cm across, in panicles :
   2. Sepals erect, with recurved tips; filaments hairy — 1. **C. connata**
   2. Sepals spreading from the base; filaments glabrous — 2. **C. grata**
1. Flowers 5-10 cm across, in axillary fascicles — 3. **C. montana**

1. **C. connata** DC., Prodr. 1 : 4, 1824; Hk.f. et Th., in Hk.f., FBI. 1 : 6, 1872. **Fig. 8**

Climbing shrub with glabrous, grooved twigs; young parts hairy. Leaves pinnate, with petiole-base simply connate or united into a flat expanse; leaflets 3-7, broad-ovate 5-15 × 3-10 cm from a cordate base, at times shallowly 3-, rarely 5-lobed, coarsely serrate. Flowers campanulate, yellowish-white, in lax axillary panicles; sepals oblong; 2-2.5 cm long, felted on either side. Achenes silky, 3-4 mm long.

Occ. : Rambling on wayside scrubs between Liddarwat and Satlanjan; also roadsides near Mundlan.

Fls. Aug.-Sept.

Gagar pathri near Satlanjan (2950 m) *B.M. Sharma* 19318 A & 19318 B; Kolahoi above Liddarwat (10,000′) *Duthie* DD no. 13523; Liddarwat-Satlanjan route (2920 m) *P.S. Jamwal* 15665.

2. **C. grata** Wall., Pl. As. Rar. 1 : 83, 1830; Hk.f. et Th., l.c. 3.

More or less pubescent, climbing shrub with deeply furrowed twigs. Leaves pinnate; leaflets 5, rarely only 3, ovate or ovate-lanceolate 2.5-7.5 cm long, from cordate base, acuminate, inciso-serrate, at times deeply lobed, villous, or above glabrate. Flowers cream-coloured 1.5-2 cm across, fragrant, many in large, leafy, usually leaf-exceeding panicles; achenes pubescent.

Occ. : Very common in open places at elevations lower to that of Pehalgam.

Fls. July-Sept.

Pehalgam (2100 m) *B.M. Sharma* 14629; 2 km below Pehalgam (2100 m) *P.S. Jamwal* 15668.

3. **C. montana** Buch.- Ham. ex DC., Syst. 1 : 164, 1817; Hk.f. et Th., l.c. 2.

Woody, climbing or rambling shrub, with terete stems; easily distinguishable from preceding species of the genus by characters of key and from its 3-foliolate leaves fascicled on arrested branchlets; leaflets pubescent to glabrate, ovate or lanceolate, 2.5-8 × 1.5-4 cm, coarsely toothed. Flowers pure white; sepals 4, spreading.

Occ. : Sporadic rambler, over shrubs and open hill sides.

Fls. May-June

Chota Khalen beyond Liddarwat (2900 m) *B.M. Sharma* 14645.

2. ANEMONE Linn.

1. Achenes embedded in dense wool — 1. **A. rupicola**
1. Achenes not embedded in wool :
   2. Primary divisions of radical leaves 3; flowers solitary or up to 3 on a stem — 2. **A. obtusiloba**
   2. Primary divisions of radical leaves 5-7; flowers many, in umbels or cymes :
      3. Leaves densely hairy on either surfare, 5-10 cm in diam.; achenes with a sharp, straight beak — 3. **A. polyanthes**

3. Leaves, on the upper surface at least, glabrous, up to 25 cm in diam.; achenes with a hooked beak **4. A. tetrasepala**

1. A. rupicola Camb. in Jacq., Voy. Bot. Inde. 5, t. 2, 1844; Hk.f. et Th., l.c. 8.

Perennial herb with 10-25 cm tall, softly silky flowering scape from a long woody rootstock. Leaves long-petioled, 5-10 cm across, 3-partite into acutely toothed or sharply incised, mostly petiolulate segments; involucral leaves, sessile, 3-lobed. Flowers white or cream, 6-7 cm across, solitary or at times 2-3.

Occ. : Common and often gregarious on open alpine slopes about Sangam valley and near Sheshnag; also along the Baltal Sangam route.

Fls . June-July.; *Frts*. Aug.

Near Sheshnag (3725 m) *B.M. Sharma* 14621; Sangam Valley (3500 m) *P.S. Jamwal* 15650. **Fig. 9**

2. A. obtusiloba D. Don, Prodr. Fl. Nep. 194, 1825; Hk.f. et Th., l.c. 8.

Perennial, softly hairy, tufted herb with 12-30 cm tall flowering scapes from a woody, fibrous rootstock clothed with remnants of sheathing leaf-bases. Leaves hairy on either surface, 3-5 cm across, with primary divisions broadly cuneate, sessile, lobed and crenately toothed; involucre of sessile, 3-fid leaves. Flowers white, with a purple-, or leaden-blue blotch on under surface of each perianth lobe, 1.6-3.5 cm across, 1-3 to each stem; perianth lobes silky without. Achenes coarsely hairy.

Occ. : Commonest of our Anemones, being distributed throughout on moist slopes and open temperate and alpine pastures. Very variable to the degree of pubescence, the latter decreasing with rising altitude.

Fls. June-Aug.

Chandanwari (2750 m) *B.M. Sharma* 14636; Badzalkhod nallah, Lidder Vy. (13000′) *Duthie* DD, s.n.; Liddarwat (2750 m) *P. S. Jamwal* 19359.

var. **potentilloides** (Camb.) Lauener, Notes R. bot. Gdn Edinb. 23 : 187, 1962; Stewart, l.c. 261. *A. potentilloides* Camb., mss. *A. rupestris* Hk.f. et Th., l.c. 9.

Much resembling the previous taxon but with yellow flowers and mostly petiolulate primary divisions of the leaves. Also, it is more slender and less hairy to almost glabrous and more commonly distributed in the upper altitudinal limits of the species.

Occ. : Mostly in wet places.

Fls. June-Aug.

Tarsar, along lake (4100 m) *B.M. Sharma* 14007; Below Dudhsar (3600 m) *B.M. Sharma* 14638.

3. **A. polyanthes** D. Don, Prodr. Fl. Nep. 194, 1825; Hk.f. et Th., l.c. 9. *A. narcissifolia* Linn. var. *polyanthes* (D. Don) Finet et Gagnep. in Bull. Soc. Fr. 51 : 75, 1904; Mukerjee, Bull. bot. Surv. India, 1 : 140, 1959. **Fig 10**

Densely hairy, perennial herb, with robust, 30-60 cm tall flowering scapes from a stout, woody rootstock crowned with old leaf remains. Leaves reniform, cordate, with cuneate, 3-fid, crenate primary divisions. Flowers white, 2.5-4 cm across, umbellate, from an involucre of sessile, cuneate, 2 5-4 cm long, 3-lobed leaves; perianth lobes softly hairy without, ovaries glabrous.

Occ. : Common in open, wet slopes about Kolahoi, Saswarian, Nangneri and similar localities in the upper reaches of the eastern Liddar valley.

Fls. July-Aug.

Below Dudhsar (3600 m) *B.M. Sharma* 14041; Sangam valley obove Liddarwat (12-13,000′) *Duthie* 13557; Near Sheshnag, Liddar vy. (12-13,000′) *Duthie* 14121; Sonsal nallah, Liddar vy. 13-14000′) *Duthie* DD. no. 13368.

4. **A. tetrasepala** Royle, Ill. Bot. Himal. 53, 1834; Hk.f. et Th., l.c. 10.

Robust, perennial herb with habit and aspect of *A. polyanthes* but distinguishable by the characters of key and almost always compound, contracted or effuse umbels; involucral and leaf segments irregularly acutely toothed or inciso-serrate; petioles and young leaves woolly. Perianth lobes generally 4, glabrous without.

Occ. : Habitats and frequency same as of the preceding species with which it is often associated.

Fls. July-Aug.

Nangneri nallah (2850 m) *B.M. Sharma* 14115; Saswarian nallah (3300 m) *B.M. Sharma* 14174; Kainmul above Pehalgam (10-11000′) *Duthie* 13080; Near Zojipal (3500 m) *P.S. Jamwal* 15651.

3. THALICTRUM Linn.

1. Leaves all radical; flowers in a simple raceme 1. T. alpinum

1. Leaves alternate; flowers in panicles :

2. Achenes flattened, laterally compressed, long–, or short-stalked;

3. Achenes membranous, shorter than their stalks, ventral suture winged 2. T. elegans

3. Achenes 3-ribbed or, —nerved on either face, longer than their stalks; ventral suture wingless :

4. Flowers red; achenes 3-5 in a head 3. T. secundum

4. Flowers greenish or greenish-white; achenes many, rarely 5 or only 4 in some heads :

5. Achenes and under-surface of leaflets glabrous 4. T. cultratum

5. Achenes and under-surface of leaflets glandular-puberulous 5. T. platycarpum

2. Achenes neither flattened nor laterally compressed, sessile 6. T. minus

1. **T. alpinum** Linn., Sp. Pl. 545, 1753; Hk.f. et Th., l.c. 12. *T. microphyllum* Royle, Ill. Bot. Himal. 51, 1839. **Fig. 11**

Slender, glabrous perennial herb with 8-12 cm high, solitary stem. Leaves pinnate, rarely bipinnate; leaflets small, 3-lobed or, -partite, glaucous beneath. Flowers greenish, 3-4 mm across, nodding, in a simple raceme; stamens variable in number, mostly 6-10; anthers beaked. Achenes few, oblong, striate.

Occ. : In wet or damp places and along the sides of stones in alpine meadows.

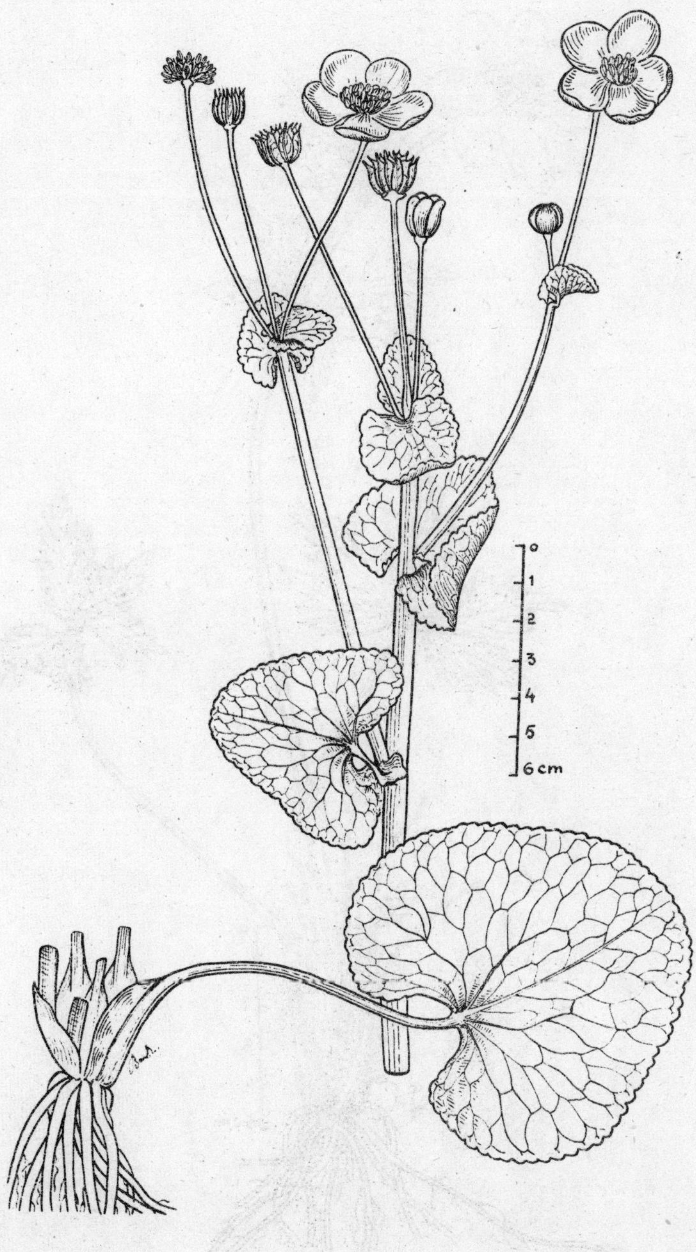

Fig. 1. *Caltha palustris* Linn. var. *alba* Hk.f. & Th.

Fig. 2. *Trollius acaulis* Lindl.

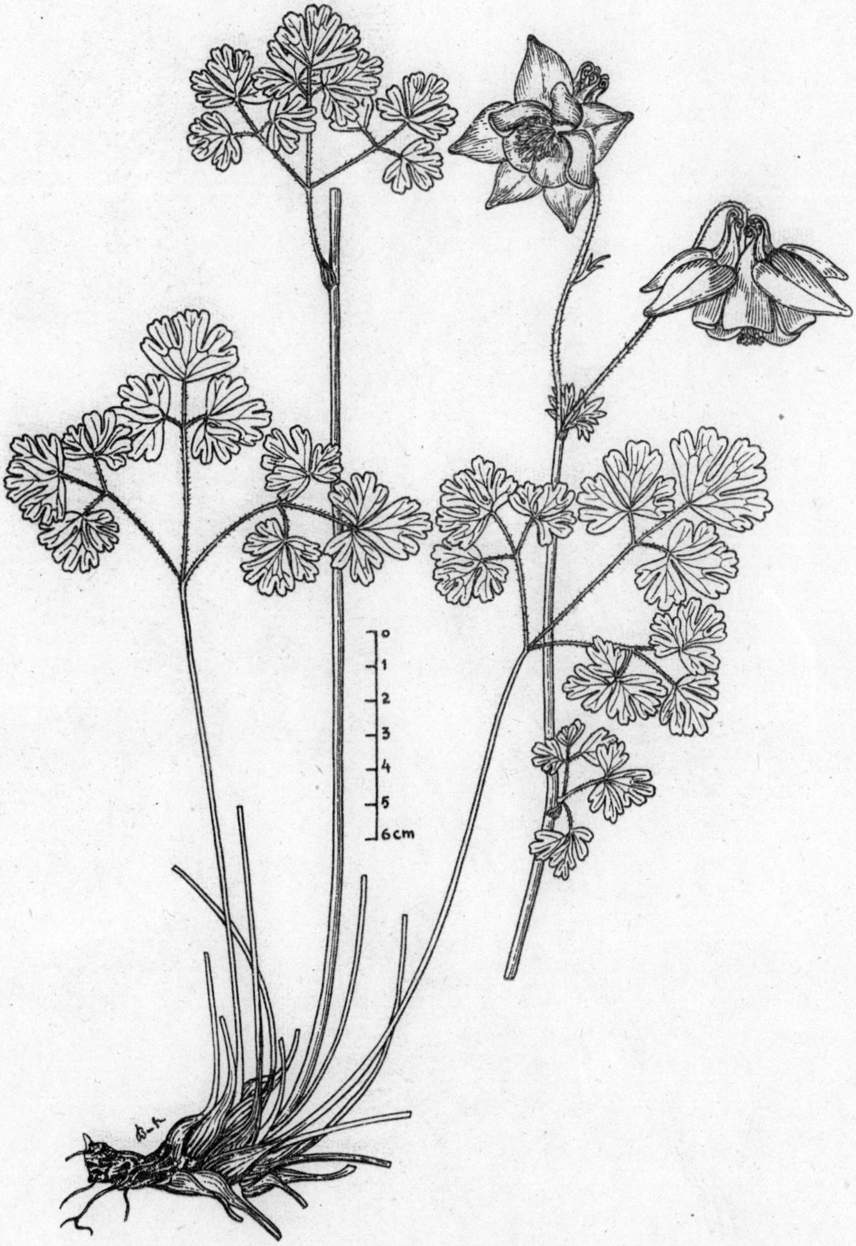

Fig. 3. *Aquilegia moorcroftiana* Wall. ex Royle

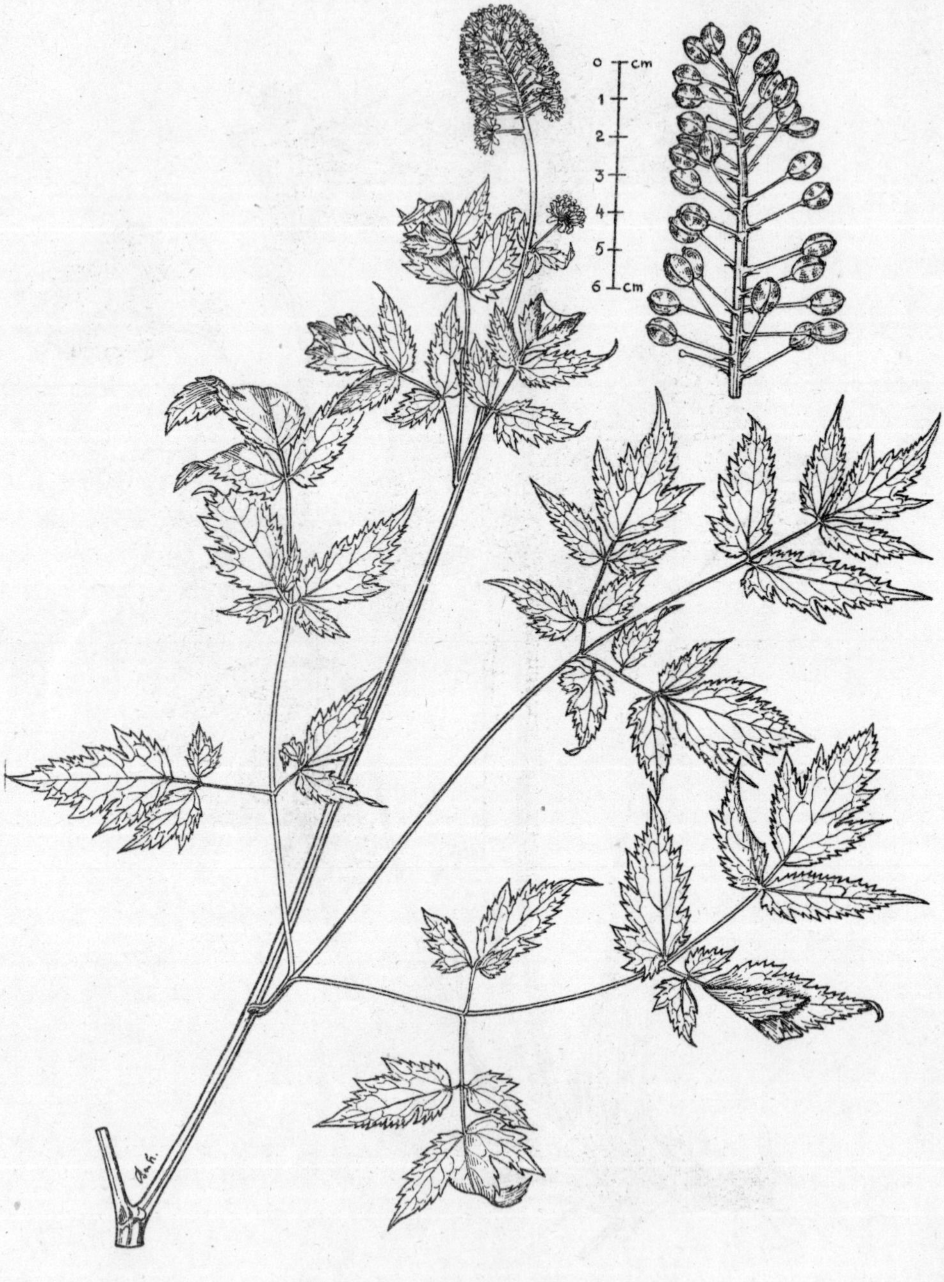

Fig. 4. *Actaea spicata* Linn.

Fig. 5. *Delphinium roylei* Munz

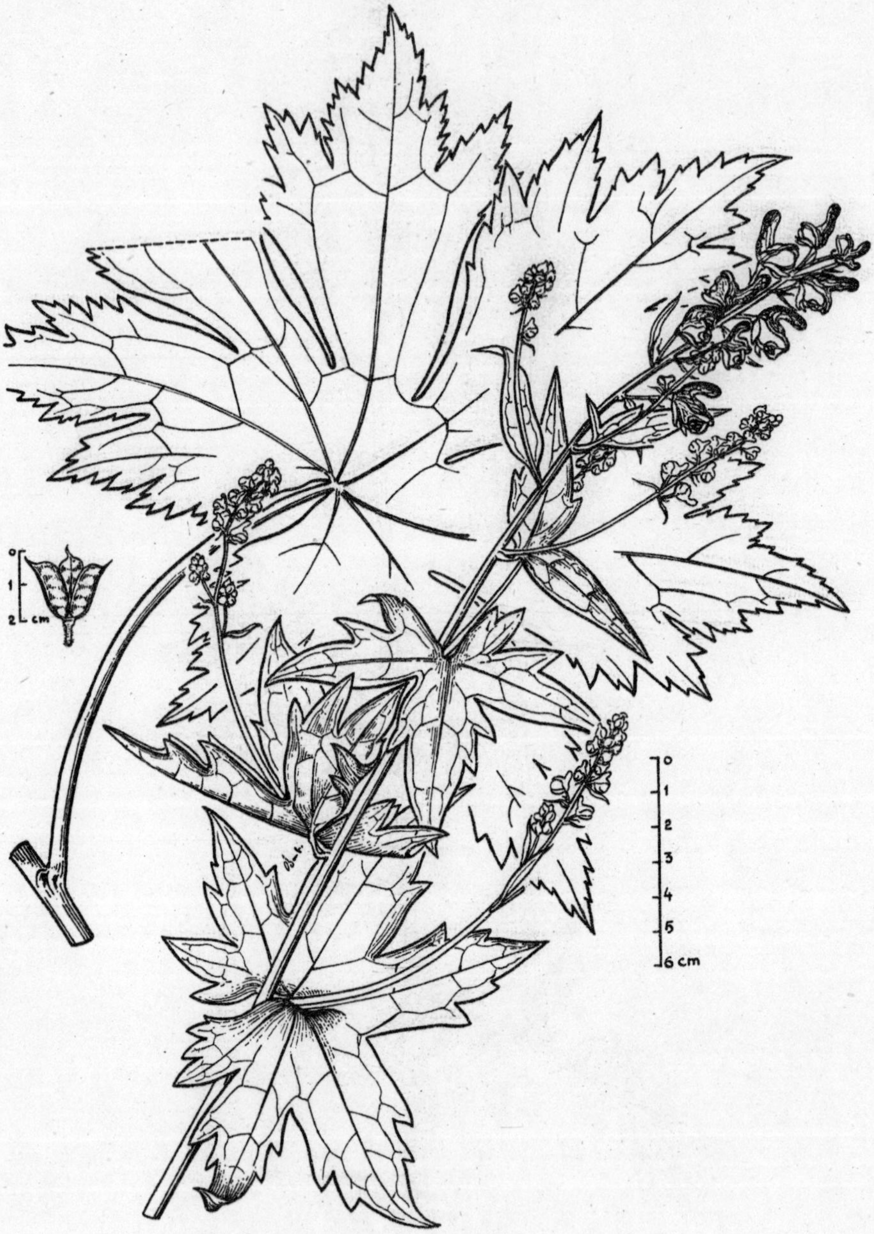

Fig. 6. *Aconitum laeve* Royle

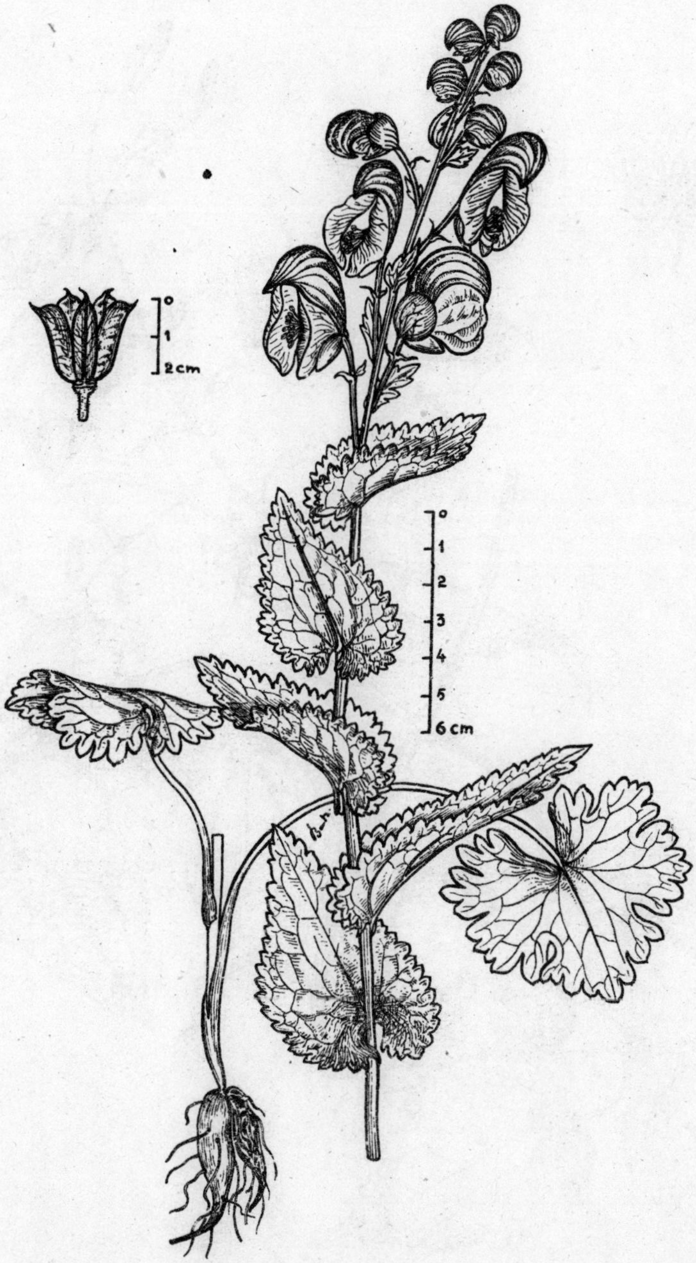

Fig. 7. *Aconitum heterophyllum* Wall. ex Royle

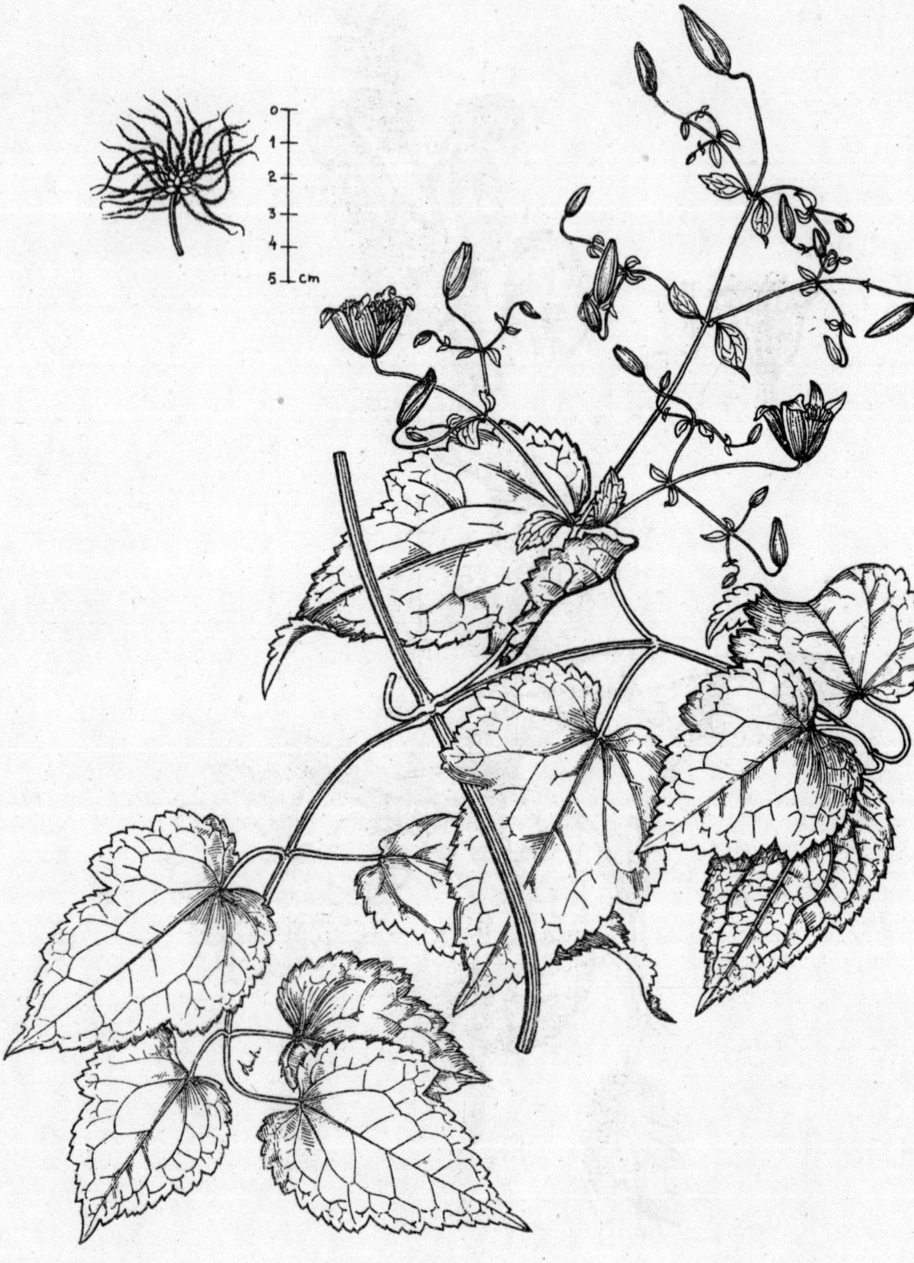

Fig. 8. *Clematis connata* DC.

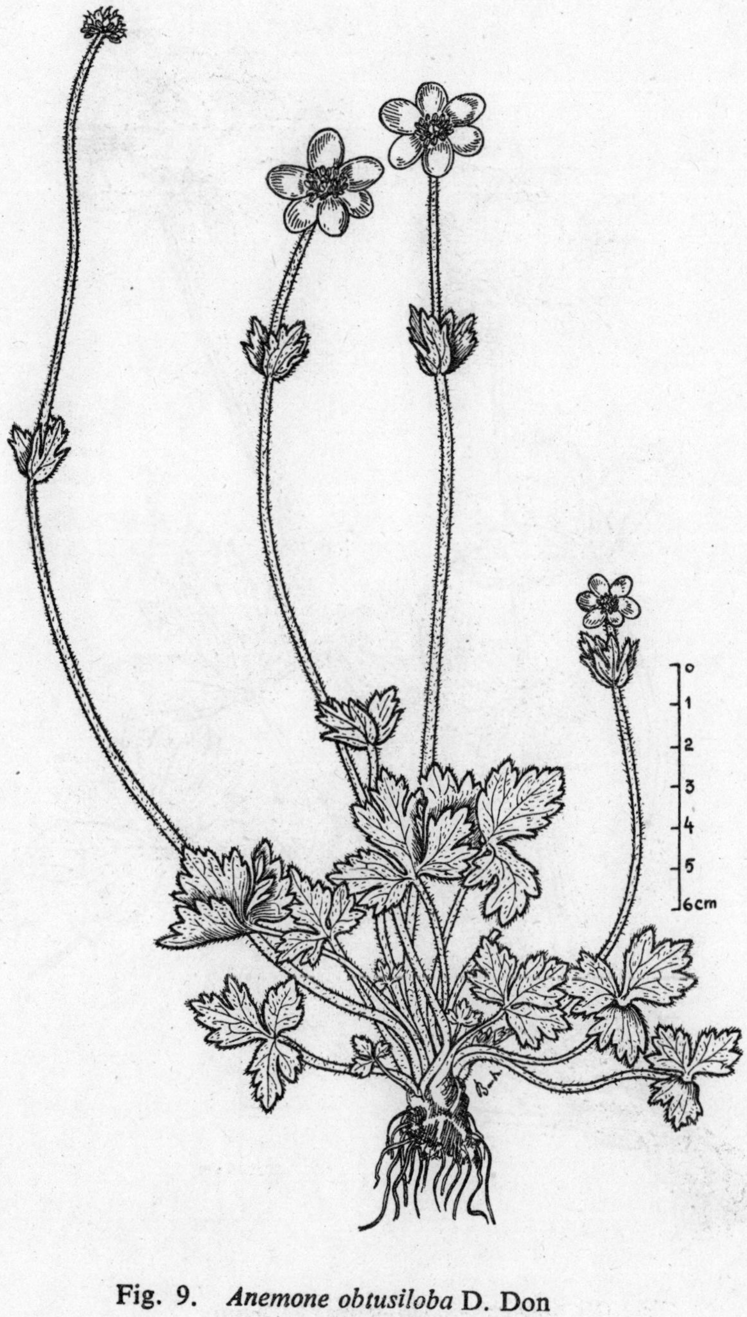

Fig. 9. *Anemone obtusiloba* D. Don

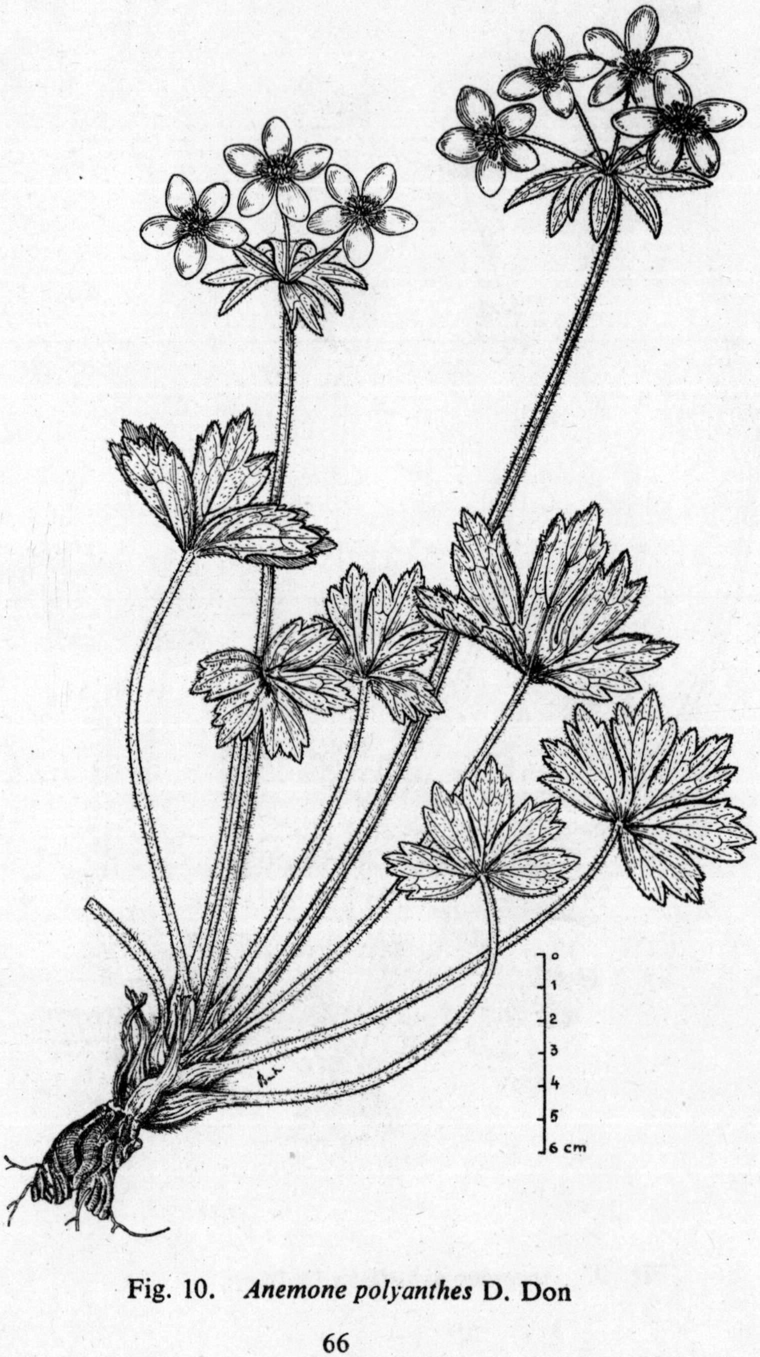

Fig. 10. *Anemone polyanthes* D. Don

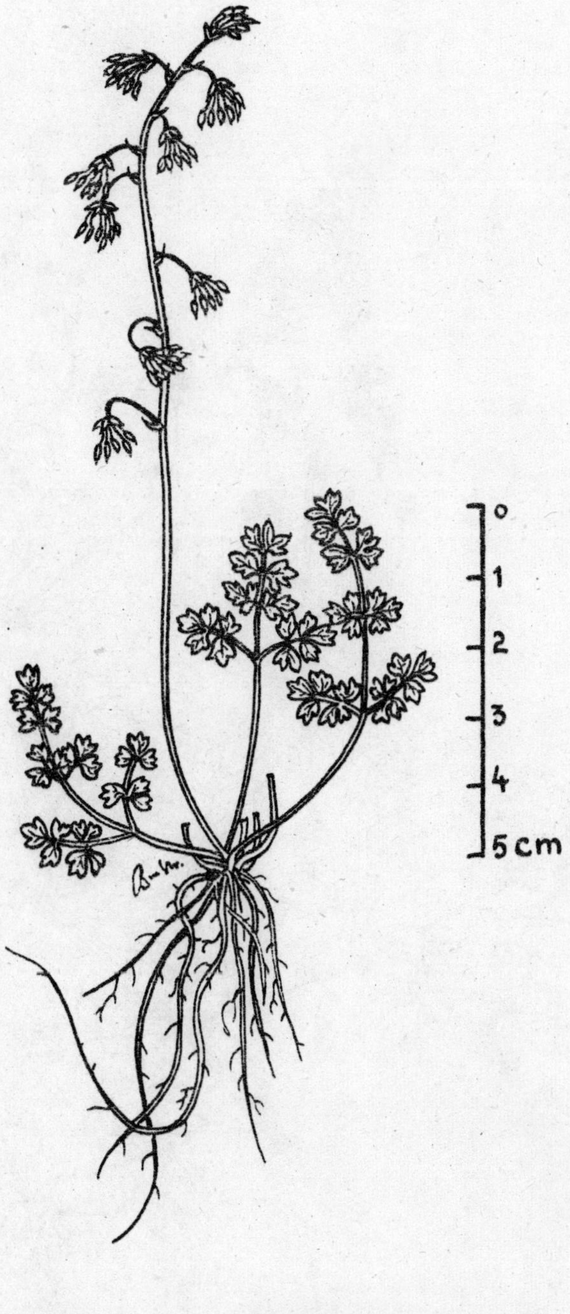

Fig. 11. *Thalictrum alpinum* Linn.

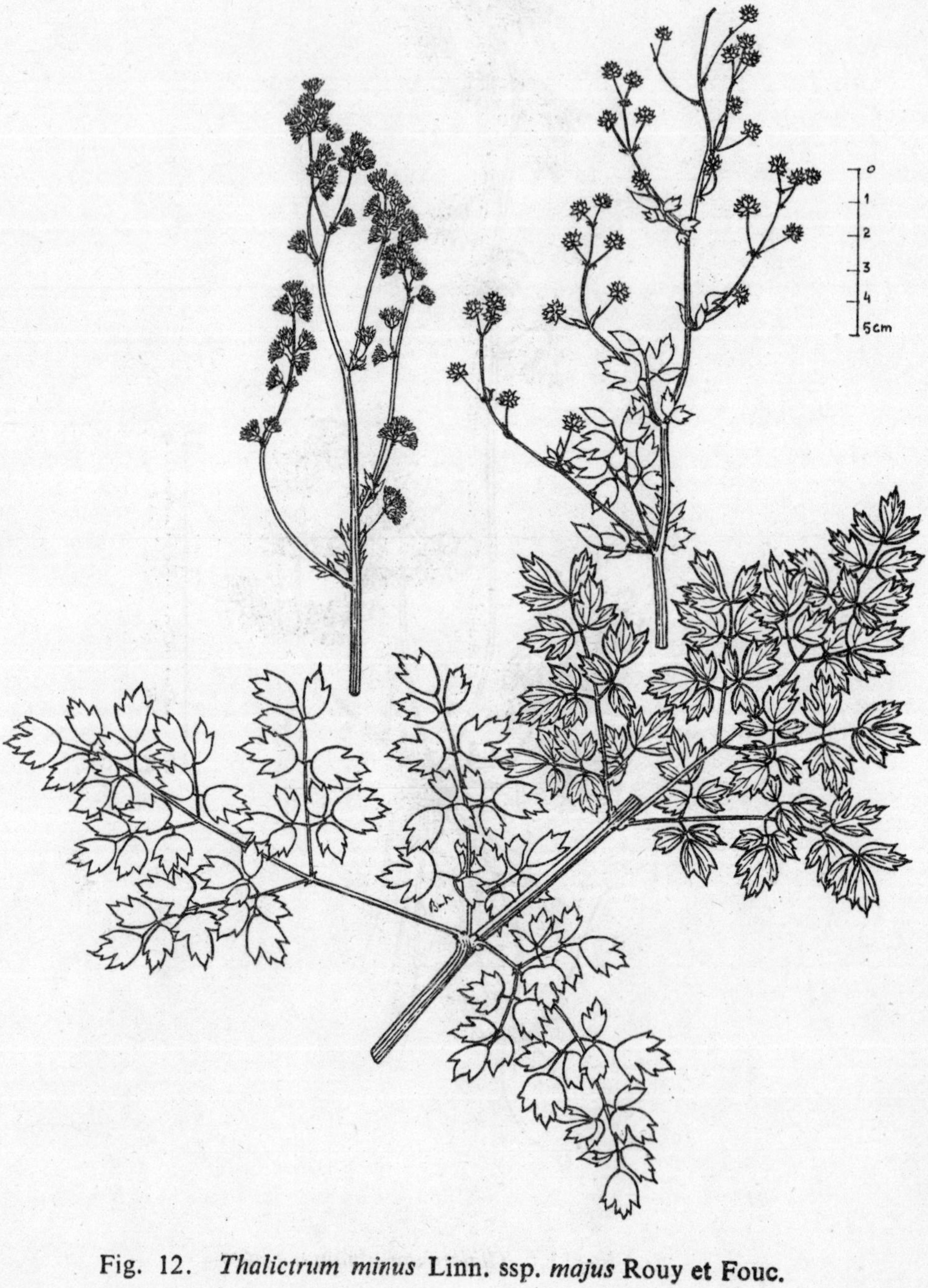

Fig. 12. *Thalictrum minus* Linn. ssp. *majus* Rouy et Fouc.

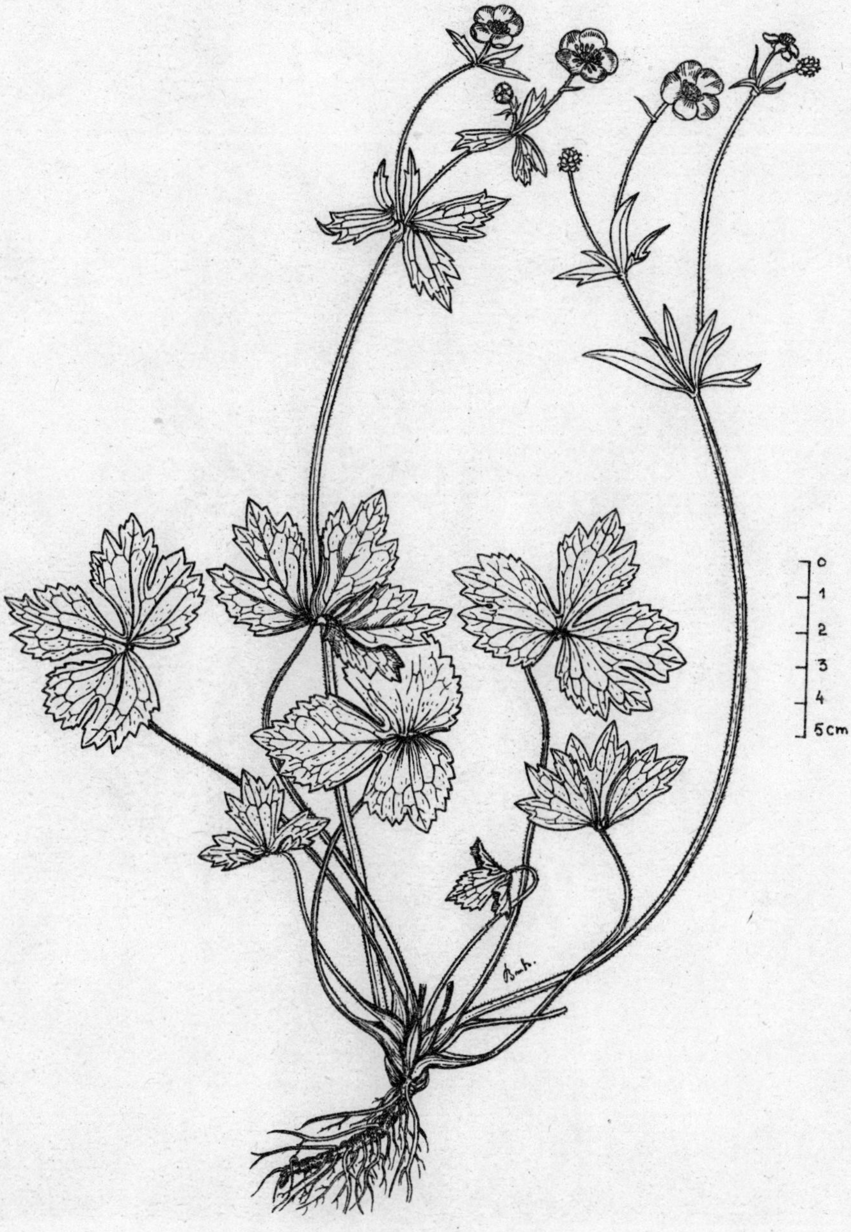

Fig. 13. *Ranunculus hirtellus* Royle

Fig. 14. *Paeonia emodi* Wall. ex Royle

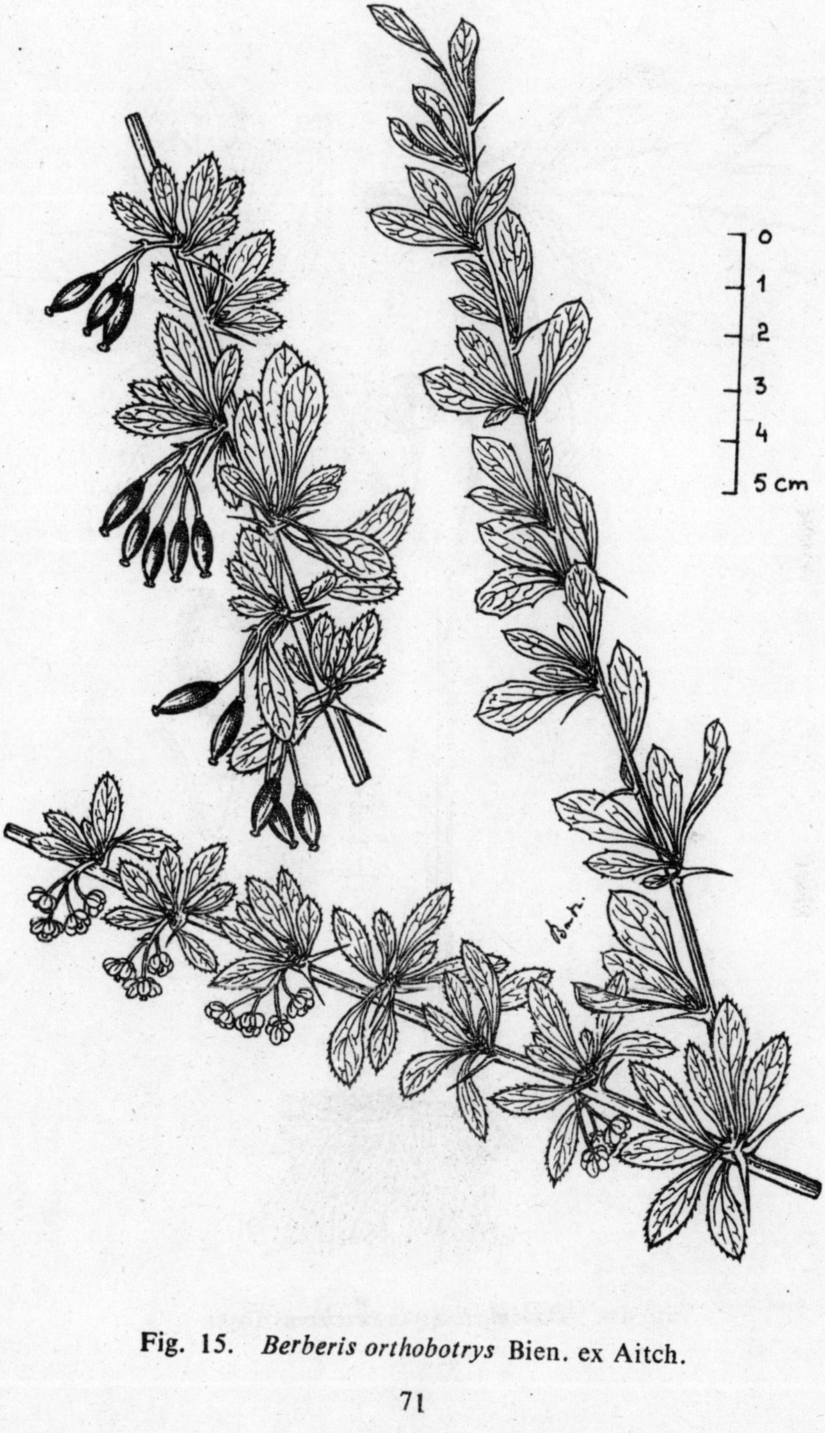

Fig. 15. *Berberis orthobotrys* Bien. ex Aitch.

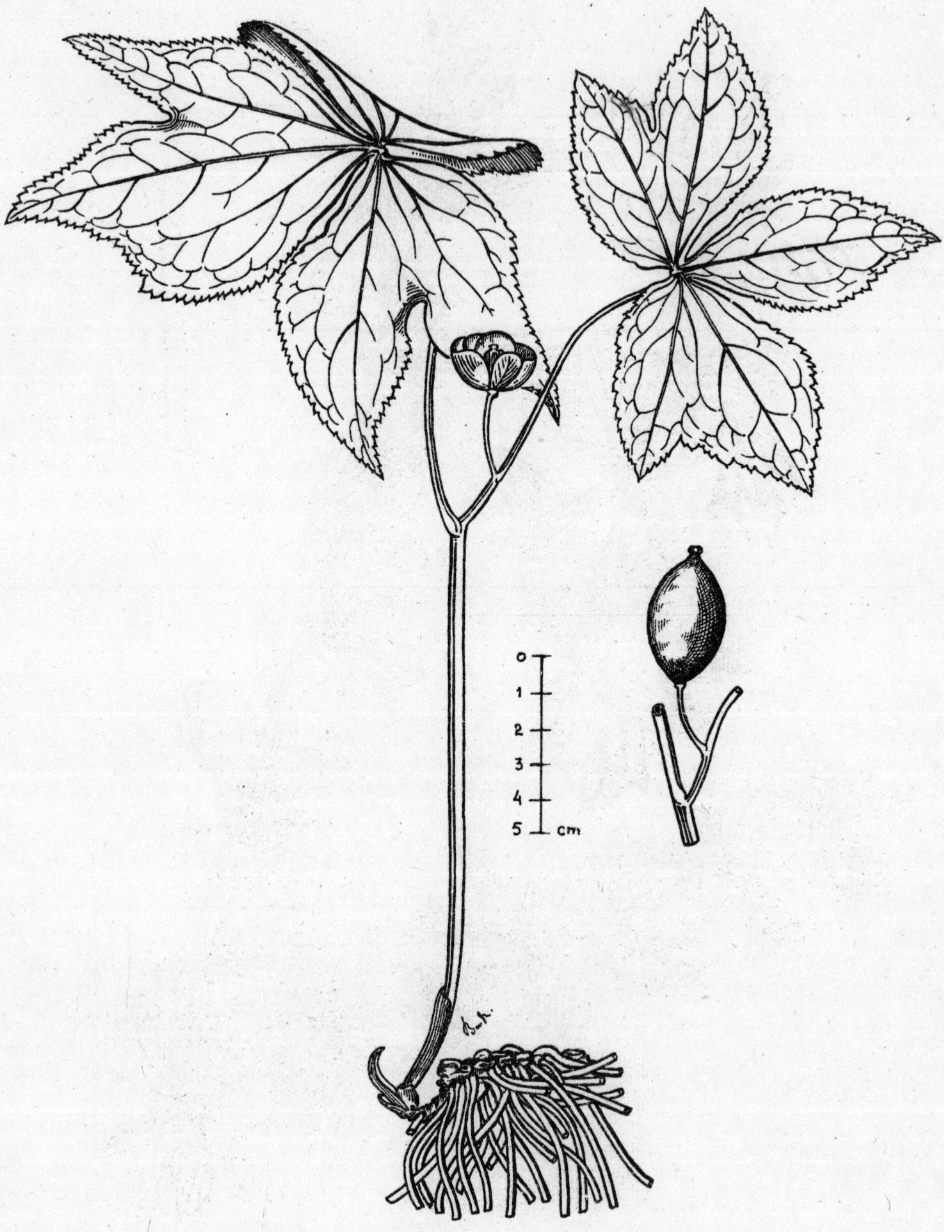

Fig. 16. *Podophyllum hexandrum* Royle

Fls. June-July.

Tarsar (4025 m) *B.M. Sharma* 14010; Dudhsar (3950 m) *B.M. Sharma* 14631; Lidder Vy. above Kainmul (11-12000′) *Duthie* 13137.

2. **T. elegans** Wall. ex Royle, Ill. Bot. Himal. 51, 1835; Hk.f. et Th., l.c. 10.

Slender, perennial, up to 30 cm tall, stiff herb with ternately decompound, 5-8 cm long leaves; leaflets 3-5 mm long, glaucous beneath, 3-lobed or, -partite. Flowers very small, green purple in a sparingly branched panicle; anthers pointed. Achenes long-stalked 6-12, obliquely obovate.

Fls. Aug.

Lidder Vy. above Kainmul (11-12000′) *Duthie* 13137/a. Cited by Stewart, l.c. 275, from Pehalgam, Aru, Shishnag etc. Some plants occurring among boulder moraines in Seikwas belong here.

3. **T. secundum** Edgew. in Trans. Linn. Soc. 20 : 26, 1846; Stewart, l.c. 277. *T. pauciflorum* Royle, Ill. Bot. Himal. 52, 1835; Hk.f. et Th , l.c. 11.

Included on the authority of Stewart's (l.c. 277) citations from Aru-Harnag and Pehalgam. It has panicles with solitary or few flowers at the ends of long branches; achenes obovate with straight dorsal suture, short-stalked, 3-nerved on either face; style straight.

4. **T. cultratum** Wall., Pl. As. Rar 2 : 26, 1832; Hk.f. et Th., l.c. 11.

Glabrous, 0.6-1.3 m tall, branched perennial herb. Leaves ternately decompound, 6-30 cm or longer even, from an auricled sheathing base; leaflets thin, obovate-cuneate, 0.5-1.7 cm long, divided into mostly 3, bluntly toothed lobes; teeth mucronate, of the upper leaves acute. Flowers greenish-white, in lax panicles; sepals 0.7-1 cm long; stamens many, longer than the sepals, with filiform, purple filaments and long pointed anthers. Achenes short-pedicelled, half-obovate straight dorsal suture, 5-6 mm long excluding the straight or hooked style.

Occ. : Very common among thickets and rock boulders throughout the areas about Aru, Liddarwat and Hamwas, also among junipers in Kolahoi and Seikwas.

Fls. July-Aug.; *Frts.* Aug.-Sept.

Kolahoi (3300 m) *B.M. Sharma* 14063; Aru (2350 m) *B. M. Sharma* 14627; Nangneri (2800 m) *P.S. Jamwal* 14128.

5. **T. platycarpum** Hk.f. et Th., Fl. Ind. 1 : 13, 1855 et in FBI. 1 : 11, 1872. *T. cultratum* Wall. ssp. *platycarpum* Bruhl, Ann. R. bot. Gdn Calcutta, 5 : 72, 1896.

Branched, 30-80 cm tall, stiff, slender perennial. Leaves similar to those of the preceding species but only up to 18 cm long and the leaflets narrower, 4-7 mm long, beneath glaucous and glandular-puberulous. Flowers small, greenish, in lax, leafy panicles. Achenes 3-4 mm long excluding styles, similar to those of *T. cultratum* but on much-longer stipes and with faces more strongly ribbed.

Occ. : In open places and among junipers on slopes about Zojipal, Sheshnag and Sangam valley.

Fls. July-Aug.; *Frts.* Aug.-Sept.

Zojipal (3450 m) *B.M. Sharma* 14660; Sangam valley (3300 m) *P.S. Jamwal* 15662.

6. **T. minus** Linn. ssp. **majus** Rouy et Fouc., Fl. Fr. 1 : 16, 1893; Stewart, l.c. 276. *T. minus* Linn. var. *majus* (Jacq.) Hk.f. et Th. in Hk.f., FBI. 1 : 14, 1872. *T. elatum* Jacq., Hort. Vindob. 3 : 49, 1776. **Fig. 12**

Glabrous, 0.6-1.2 m tall, rather stout, glaucous perennial. Leaves pinnately decompound, sheathing at base; leaflets obovate to oblong from a cuneate or truncate base, up to 2.5 m long, 3-4-lobed at apex, rather prominently veined beneath; lobes sharply toothed. Flowers yellowish or pale-green to whitish, in a nearly leafless panicle. Achenes generally up to 20 in a head, ovoid, pointed at either end, strongly ribbed, *c.* 3 mm long excluding the beak.

Occ. : On humus rich scrubby slopes in Aru - Pehalgam - Chandanwari area. Common.

Fls. July.; *Frts.* Aug.-Sept.

Aru (2350 m) *P.S. Jamwal* 19369; Aru (2300 m) *B.M. Sharma* 14622; Chandanwari (2700 m) *B.M Sharma* 14660.

4. **CALLIANTHEMUM** C.A. Meyer

1. **C. pimpinelloides** (D. Don ex Royle) Hk.f. et Th., Fl. Ind. 26, 1855; Stewart, l.c. 263. *Ranunculus pimpinelloides* D. Don ex Royle, Ill. Bot. Himal. 53, 1834. *C. cachemiricum* Camb. in Jacq. Voy. Bot. Inde. 5, t. 3, 1844; Hk.f. et Th. in Hk.f., FBI. 1 : 14, 1872.

Small, fibrous-rooted, rather glaucous, tufted, scapose, perennial herb. Leaves all radical, long-petioled, imparipinnate; segments orbiculate, 2-3 sect into deeply divided pinnules. Flowers creamy white, 2-4 cm across, solitary on 3.5-10 cm long scapes; petals oblong-cuneate, 6-12, 2-3 times longer than the often pinkish-tinged, broadly rounded 5 sepals. Achenes ovoid, tipped by a short style.

Occ. : On wet rocks and grassy slopes between Mundlan and Aru. Common.

Fls.. April-May.

Roadside slope beyond Mundlan (2350 m) *B. M. Sharma* 14611; 2 km beyond Mundlan (2300 m) *P.S. Jamwal* 15643.

5. **ADONIS** Linn.

1. A. **chrysocyathus** Hk.f. et Th. in Hk.f., FBI. 1 : 15, 1872.

Tufted, perennial herb with leafy, 15-25 cm tall flowering stems, elongating to 40-50 cm in fruit; rootstock scaly. Leaves pinnately much dissected into narrow, linear segments, radical and lower cauline petiolate. Flowers golden yellow, 2-5 cm across, solitary, terminal; sepals 7-8, petaloid, about half the length of the 16-24 petals. Achenes glabrous, 5 mm long excluding their stout, hooked beaks, convex and obtusely keeled dorsally, numerous in a 1.5 cm thick head.

Occ. : Forming dense clumps in open alpine pastures in E. Liddar valley; often associated with *Iris hookeriana.*

Fls. June-Aug.; *Frts.* Aug.-Sept.

Santsingh Pari (3000 m) *B.M. Sharma* 14619; Near Sheshnag (12-13000′) *Duthie* DD. no. 13326; Zojpal (11000′) Coll. probably *R.L. Badhwar*, RL. Herb. nos. 153 & 154; Amarnath, *L.D. Kapoor* 229; Zojipal-Sheshnag route (3550 m) *P. S. Jamwal* 15647.

Zojpal sheets carry the remark, "Tried about 1 lb root, leaves, and fruits on a lamb but it did not die although the plant is said to be poisonous".

6. CERATOCEPHALUS Pers.

1. **C. falcatus** (Linn.) Pers., Syn. Pl. 1 : 341, 1805; Kitam., Fl. Afgh. 122, 1960; Stewart, l.c. 263; For. Fl. Sgr. 161. *Ranunculus falcatus* Linn , Sp. Pl. 556, 1753; Hk.f. et Th., l.c. 16.

Small, woolly annual, up to 15 cm high in fruit. Leaves all radical, pinnate or 3-fid, with lobes divided into narrow, linear segments. Flowers small yellow, solitary on 1-several, 2-8 cm long scapes. Achenes numerous on an elongated receptacle, 5-7 mm long including their sharp, broad, falcate beaks; empty cells far apart.

Occ. : Common and often in tufts on dry grassy places along the roads in Pehalgam.

Fls. April; *Frts.* April-May.

Pehalgam-Mundlan road (2300 m) *B.M. Sharma* 14613; Pehalgam (2250 m) *P.S. Jamwal* 15644.

7. RANUNCULUS Linn.

1. Submerged plants; leaves multi-dissected into capillary segments; flowers white **1. R. trichophyllus**
1. Terrestrial or marshy herbs; leaves not as described above; flowers yellow:
    2. Petals 10-15 :
        3. Leaves orbicular, crenate-lobed; scapes usually several **R. polypetalus**
        3. Leaves elliptic-oblong, entire or crenate; scape solitary **R. kamchaticus**
    2. Petals 5 :
        4. Radical leaves undivided **R. lobatus**
        4. Radical and cauline leaves both lobed or divided :
            5. Achenes compressed, intramarginally ribbed :
                6. Stems producing runners from among the radical leaves; achenes minutely dotted **2. R. diffusus**
                6. Stems not producing runners; achenes not dotted **3. R. laetus**
            5. Achenes turgid or subinflated, not intramarginally ribbed :
                7. Sepals appressedly hairy; achenes 2-2.5 mm excluding beak; mostly hairy **4. R. hirtellus**
                7. Sepals glabrous; achenes 1 mm excluding beak, glabrous **5. R. munroanus**

Ranunculus polypetalus Royle *(Oxygraphis polypetala* Hk.f. et Th.), **R. kamchaticus** DC. (*O. glacialis* Bunge) and **R. lobatus** Jacq. ex Camb. have been keyed out because of their having been cited by Stewart (l.c. 271, 272) from Mt. Kolahoi. We have not been able to collect these, nor is there any specimen from our area in the Forest Research Institute Herbarium.

1. **R. trichophyllus** Chaix in Vill., Hist. Pl. Dauph. 1 : 335, 1786; Cook. in Fl. Eur. 1 : 237, 1964; Stewart, l.c. 273;

For Fl. Sgr. 162. *R. aquatilis* Linn. var. *trichophyllus* (Chaix) Hk.f. et Th., l.c 16.

Much branched. 30-70 cm long, aquatic herb, rooting only at the lower joints; leaves capillaceo-multifid. Flowers 0.8-1 cm across, on 0 6-2.5 cm pedicels, the latter up to 4.5 cm in fruit. Achenes transversely wrinkled, *c.* 1.5 mm long, in globose heads.

Occ. : Abundant in water channels running across the lawns of Pehalgam Club.

Fls. & Frts. June-Aug.

2. **R. diffusus** DC., Prodr. 1 : 38, 1824; Hk.f. et Th., l.c. 19.

Decumbent or ascending, branched, hairy perennial; stems 30-70 cm long, clothed with soft, mostly patent hairs and rooting at the lower nodes. Radical and lower cauline leaves long petioled, 5-14 cm across, deeply 3-lobed into cuneate-obovate, sharply inciso-dentate segments, upper leaves smaller and with shorter petioles, the uppermost sessile or nearly so. Flowers long-stalked, 1.2-2 cm across; sepals hairy. Fruit a globose head of 3-4 mm long achenes with straight or hooked styles.

Occ. : In shady, wet places along the water channels in Aru.

Fls. June-Aug.; *Frts.* July.-Aug.

Aru (2400 m) *B.M. Sharma* 14663; Mundlan (2350 m) *P.S. Jamwal* 15664.

3. **R. laetus** Wall. ex Royle, Ill. Bot. Himal. 53, 1834; Hk.f. et. Th., l.c. 19.

General aspect of the preceding species but stems erect, mostly appressedly hairy, not rooting from the joints. Flowers 1.5-2.5 cm across; sepals villous with long hairs; styles straight.

Occ. : Very common and often gregarious in damp, open or shaded places up to *c.* 2850 m throughout the area.

Fls. & Frts. June-Aug.

Liddarwat (2800 m) *B. M. Sharma* 14383; Pehalgam (7200′) *H.D. Thapliyal* 26486; Aru (2450 m) *P.S. Jamwal* 15700.

4. **R. hirtellus** Royle, Ill. Bot. Himal. 53, 1834; Hk.f. et Th., l.c. 18. **Fig. 13**

Variable, appressedly pubescent, perennial herb from a fibrous rootstock; stems erect or decumbent, often tufted, 8-45 cm long, 1-many-flowered. Radical leaves long-petioled, cordate, 1.5-7 cm across, deeply 3-parted into obovate-cuneate, upwards inciso-dentate segments, or the latter oblong to elliptic and entire to upwards 1-2-toothed; cauline leaves 3-5-partite. Flowers 1-2 cm across, achenes many, in a globose or ovoid head.

Occ. : Usually in shaded moist places along banks in forested areas and in open, grassy, damp habitats on alpine meadows.

Fls. & Frts. July.-Aug.

Liddarwat nallah (3000 m) *B. M. Sharma* 14184; Dudhsar (3850 m) *B M. Sharma* 14045; Seikwas (3350 m) *B M. Sharma* 14664; Amarnath (12500′) common up to 14500′, *K.M. Vaid* DD. Herb no. 117532; Dudhsar (3900 m) *P.S. Jamwal* 15666.

5. **R. munroanus** J. R Drum. ex Dunn Kew Bull. 279, 1925; Stewart, l. c. 271. *R. munroanus* var. *minor* Kitamura, Pl. W. Pak. 63, 1964.

Low perennial herb with slender, lanuginose, 4-8 cm long, erect or decumbent, few-many-flowered. usually tufted stems; roots a sessile cluster of fusiform tubers. Radical leaves long-petioled, reniform, 0.8-2 cm across, from a cordate base, deeply divided into usually 3, obovate to suborbiculate, broadly and remotely crenate lobes; cauline leaves similar, smaller, with shorter petioles. Flowers 6-9 mm across, on long or short, axillary pedicels; achenes ovoid, in a globose, 4 mm wide head.

Occ. : Common and often forming small, dense patches in shady nooks and rock crevices in open alpine zone; also along moist banks.

Fls. & Frts. July-Sept.

Seikwas (3350 m) *B.M. Sharma* 14018; Below Dudhsar (3600 m) *B.M. Sharma* 14628; Sangam Vy. above Lidderwat (12-13000′) *Duthie* 13555 (under *R. hyperborus* var. *radicans*); Sheshnag (3650 m) *P.S Jamwal* 15667.

VI. Paeoniaceae

1. PAEONIA Linn.

1. **P. emodi** Wall. ex Royle, Ill. Bot. Himal. 57, 1834; Hk.f in Bot. Mag. t. 5719, 1868; Hk.f. and Th , l.c. 30. **Fig. 14**

Stout, glabrous, perennial herb with thick, fleshy roots; stems leafy, 30-70 cm tall, rather woody. Leaves alternate, 15-30 cm long including petiole, 1-2-ternate; leaflets entire or deeply incised into lanceolate, acuminate ultimate segments. Flowers white, 6-10 cm across, solitary, mostly from the upper leaf-axils; sepals persistent, orbicular, concave; petals 5-8. Follicles solitary, rarely 2, ovoid, 2.5-3 cm long, somewhat curved.

Occ. : Common and more often gregarious on open hillsides near Pehalgam; also in wooded places near Chandanwari. Rare in W. Liddar Valley.

Fls. April-May; *Frts.* June.

Pehalgam-Gurkum route (2225 m) *B M. Sharma* 14611; Chandanwari, *H.D. Thaplival* DD no. 26528/128443; Kinimola nallah, Lidder Vy., *Inayat* 25462; above Pehalgam (8000′) *K.M. Vaid* DD no. 117529.

Seeds are emitic and purgative; roots are stated to be useful for uterine diseases.

VII. Berberidaceae

Shrubs; leaves simple, fruit a berry **1. Berberis**
Herbs; leaves compound; fruit a dry, dehiscent capsule **2. Epimedium**

1. BERBERIS Linn.

1. Inflorescence a simple raceme; leaves usually 3-7 cm long **1. B. pachycantha**
1. Inflorescence umbellate, subumbellate or racemose umbellate; leaves below 3 5 cm long :
   2. Ovary, 3-5-ovuled; berries pruinose **2. B. orthobotrys**
   2. Ovary 2-ovuled; berries epruinose **3. B. jaeschkeana**

1. B. **pachycantha** Koehne, Deutsche Dendr. 170, 1893; Ahrendt, J. Linn. Soc. Bot. 57:191, 1961; Jafri in Fl W. Pak. no. 87 : 21, 1975, *B. zabeliana* C.K. Schneid., Bull. Herb. Boiss. ser. 2, 5 : 667, 1905. *B. petiolaris* auct. non Wall. ex G. Don 1831; Parker, For. Fl. 12, 1918.

Deciduous, 2-3-m tall, glabrous; twigs dark-red to pale, unarmed or with slender, usually simple, 0.5-1 cm long spines. Leaves obovate to elliptic-oblong, 3-7 × 1-3.2 cm, prominently reticulate, cuneate into a short slender petiole; margin with numerous close-set, irregular serrations. Flowers yellow, 5-6 mm across, in 4-10 cm (excl. peduncle), often pendulous racemes. Berries epruinose, oblong-ovoid, 0.8-1 cm long, red, estylose.

Occ. : In moist, shaded places.

Fls. May-June; *Frts.* Aug.-Sept.

Pehalgam, Liddar range (7000′) *W.J. Lambert* DD. nos. 44707, 44708 & 44711; Pehalgam (8000′) *W.J. Lambert* DD. no. 46229.

2. **B. orthobotrys** Bien. ex Aitch. J. Linn. Soc. Bot. 19 : 150, 1882; Ahrendt, l.c. 142; Stewart, l.c. 279. **Fig. 15**

Glabrous, 1.5-2 m high; twigs sulcate, dark-, or pale-brown or yellowish, armed with 1-1.5 cm long, 1-3-fid spines. Leaves obovate, 1-3.5 × up to 1.2 cm, cuneate into 0.5-1.3 cm long, slender petiole, pale beneath. Flowers yellow, 0.7-1 cm across, subumbellate or racemose-umbellate. Berries oblong or ellipsoid, 1-1.2 cm long, red, coated with greyish-white bloom.

Occ. : Open places along stream beds in Pehalgam and Aru.

Fls. April-May; *Frts.* June-Aug.

Aru (2300 m) *B. M. Sharma* 14270; Pehalgam (2100 m) *P.S. Jamwal* 15673.

3. **B. jaeschkeana** C.K. Schneid , Bull. Herb. Boiss. ser. 2, 5 : 399, 1905; Ahrendt, l.c. 238; Stewart, l.c. 279.

About 1 m high; twigs armed with 3-fid spines. Leaves membranous, oblanceolate or oblong-elliptic, up to 2.5 × 0.8 cm, nearly sessile; margin 2-5-spinose serrulate. Flowers 3 to 5 in an umbellate or subcorymbose inflorescence. Berries red, oblong-ovoid, 0.8-1 cm long, nearly estylose.

Occ. : Open places on the slope to Pisu top.

Fls. June; *Frts.* July-Aug.

Pisu Ghati (3150 m) *B.M. Sharma* 16673;, Liddar Vy. above Kainmul (11-12000′) *Duthie* 13150; Pisu Ghati (3150 m) *P.S. Jamwal* 16363.

2. EPIMEDIUM Linn.

E. elatum Morr. et Decne, Annls Sci. nat. ser. 2(2) : 356, 1834; Hk f. et Th., l.c. 112.

Glabrous, erect perennial, 0.3-0.8 m high from a stout, creeping, scaly rhizome; stems, branches and petioles stramineous. Leaves 2-3-ternate, with each ultimate pinna of 3, membranous, petioluled, ovate-cordate, spinulose-serrate leaflets 2-9 × 1.5-7.5 cm; lateral leaflets usually oblique. Flowers yellowish-white, 6-7 mm across, in irregular, up to 25 cm long panicles;

sepals petaloid, 8; petals 4, the inner two slipper-like; stamens 4, antipetalous. Capsule elongate, 6-10 cm long (excl. style), 2-3-seeded; seeds reniform, black, *c.* 2 mm diam.

Occ. : Sporadically met with among scrubs and as undergrowth of fir forests about Aru and along the route to Chhumanaisar.

Fls. June-July; *Frts.* Aug -Sept.

VIII. PODOPHYLLACEAE

1. PODOPHYLLUM Linn.

1. **P. hexandrum** Royle, Ill. Bot. Himal. 2(1) : 64, 1834; Rao, Rec. bot. Surv. India, 18(2) : 14, 1960; For. Fl. Sgr. 163; Soejarto et al., Taxon, 28 : 549-51, 1979 and 30 : 652-56, 1981. *P. emodi* Wall. ex Honig., Thirty five years in the East 2 : 35, pl. 20, 1852. Hk. f. & Th., l.c. 64. *P. emodi* var. *hexandrum* (Royle) Chatt. & Muker., Rec. bot. Surv. India, 16(2) : 45, 1953. *P. hexandrum* var. *emodi* (Honig) Selivan., Bot. Zuhrn. 54 : 1605, 1969. **Fig. 16**

Glabrous, 15-50 cm tall, more or less succulent, perennial herb from a creeping 2-5 cm long knotty rhizome bearing thick, fibrous roots; stems 1 or 2, simple, leafy about the top. Leaves 2, or sometimes 3, alternate, palmate, 10-25 cm across, deeply 3-parted into sharply toothed to almost entire, simple or 2-3-lobulate segments; terminal segments broadly elliptic, the lateral ones obliquely ovate. Flowers white to pinkish, 2.5-4 cm across, terminal, ultimately becoming supra-axillary; sepals 3, petaloid; stamens usually 6. Berry pulpy, 2.5-5 cm long, ovoid or oblong-ellipsoid, many-seeded, red on ripening.

Occ. : Abundant on humus rich forest floor in Liddarwat area but not uncommon in open, stony places about Satlanjan.

Fls. May-Aug.; *Frts.* June-Sept.

Satlanjan (3150 m) *B.M. Sharma* 19357; Chandanwari, *H. D. Thapliyal*, DD no. 26536/127839; Pisu Ghati (11500-13000′) *K. M. Vaid* 117534; Chandanwari (2700 m) *P. S. Jamwal* 15672.

Rhizomes and root are the source of a medicinal resin whose major active constituent is podophyllotoxin. On account of its tumour necrotizing properties the drug has regained consideration as a possible cure for some forms of cancer. Ripe fruits are edible.

IX. PAPAVERACEAE

1. Stigmas forming a radiating disc adnate to the top of the ovary; capsule opening by pores beneath the stigmatic disc **1. Papaver**
2. Stigmas forming a capitate head on a distinct columnar style; capsule opening by short valves **2. Meconopsis**

1. PAPAVER Linn.

P. nudicaule Linn., Sp. Pl. 507, 1753; Hk.f. et Th., l.c. 117. **Fig. 17**

Perennial, usually tufted herb, rough with dark brown, stiff, setose hairs, rootstock clothed with remains of old leaf-bases. Leaves all radical, long-stalked; 5-15 (incl petiole) x 1.5-6 cm, bristly or ciliate, pinnately lobed lobes pinnatifid into oblong, acute or obtuse segments. Flowers yellow, orange or reddish-orange, 2.5-4 cm across, solitary on 15-40 cm tall,

1-few leafless scapes; sepals caducous, stiffly setose. Capsule ablong to oblong-globoid, 1.2-15 × 0.7-1 cm, clothed with stiff, black or blackish-brown setae.

Occ. : In stony ground beside the streams and on open alpine slopes. Not common.

Fls. Aug.-Sept.

Santsingh Pari (3900 m) *B.M. Sharma* 14620; Kolahoi, Liddar Vy. *Inayat* DD. no. 25482; near Mahagunus Pass (13500′) *K. M. Vaid*, DD s.n.; Panchtarni-Yangan-Shishnag, *K.S. Ahluwalia* RL. Herb. nos. 667-670; Amarnath (3750 m) *P.S. Jamwal* 15648.

2. **MECONOPSIS** Vig.

M. aculeata Royle, Ill. Bot. Himal. 67, t. 15, 1833; Hk.f. et Th., l.c. 118. **Fig. 18**

Prickly, 30-75 cm tall; leafy herb from a long, tapering rootstock; sap yellow. Leaves petiolate, irregularly pinnatifid, the uppermost sessile, radical ones 10-20 × 2.5-6 cm, with obtuse or acute lobes, the cauline oblong or lanceolate, with variously shaped lobes and more or less sheathing petiolar-bases. Flowers deep-, or sky-blue to purplish blue, 5-7 cm across, racemed or in pseudoracemes; sepals suborbicular-ovate, caducous. Capsule prickly, 1.2-1.8 (excl. style) × 0.6-0.6 cm, tapering at both ends.

Occ. : Common, but not abundant, among moraines and on rocky, cliffs throughout the area at elevations above 3200 m.

Fls. July.-Aug.

Above Satlanjan (3300 m) *B.M. Sharma* 19356; Kainmul, on wet rock, Liddar Vy. (10-11000′) *Duthie* 13083; Kolahoi (3600m) *P.S. Jamwal* 15679; Mahagunus (*c.* 11300′) *R. L. Badhwar*, R. L. Herb. no. 682

Roots powdered and eaten with water as stomachic.

X. Fumariaceae

CORYDALIS Vent.

1. Radical leaves absent — **1. C. diphylla**
1. Radical leaves few or many :
    2. Rootstock tuberous, branched with small bulbous structure above the scale leaves and emitting tuberous roots below; flowers blue — **2. C. cashmeriana**
    2. Neither the rootstock as described above nor the flowers blue :
        3. Leave simply 3-5-lobed or, -partite; flowers variegated with purple against white or yellowish background; capsule 1.2-2.5 cm diam., bladder-like and much inflated — **3. C. crassifolia**
        3. Leaves multisegmented; flowers yellow or yellow with purple or green tips; capsule below 1 cm diam., neither bladder-like nor so inflated :
            4. Annual or biennial, or sometimes perennating through a very short subterranean rhizomatous stock; seeds opaque, tubercled — **4. C. stewartii**

4. Perennials, with thickened, fusiform or slender rootstock, the latter often branched and covered with withered, sheathing leaf bases; seeds (*sikkimensis ?*) shining, smooth :
 5. Leaves equally twice ternate; posticous petal dorsally unwinged **5. C. longipes**
 5. Leaves pinnatisect or (rarely) very unequally ternate; posticous petal winged :
  6. Stems branched :
   7. Stems naked or 1-leaved; plants glaucous, usually glandular-pilose **6. C. tibetica**
   7. Stems leafy; plants subglaucous, glabrous :
    8. Racemes simple; flowers yellow; capsules obovate–oblong, obtuse **7. C. ramosa**
    8. Racemes ± thyrsiform; flowers with purple tips; capsules elliptic, acute **8. C. thyrsiflora**
  6. Stems simple :
   9. Petals tipped with purple, brown or green :
    10. Diffuse; stems naked or 1-leaved; spur straight **6. C. tibetica**
    10. Erect; stems bearing a pair of opposite leaves above its middle; spur slightly hooked **9. C. falconeri**
   9. Petals yellow, not tipped as above :
    11. Stems almost naked or the cauline leaves borne only low down on the stem and solitary or very often 2 and ± opposite; lower bracts broadly cuneate **10. C. govaniana**
    11. Stems bearing 2-4 leaves disposed throughout in opposite or subopposite pairs or sometimes merely alternate; lower bracts lanceolate or linear-, to broad-elliptic :
     12. Ultimate leaf-lobules 3-5×1-1.5 mm, discrete; bracts, at least the lower, linearily segmented or incised; capsules *c.* 1.5 cm long **11. C. sikkimensis**
     12. Ultimate leaf-lobules up to 2 (2.5)×0.3-0.5 cm, confluent; bracts almost entire; capsules 0.6-1 cm long, broadly ovate or obovate **12. C. clarkei**

1. **C. diphylla** Wall., Tent. Fl. Nep. 54, 1824; Prain, J. Asiat. Soc. Beng. 65(2) : 20, 1896; Stewart, l.c. 289; For. Fl. Sgr. 164. *C. rutaefolia* Hk.f & Th., Fl. Ind. 262, 1855; Hk.f., FBI. 1 : 122, 1872 (non Sibth.).

Glabrous, subglaucous, perennial from a deeply buried, globose tuber; stem usually simple, slender, ± erect, 5-25 cm above the ground level. Leaves 2, (rarely 1) opposite, 3-15 cm long, 2-3-ternately cut, with petioles very often longer than petiolules; ultimate leaf-lobules ovate, oblong or elliptic-oblong, 3-5 to each pinnule, variable in size, the median largest and 1-4×0.3-1 cm. Flowers purple with darker tips, 1.5-2.7 cm long, lax, in a mostly 5-10-flowered raceme; spur equalling or longer than the emarginate lamina of the posticous petal, infundibuliform, recurved with mostly down curved, blunt tip. Bracts foliaceous, ovate-, or oblong-lanceolate, entire. Capsule ovate-oblong, 1-1.5×0.4-0.5 cm.

Occ. : Very common on humus rich, open and wooded slopes.

Fls. April-May.

Mundlan (2300 m) *B.M. Sharma* 15611; Aru (2400 m) *B. M. Sharma* 16690; Pehalgam (2100 m) *B.M. Sharma* 16692; Kinimola nala, Lidder vy. *Inayat* 25469 & 25471; Lidderwat, Lidder vy. *Inayat* 25472 & 25473; Pehalgam-Chandanwari route (2400 m) *P.S. Jamwal* 16325.

2. **C. cashmeriana** Royle, Ill. Bot. Himal. 69, t. 16, f. 1, 1839; Hk.f., l.c. 123. **Fig. 19**

Perennial, 8-20 cm tall; stems ± erect, simple, 1-few. Leaves glaucous, the radical ones 1-few, long stalked, 6-15 cm long incl. petiole, palmately 3-partite into petiolulate, deeply 3-lobed pinnules, the lobes toothed or incised into linear or lanceolate, mucronulate, 2-3 mm broad, ultimate lobules; cauline leaves mostly solitary, sessile or subsessile, more deeply partite and disposed about or above the middle of the stem. Flowers bright blue with darker tips, 1.2-2 cm long, 5-8 in a subumbellate raceme; spur slightly down curved, blunt ± equalling the concave, acute or acuminate lamina of the posticous petal. Bracts foliaceous, at least the lower 3-fid. Capsule linear-oblong, 1.5-1.8 × 0.25-0.3 cm, deflexed.

Occ. : Frequent in open stony or grassy alpine and subalpine ground.

Fls. June-Aug.; *Frts.* Aug.

Nangneri nallah (2850 m) *B.M. Sharma* 14129; Dudhsar (3900 m) *B.M. Sharma* 14226; Lidder vy. above Kainmul (12000′) *Duthie* 13114; Sangam vy. above Lidderwat, *Duthie* s.n. DD; Budzalkhod nallah, Lidder vy. (13000′) *Duthie* 14132; Pisu Ghati (3000 m) *P.S. Jamwal* 16314.

3. **C. crassifolia** Royle, Ill. Bot. Himal. 69, 1839; Hk.f., l.c. 127; Jafri in Nasir et Ali, Fl. W. Pak. 73 : 38, 1974. **Fig. 20**

Glabrous, glaucous perennial from a prostrate, slender rootstock covered above with withered leaf-bases; stems 1-2; simple, 10-28 cm tall. Leaves fleshy, the radical ones few, long-stalked, 12-30 cm long from base of petiole to tip of the reniform, deeply 3 (-5)-lobed or, -partite, 5-12 cm wide lamina; lobes hemiorbicular from a cuneate base, crenate or lobulate. Cauline leaves usually 3, subopposite or whorled. Flowers 1.5-2.5 cm long, in a dense, 3-6 cm long, mostly simple raceme; spur slightly down curved, blunt, shorter than the broad, obtuse lamina of the posticous petal. Bracts foliaceous, spathulate or obovate-lanceolate, the lower 1-2 × 0.5-1.2 cm, entire. Capsule ellipsoid or subglobose, 1.5-3.2 × 1.2-2.5 cm.

Occ. : Common denizen of morainal deposits in the alpine zone.

Fls. July-Sept.; *Frts.* Aug.-Sept.

Amarnath, near cave (±3900 m) *B.M. Sharma* 16679; Mt. Kolahoi (4500 m) *B.M. Sharma* 14210; Amarnath (12500-14000′) *K.M. Vaid* DD. no. 117531; near Amarnath cave (13000′) DD. no. 44963; Sangam vy. (3900 m) *P.S. Jamwal* 16395.

Leaves are stated to be consumed as spinach by the nomads.

4. **C. stewartii** Fedde, Repert. 19 : 119, 1923; Stewart, l.c. 291.

Glabrous, subglaucous, with branched, elongate, straggling or semi erect, rather fleshy, weak stems. Leaves 2-3-pinnate, 10-20 (incl. ± half as long petiole) × 8-17 cm; gradually short stalked and ultimately almost sessile upwards; pinnules like the pinnae petiolulate, 0.5-1.5 cm long and ± as broad, mostly 3-parted into obovate-oblong, cuneate, 0.5-1.2 cm long, rounded-obtuse, truncate, subemarginate or mucronulate ultimate lobules. Flowers yellow with brown tips, 0.9-1.3 cm long, in terminal, 5-10-flowered racemes; spur slender, obtuse, curved, slightly longer than the acute or subacuminate lamina of posticous petal. Bracts incised, upwards gradually entire and linear. Capsule linear-oblong, 1.2-1.5 × 0.15-0.2 cm, subtorrulose.

Occ. : Common among scrubs along waysides and forest paths, usually up to 3000 m.

Fls. July-Sept.; *Frts.* Aug.-Sept.

Armeun nalla, Aru (2500 m) *B. M. Sharma* 19266; Aru-Chummanaisar route (2600 m) *B. M. Sharma* 16635; below Chandanwari (2650 m) *P.S. Jamwal* 16355.

5. **C. longipes** DC., Prodr. 1 : 128, 1824; Prain, l.c. 28; Stewart, l.c. 290. *C. sibirica* auct. non Pers.; Hk.f., l.c. 125.

Glabrous, procumbent, much branched, subglaucous herb with 30-45 cm long, weak, leafy stems and slender rootstock. Leaves 2-ternate, membranous, the lower long-stalked, 10-20 × 3-6 cm, upwards smaller, shortly petioled and ultimately almost sessile; pinnae 3(-5), petiolulate like the pinnules, the latter broad-ovate, 1-2 cm wide and deeply segmented into oblong or obovate-oblong, ultimate lobules 0.3-1 × 0.1-0.3 cm and subrotund at apex. Flowers yellow with brown tips, *c.* 8 mm long, in 3-10-flowered, short terminal racemes elongating in fruit; spur blunt or subobtuse, straight or slightly upcurved, shorter than the hood-like lamina of the posticous petal. Bracts leafy, segmented, upwards smaller, simply toothed to almost entire. Capsule linear or linear-obovate, 1-1.7 × 0.2-0.35 cm, subtorulose.

Occ. : In habitats similar to those for the preceding species.

Fls. July-Aug.; *Frts.* Aug.

Gha Shangan (2900 m) *B.M. Sharma* 19263; Armium Glen (2700 m) *B. M. Sharma* 16634; Pisu Ghati (3000 m) *P.S. Jamwal* 16318. Closely resembling and perhaps indistinguishable from Dehradun specimen ticketed as "*C. cashmiriana* Duthie & Prain; Kolahoi valley near Lidderwat (11-12000′) *Duthie* 13521".

6. **C. tibetica** Hk.f. & Th., Fl. Ind. 265, 1855; Hk.f., l.c. 124.

Included after Jafri, l.c., who has cited its gatherings from above Pehalgam etc. by *Stewart et al.*

7. **C. ramosa** Wall. ex Hk.f. & Th., Fl. Ind. 267, 1855; H.k.f., l.c. 125.

Much branched, straggling or semi-erect, 20-40 cm high, glabrous, subglaucous herb from a slender, simple or branched rootstock; stems leafy. Leaves 2-3-pinnate or decompound from sheathing bases, 6-12 (incl. ± half as long petiole) × 3-9 cm; pinnules like the pinnae petiolulate, the ultimate ones deeply segmented into linear-, to obovate-oblong, acute or obtuse, mucronulate lobules 0.3-1.2×0.15-0.4 cm. Flowers yellow, 1-1.3 cm long, in simple, 10-20 flowered, terminal, 3-12 cm long, in simple, 10-20 flowered, terminal, 3-12 cm long racemes; spur straight or curved, blunt, equalling or longer than the hooded lamina of the posticous petal. Bracts leafy, the lower 3-sect, upwards dissected to subentire or entire. Capsules 0.7-1×0.3-0.5 cm.

Occ. : Usually among the scrubs up to *c.* 3600 m.

Fls. June-Aug.; *Frts.* July-Aug.

Aru (2400 m) *B.M, Sharma* 19265 A & 19265 B; Lidder vy. near Shishnag (12000′) *Duthie* 14131; Pisu Ghati (3000 m) *P.S. Jamwal* 16319.

8. C. **thyrsiflora** Prain, Asiat. Soc. Beng. 65(2) : 35, 1896; Stewart, l.c. 292. *C. gortschakovii* Hk.f. & Th., Fl. Ind. 267, 1855; Hk.f, l.c. 125, non Schrenk.

Diffuse or ascending, 10-40 cm high, glabrous, subglaucous herb from a thickened, often branched rootstock. Leaves 2-3-pinnatisect, the radical ones long stalked from sheathing bases; pinnae alternate or subopposite, petiolulate to subsessile, 1-4×1-2.5 cm; pinnules 3-5 to each pinna, 0.5-1.5 cm long and ± as broad, shallowly or deeply segmented into 3-5 oblong-obovate to almost linear, acute, mucronulate lobules 0.3-1.2×0.1-0.4 cm. Cauline leaves similar, alternate or subopposite, smaller upwards and ultimately stalkless. Flowers yellow with purplish tips, 1.5-1.7 cm long, in dense, ± thyrsiform, branching inflorescence; spur broad, straight or slightly down curved, blunt, about as long as the obtuse lamina of the posticous petal. Bracts leafy, linearily incised, upwards linear and entire. Capsule 5-7×*c.* 4 mm, 3-4-seeded.

Occ. : On wet islands in stream beds and along banks in alpine and subalpine zone.

Fls. July-Sept.; *Frts.* Sept.

Liddarwat gagarpathri (3000 m) *B.M. Sharma* 14140; Kolahoi (3450 m) *B.M. Sharma* 16612; Sonsal nallah, Lidder vy. by the lake (13-14000′) *Duthie* 13347; Sangam vy. (3900 m) *P.S. Jamwal* 16394; Satlanjan (3000 m) *P.S. Jamwal* 16348.

9. **C. falconeri** Hk.f. & Th., FBI. 1 : 124, 1872.

Cited from Pehalgam by Stewart, l.c. 289.

10. **C. govaniana** Wall., Tent. Fl. Nep. 55, 1824; Hk.f., l.c. 124. **Fig. 21**

Glabrous, more or less glaucous; stems erect, tufted, 20-60 cm tall from a stout, elongated rootstock crowned with withered leaf bases. Leaves oblong, 2-pinnate, the radical ones nearly as long as the stems, with petioles sheathing at bases and about equalling or longer than the lamina; pinnae opposite or subopposite, obovate-oblong to suborbiculate, 2-5 cm long, deeply segmented into 3-5, petiolulate, subsessile or sessile, cuneate pinnules incised into linear, oblong or obovate-oblong, acute, obtuse or subobtuse, mucronulate lobules 0.8-1.8×0.1-0.5 cm. Cauline leaves usually 2, opposite or subopposite, down below the middle of the stem. Flowers yellow, 2-2.5 cm long, in dense, 5-10 cm long racemes; spur straight or slightly down curved, about equalling to slightly longer than the convex lamina of posticous petal. Bracts much incised.

Capsule oblong, 1.2-1.5×0.3-0.5 cm, narrowed to the base, rounded to subobtuse at apex.

Occ. : Common in habitats similar to those of *C. thyrsiflora*.

Fls. June-Aug.; *Frts.* July-Aug.

Liddarwat, gagarpathri (3000 m) *B.M. Sharma* 14142; Satlanjan-Kolahoi route (3150 m) *B.M. Sharma* 16613; Lidder vy. above Kainmul (12000′) *Duthie* 13126; Sonsal nallah, Lidder vy., (13-14000′) *Duthie* 13367; Liddarwat, Liddar vy., *Inayat* 25479; Tolian, Lidder vy. *Inayat* 25480; Yamher Pass (13400′) *M.V. Laurie* 16794; Glacier snout, Kolahoi (3700 m) *P.S. Jamwal* 16347.

11. **C. sikkimensis** (Prain) Fedde, Repert. 17 : 20, 1921; Stewart, l.c. 291. *C. duthei* Maxim. var. *sikkimensis* Prain, l.c. 33.

Medium, glabrous herb with simple stems about equalling the length of radical leaves; rootstock covered with sheathing leaf bases. Radical leaves oblong, 2-3-pinnatisect, with 4-5 pairs of petiolulate to subsessile lateral pinnae and a terminal one, ultimately segmented into numerous, linear, oblong or oblong-obovate, acute or obtuse lobules. Cauline leaves 2-4, alternate to subopposite. Flowers yellow, 1.5-1.8 cm long (incl. straight, conical spur) in many flowered, simple, ovoid racemes.

A solitary specimen in Dehradun Herbarium, ticketed as "Sorus above Pehalgam (13000′) *R.R. Stewart* 8058"; cited also from Yamher Pass by Stewart, l.c.

12. **C. clarkei** Prain, J. Asiat. Soc. Beng. 65(2) : 34, 1896; Stewart, l.c. 288. *C. moorcroftiana* Wall. ex Hk.f. & Th., Fl. Ind. 266. 1855 p.p.; Hk.f., l.c. 125 p.p.

Glabrous, tufted, robust, glaucescent herb; rootstock crowned with sheathing leaf bases, thick, with branches elongated and twisted together. Radical leaves 4-5-jugately pinnate, nearly as long as the stem; cauline leaves 4, in 2 opposite or subopposite pairs. Flowers light yellow, in dense, conspicuously large-bracteate racemes; bracts veined.

Cited from Yamher Pass by Stewart, l.c.; also reported in Jafri (l.c., 23) to have been gathered by Stewart from near Mt. Kolahoi and in Nafran, upper Liddar vy. (4500 m).

XI. BRASSICACEAE

1. Fruit a silicula, i.e. less than 3 times as long as wide :
   2. Siliculae laterally compressed (i.e., at right angles to the septum); septum narrow :
      3. Siliculae indehiscent, breaking up longitudinally into two nearly orbicular, 2.2-3 cm wide halves ... **16. Megacarpaea**
      3. Siliculae dehiscent throughout their length, their halves not more than 1 cm wide :
         4. Siliculae completely winged or at least their tops winged ... **17. Thlaspi**
         4. Siliculae wingless or obscurely winged :
            5. Siliculae triangular; radicle incumbent ... **15. Capsella**

5. Siliculae oblong, elliptic-oblong or oblong-lanceolate; radicle accumbent **17. Thlaspi**
2. Siliculae dorsally subcompressed or compressed (i.e. parallel to the septum) or not compressed; septum broad and equalling the silicula width or rudimentary or wanting :
6. Leaves 5 cm and more in length, at least the basal ones pinnatifid **2. Rorippa**
6. Leaves 2.5 cm and less in length, entire or toothed at the most, not pinnatifid:
7. Petals deeply 2-fid or absent **9. Erophila**
7. Petals entire or subemarginate :
8. Leaves hairy; cotyledons accumbent mostly **8. Draba**
8. Leaves glabrous; cotyledons incumbent **13. Aphragmus**
1. Fruit a siliqua, i.e. at least 3 times as long as wide :
9. Siliquae indehiscent, irregularly deeply torulose, breaking up on each side into 1-celled, 1-seeded segments **18. Chorispora**
9. Siliquae dehiscent throughout their length, terete, compressed or tetragonous, rarely subtorulose but then not breaking up as described :
10. Leaves all or at least the lower pinnatifid or pinnatisect :
11. Flowers yellow or yellowish :
12. Flowers 2-2.5 mm across; petals equalling to or shorter than the sepals **2. Rorippa**
12. Flowers 5 mm or more across; petals longer than the sepals :
13. Flowers 8 mm or more across; siliquae appressedly hairy **11. Erysimum**
13. Flowers 5-7 mm across; siliquae glabrous :
14. Leaves glaucous, the cauline undivided; seeds 2-seriate **4. Turritis**
14. Leaves green, all or at least some of the lower cauline pinnately divided; seeds 1-seriate **3. Barbarea**
11. Flowers white, lilac or pink :
15. Cauline leaves glaucous, undivided **4. Turritis**
15. Cauline leaves green, pinnately divided :
16. Siliquae slender, strongly compressed, opening from below upwards; seeds 1–seriate **7. Cardamine**
16. Siliquae tumid, subcylindric, opening uniformly; seeds 2-seriate **1. Nasturtium**
10. Leaves simple :
17. Cauline leaves absent **6. Drabopsis**
17. Cauline leaves many or few :
18. Siliquae unilocular or imperfectly bilocular :
19. Plants annual; siliquae linear oblong, 1-1.7 cm long; fruiting pedicels 0.6-1.2 cm long **14. Lignariella**
19. Plants perennial, with somewhat thickened rootstock; siliquae sublanceolate to elliptic, 0.5-1 cm long; fruiting pedicels 4-6 mm long **13. Aphragmus**
18. Siliquae completely bilocular :
20. Radicle incumbent *(obliquely in Arabidopsis thaliana)* ;
21. Flowers 1 cm or more across yellow **11. Erysimum**
21. Flowers small, white, pink or lilac :
22. Seeds 4-6 in each loculus, 2-seriate **13. Aphragmus**

22. Seeds many in each loculus, 1-seriate **10. Arabidopsis**

20. Radicle accumbent :

23. Plants fruiting after the basal leaves perish; siliquae erect, ± appressed to stem **4. Turritis**

23. Plants fruiting while still with basal leaves; siliquae erect, ascending or spreading but not appressed to stem :

24. Basal and lower cauline leaves, some at least, 3-5 (—10)-lobed or toothed at top only; seeds sub 2-seriate **12. Christolea**

24. Leaves entire, dentate or sinuate-dentate but not as described above; seeds 1-, or 2-seriate :

25. Seeds 2-seriate; siliquae up to 2 cm long **8. Draba**

25. Seeds 1-seriate; siliquae 2.5 cm and more long **5. Arabis**

1. NASTURTIUM

N. officinale R. Br. in Ait., Hort. Kew. ed. 2, 4 : 111, 1812; Hk. f. & T. Anders. in Hk.f., l.c. 133. **Fig. 22**

Rather succulent, perennial herb, glabrous or sometimes scattered with simple hairs; stems 10-60 cm long, creeping and rooting in mud from lower nodes, then ascending or floating. Leaves imparipinnate, up to 10 cm long, lower 1-3-foliolate, the upper mostly 3-7-foliolate with auricled base; leaflets up to 2×1.8 cm, entire or sinuate, terminal one ovate, suborbicular or broadly cordate, the lateral ovate or ovate-oblong. Flowers white, 4.5-5 mm diam., in short racemes. Siliquae straight or upcurved, 1-1.8 cm long, horizontal or ascending; seeds reddish, ovoid, *c.* 1 mm long, netted with polygonal pits.

Occ. : Common and often gregarious in standing or slow running shallow waters.

Fls. May-Sept.; *Frts.* June-Sept.

Aru (2400 m) *B. M. Sharma* 14370; Pehalgam (2150 m) *B.M. Sharma* 14691; Aru (2425 m) *P.S. Jamwal* 16306.

2. RORIPPA Scop.

R. islandica (Oeder) Borbas, Balaton Tav. Pt. 2 : 392, 1900; Kitam., Fl. Afgh. 159, 1960; Stewart, l.c. 326. *Sisymbrium islandicum* Oeder, Fl. Dan. 3, 7 : 8, t. 409, 1768. *Nasturtium palustre* DC., Syst. Veg. 2 : 191, 1821; Hk.f. & T. Anders., l.c. 133. **Fig. 23.**

Branched, erect or nearly so, 15-60 cm high, annual or biennial, almost glabrous herb. Leaves coarsely toothed, the lower petiolate, like the auricled middle cauline one slyrate-pinnatifid and 5-15× 1-3 cm, with end lobe largest, ovate or oblong-ovate; upper leaves smaller, sessile undivided or pinnatifid. Flowers pale yellow, laxly racemed; sepals ovate, almost glabrous. Fruit glabrous, oblong-ellipsoid, 5-10×2-3 mm, spreading; seeds irregularly 2-seriate, *c.* 0.8 mm long, reddish, pitted.

Occ. : Common in muddy places, wet ditches and along drains and streams.

Fls. May-Sept.; *Frts.* June-Sept.

Aru (2425 m) *B. M. Sharma* 19349; Liddarwat (3000 m) *B.M. Sharma* 19342; Aru Ghasan (2450 m) *P.S. Jamwal* 16307; Chandanwari (2700 m) *P. S. Jamwal* 16353.

3. BARBAREA R. Br.

1. Upper cauline leaves simple or with the terminal lobe much larger and wider than the lateral ones; pods long pointed **1. B. vulgaris**
1. Upper cauline leaves pinnatifid or pinnatisect, with terminal lobe not much larger than thelateral ones; pods with a short point **2. B. intermedia**

1. B. vulgaris R. Br. in Ait., Hort. Kew. ed. 2, 4 : 109, 1812; Busch in Kom., Fl. U.R.S.S. ed. 2, 8 : 132, 1939; P.W. Ball, Fl. Eur. 1 : 281, 1964. *B. vulgaris* var. *taurica* Hk.f. & T. Anders., l.c. 134.

Erect, glabrous, sparingly branched, 56 cm tall, stiff perennial herb. Basal leaves petiolate, like the lower and middle cauline 7-19×2-4.5 cm and pinnately divided in 3-5 pairs of small lateral and a large broadly ovate, often cordate, sinuate-dentate to subentire end lobe; upper leaves smaller, oblong ovate, dentate, pinnatifid towards the base. Flowers yellow, 6-7 mm diam., laxly racemed; sepals *c.* 3 mm long, glabrous. Siliquae linear, glabrous; ascending, slightly up-curved, 1.5-3 cm long, on almost spreading pedicels; seeds ovoid, *c.* 1.3×1 mm, dark brown, netted.

Occ. : Rare; collected from wet place along a water channel.

Fls. & Frts. June.

Aru (2425 m) *B. M. Sharma* 19343.

2. B. intermedia Boreau, Fl. Centr. Fr., ed. 1, 2 : 48, 1840; P.W. Ball, l.c. 282; Stewart, l.c. 302. *B. vulgaris* var. *sicula* Hk.f. & T. Anders., l.c. 134. *B. sicula* auct. non Presl 1822. **Fig. 24**

Erect, leafy, usually branched, 15-60 cm tall, biennial herb, glabrous or sparingly pubescent with simple hairs. Leaves pinnate, the basal ones 5-12 cm long, with 3-5 pairs of small lateral lobes and a large ovate, suborbicular or cordate end lobe; upper leaves sessile, with basal segments auricled and amplexicaul. Flowers yellow, 5-6 mm diam., racemed; sepals erect, 3-3.5 mm long, glabrous. Siliquae linear, straight, compressed, 1.5-3 cm long; pedicels up to 5 mm long, ascending.

Occ. : Common along water courses at altitudes 2700-3600 m in both eastern as well as western Liddar valleys.

Fls. May-June; *Frts.* June-Aug.

Satlanjan (3100 m) *B. M. Sharma* 14308; Kolahoi valley near Lidderwat (11-12000′) *Duthie* 13507; Kolahoi (3400m) *P.S. Jamwal* 14222.

4. TURRITIS Linn.

T. glabra Linn., Sp. Pl. 666, 1753; Busch in Kom., l.c. 171; Kitam., l.c. 166; Stewart, l.c. 330. *Arabis glabra* (Linn.) Bernh., Syst. Erfurt. 1 : 195, 1800; Hk.f. & T. Anders, l.c. 135.

Erect, unbranched, 0.3-1 m tall, glaucous herb; stem strict, pubescent at base alone. Radical leaves stellately hairy, 5-12 cm long, entire or sinuate-toothed to lyrate, soon withering away; cauline leaves glaucous, oblong, or ovate-lanceolate, entire, 2-8×0.5-2.5 cm, auricled, amplexicaul, glabrous or with margins and underside of midrib sparsely hairy. Flowers white or pale yellow, *c.* 5 mm diam., in

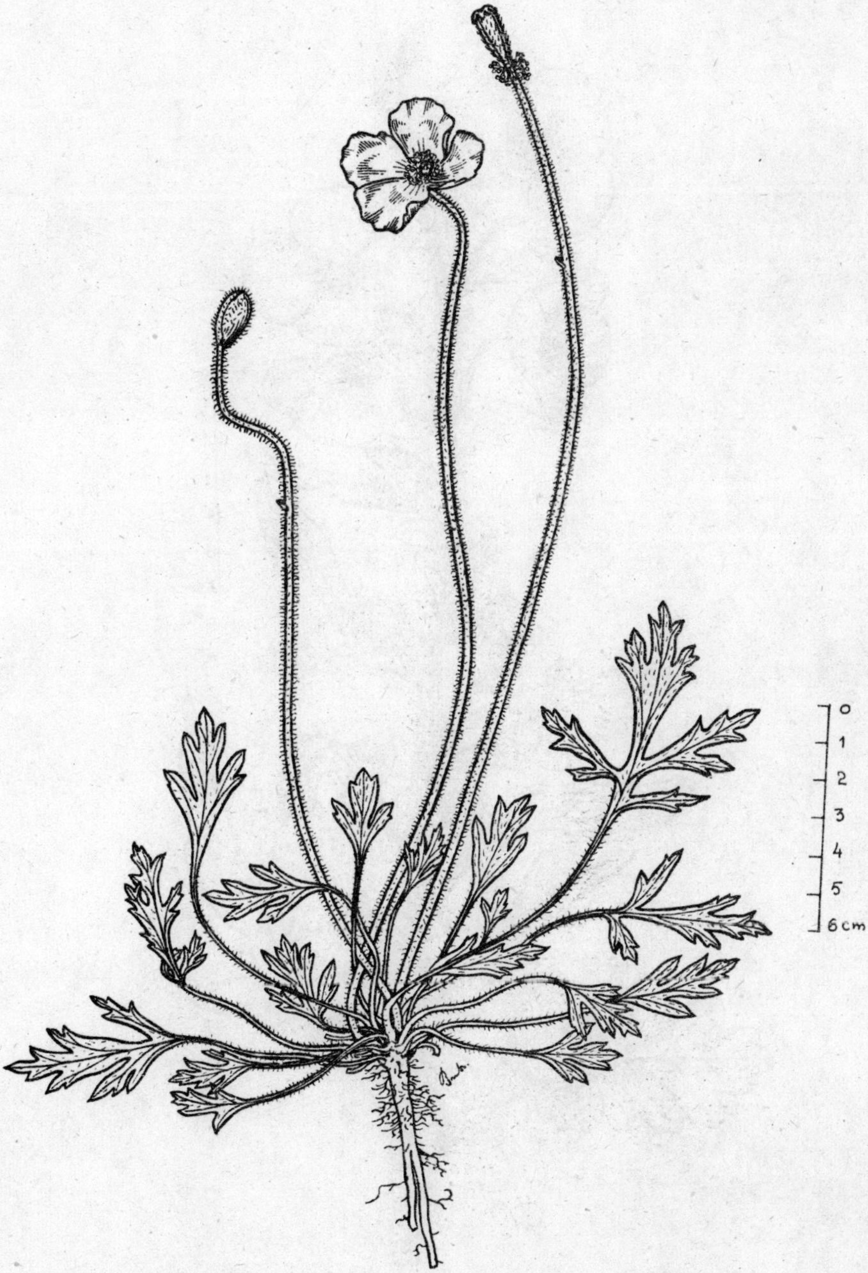

Fig. 17. *Papaver nudicaule* Linn.

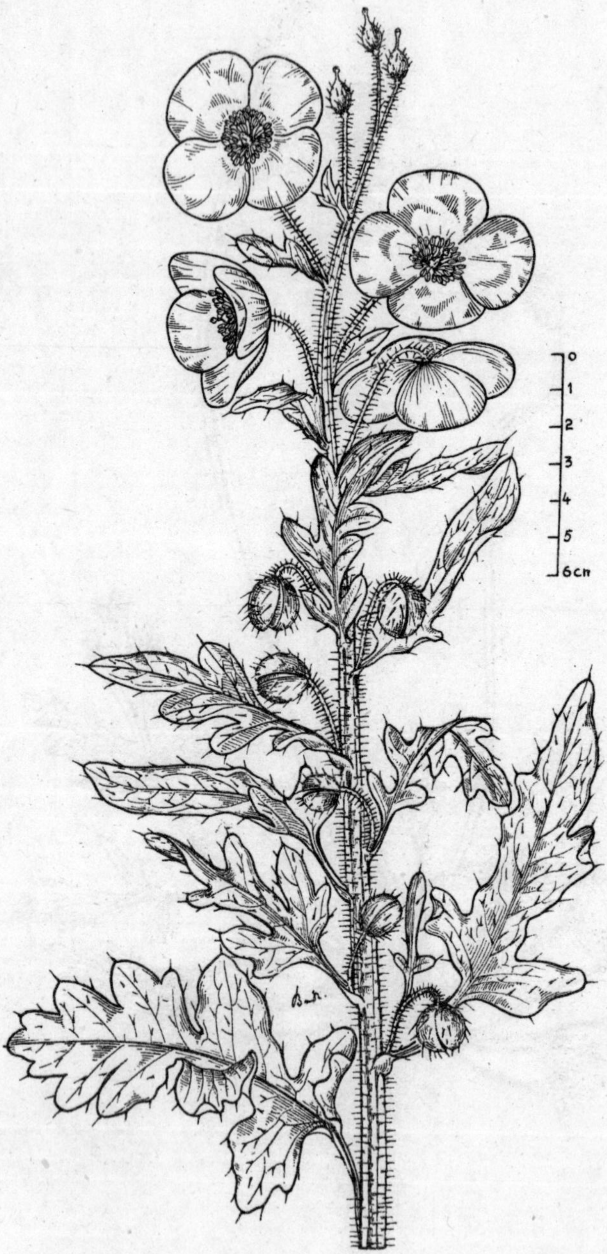

Fig. 18. *Meconopsis aculeata* Royle

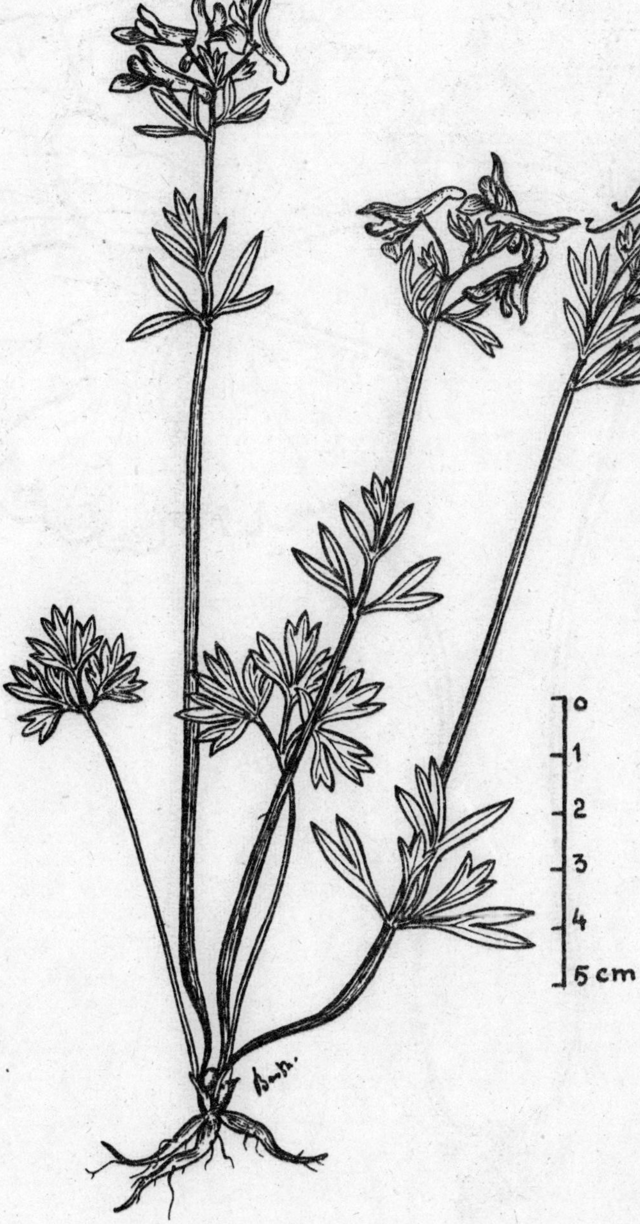

Fig 19. *Corydalis cashmeriana* Royle

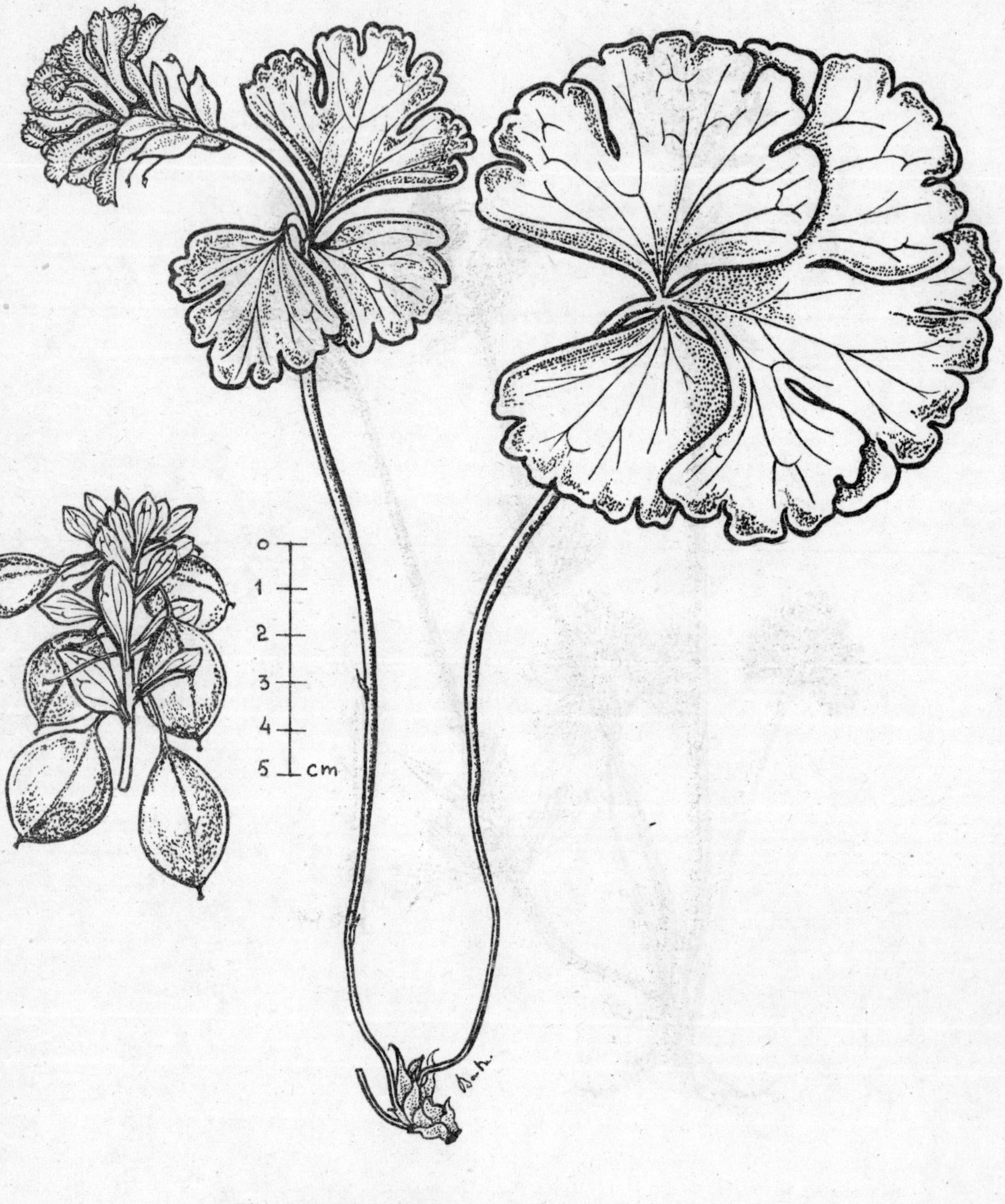

Fig. 20. *Corydalis crassifolia* Royle

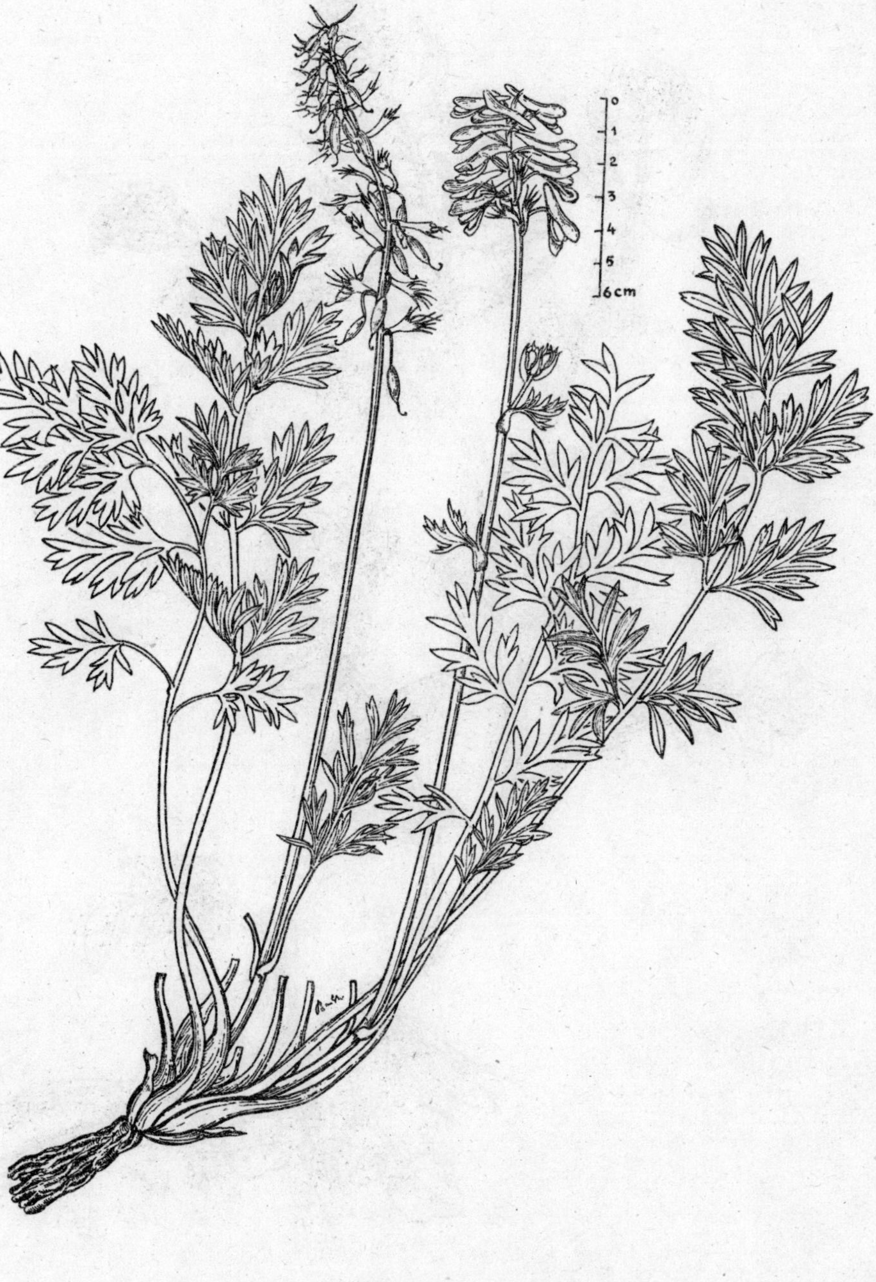

Fig. 21. *Corydalis govaniana* Wall. ex Tent.

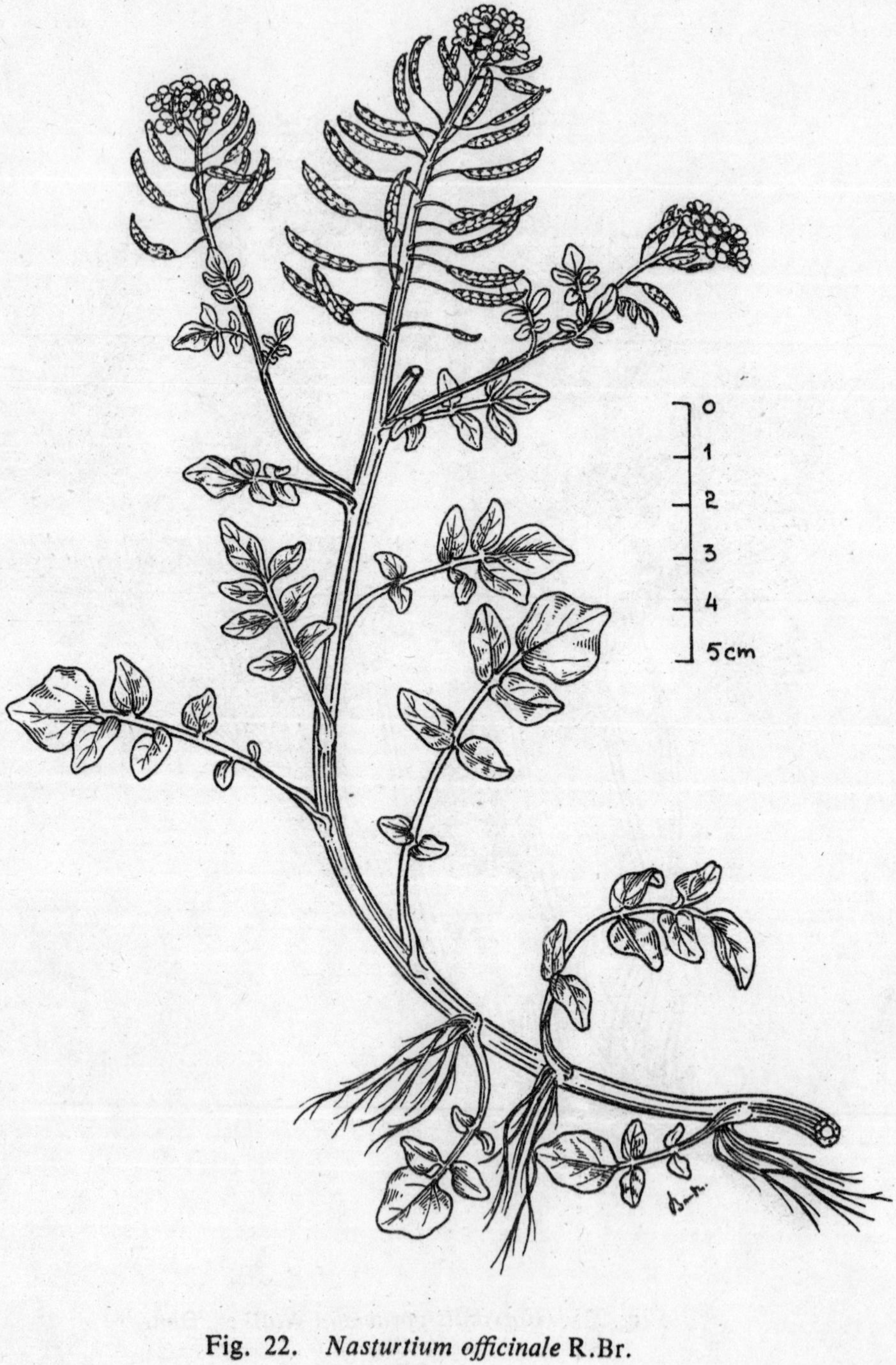

Fig. 22. *Nasturtium officinale* R.Br.

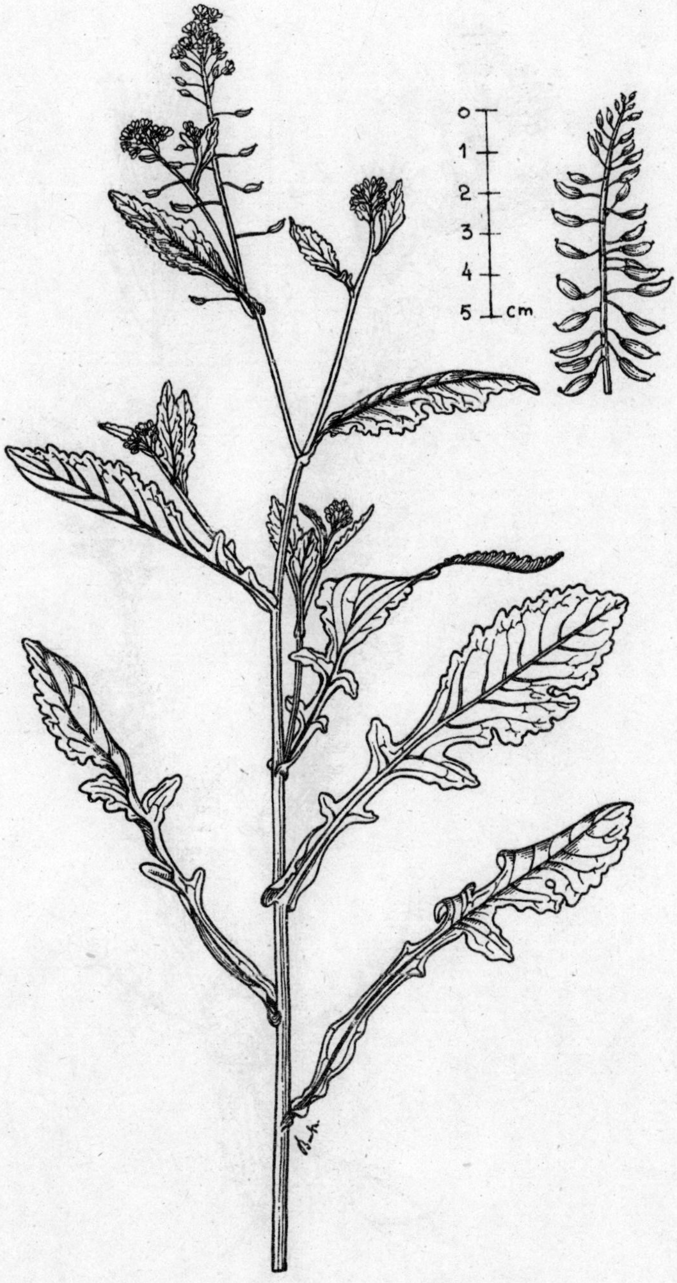

Fig. 23. *Rorippa islandica* Borbas

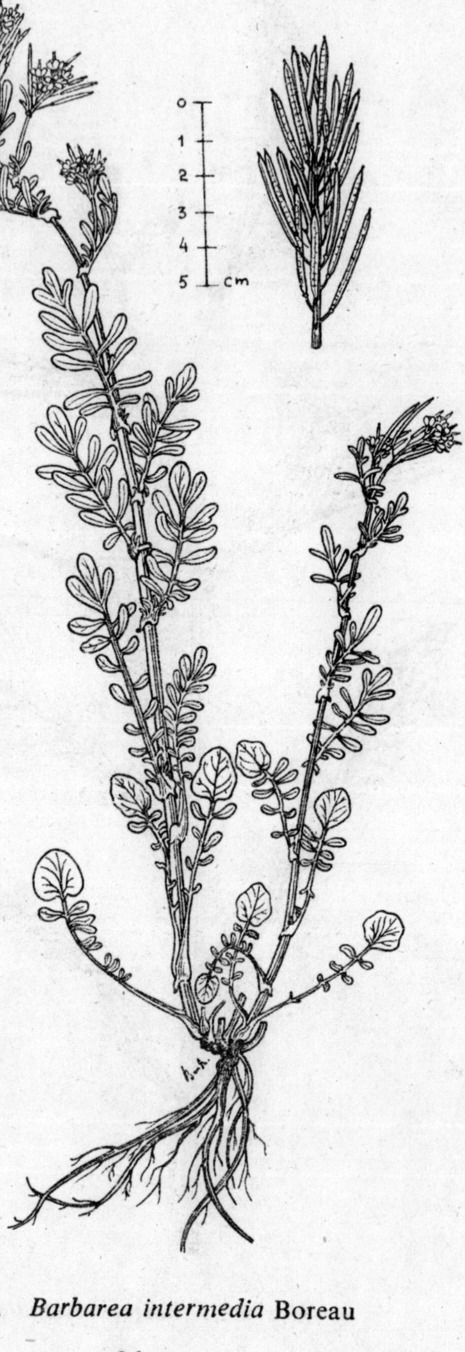

Fig. 24. *Barbarea intermedia* Boreau

terminal racemes elongating up to 30 cm in fruit. Siliquae slender, erect, linear, 4-8 cm long, crowded.

Occ. : Common in cultivated fields and on scrubby slopes about Aru and Pehalgam.

Fls. May-July; *Frts.* Aug.

Aru (2400 m) *B. M. Sharma* 14221; Pehalgam (2160 m) *B.M. Sharma* 14675; Pehalgam-Chandanwari route (2250 m) *P.S. Jamwal* 16340.

5. ARABIS Linn.

1. Cauline leaves cuneate or narrowed at the base, not auricled — **1. A. tibetica**
1. Cauline leaves sessile, amplexicaul, with obtuse basal auricles :
   2. Fruiting pedicels 0.3-1.2 cm long; style of siliquae short, below 1 mm long — **2. A. pterosperma**
   2. Fruiting pedicels 1.5-2 cm long; style of siliquae slender, elongate, 1-1.5 mm long — **3. A. amplexicaulis**

1. **A. tibetica** Hk.f. & Th., J. Linn. Soc. Bot. 5 : 143, 1861; Hk.f. & T. Anders., l.c. 136.

Perennial herb with several, 5-20 cm long, subspreading, ascending or arcuate flowering stems covered with forked hairs. Radical leaves stellately hairy, spathulate to obovate, 0.8-4 x 0.25-1 cm, sinuate to entire, the cauline oblong-oblanceolate to spathulate, 0.8-2 x 0.2-0.4 cm, entire to sparingly-toothed or,-lobulate. Flowers white or creamy, 5-6 mm diam., in terminal racemes; sepals ± half the length of 5-7 mm long petals, stellately hairy. Siliquae (immature) narrow-linear, glabrous, acute.

Occ. : Common among moraines in axposed places at high altitudes.

Fls. May.-July.

Kolahoi (3500 m) *B. M. Sharma* 19355; Kolahoi, near glacier snout (3700 m) *P. S. Jamwal* 16342.

2. **A. pterosperma** Edgew. in Trans. Linn. Soc. 20 : 33, 1851; Schulz in Notizbl. 9 : 1063, 1927; Stewart, l.c. 301. *A. alpina* auct. (non Linn.); Hk.f. & T. Anders., l.c. 135. **Fig. 25**

Simple or sometimes branched, 15-50 cm tall, perennial herb; stems coarsely hairy with forked and stellate pubescence, specially so in lower part. Radical leaves spathulate or obovate-oblong, attenuate-petiolate, 2-8.5 x 0.5-2.5 cm, stellately hairy, coarsely dentate, the cauline oblong, with usually elongate, entire apex and toothed lower part. Flowers white, *c.* 8 mm diam., laxly racemed; sepals glabrous. Siliquae narrowly linear, 3-6 cm long, glabrous, compressed, spreading or ascending.

Occ. : Common along banks throughout the Liddarwat-Kolahoi route.

Fls. June-Aug. ; *Frts.* July-Sept.

Saswarian (2950 m) *B. M. Sharma* 14209 & 19352; Chandanwari (9500′) *H. D. Thapliyal* DD. no. 26543; Satlanjan (3100 m) *P. S. Jamwal* 14208.

3. **A. amplexicaulis** Edgew. in Trans. Linn. Soc 20 : 31, 1851; Hk.f. & T. Anders., l.c. 136.

Perennial herb, differing from *A. pterosperma* in characters of key and rootstock more robust and frequently branched at crown; stems hispidly pilose with simple and forked hairs, especially so below; leaves rather thin, sinuate-toothed to almost entire, more or less hispid with forked and stellate hairs, the cauline ovate to oblong; flowers 0.8-1 cm diam.

Occ. : Often under scrubs on humus laden slopes.

Fls. June-July ; *Frts.* July-Aug.

Mundlan (2325 m) *B. M. Sharma* 14211; Aru (2425 m) *B. M. Sharma* 16625; Aru Ghasan (*c.* 2450 m) *P. S. Jamwal* 14388.

6. DRABOPSIS C. Koch

D. verna C. Koch in Linnaea, 15 : 253, 1841; Hedge in Rech. f., Fl. Iran. 57 : 212, 1968; Stewart, l.c. 312. *Arabidopsis verna* (C. Koch) N. Busch in Kom., l.c. 80. *Sisymbrium nudum* (Belang. ex Boiss.) Boiss., Fl. Or. 1 : 214, 1867. *D. nuda* (Belang. ex Boiss.) Stapf in Denkschr. Akad. Wiss. Wien. Math.-Nat. K. 51, 2 : 298, 1886; For. Fl. Sgr. 167. *Arabis nuda* Belang. ex Boiss. in Annls Sci Nat. ser. 2, 17 : 54, 1842; Hk.f. & T. Anders., l.c. 137.

Annual herb with erect, leafless, up to 15 cm long flowering stems, more or less hairy below and usually glabrous above. Leaves rosulate, spathulate or oblong-obovate, 0.5-1-5 x 0.3-0.6 cm, entire or sparsely toothed, hairy; hairs branched. Flowers yellow, 2.5-3 mm diam. ; sepals obtuse, glabrous, about half the length of the petals. Siliquae flattened, erect, linear, 2.5-3.8 cm long, glabrous, straight or slightly curved.

Occ. : Common along roadsides and other open places about Pehalgam.

Fls. April-June.

Pehalgam (2160 m) *B. M. Sharma* 14688 & 14689; Pehalgam (2160 m) *P. S. Jamwal* 16326.

7. CARDAMINE Linn.

1. Flowers 1-1.5 cm across; petals spreading — **1. C. macrophylla**
1. Flowers 3-4 mm across; petals nearly erect:
   2. Cauline leaves lobed at base — **2. C. impatiens**
   2. Cauline leaves not lobed at base — **3. C. hirsuta**

1. **C. macrophylla** Willd., Sp. Pl. 3 : 484, 1800; Hk.f. & T. Anders., l.c. 139. *C. polyphylla* D. Don, Prodr. Fl. Nep. 201, 1825. *C. macrophylla* ssp. *polyphylla* (D. Don) O.E. Schulz, Notizbl. Bot. Gart. Berl. 9 : 1071, 1927; Stewart, l.c. 304. **Fig. 26**

Perennial 30-75 cm tall, erect, leafy herb from creeping rootstock; stem glabrous, striate. Leaves 5-15 cm long, imparipinnately 7-15-foliolate mostly; leaflets lanceolate, ovate or ovate-lanceolate, 1.2-7 x 0.7-2.5 cm, pubescent on margins and sometimes nerves beneath as well, irregularly deeply toothed; mostly acuminate; teeth sharp or blunt. Flowers white or pink, 1-1.5 cm diam., in terminal racemes; sepals 4-4.5 mm long, scattered with simple hairs. Siliquae linear, glabrous, straight, 2-4.5 cm long, compressed.

Occ. : Common in stony places along water courses up to *c.* 3500 m.

Fls. June-July; *Frts.* July-Aug.

Saswarian nallah (3000 m) *B. M.*

Sharma 14173 & 14681; Liddar vy. above Pehalgam (8–9000′) *Duthie* 13477; Kinimola nallah, Liddar vy. *Inayat* 25509; Liddarwat–Tarsar route (2980 m) *P. S. Jamwal* 16311.

2. **C. impatiens** Linn., Sp. Pl. 655, 1753; Hk.f. & T. Anders., l.c. 138.

Annual or biennial, 15-50 cm high, simple or branched, erect herb; stem glabrous. Leaves usually 5-10 cm long, imparipinnately 7-17-foliolate mostly, radical ones few or many, the cauline with auricled, amplexicaul, ciliate petiolar bases; leaflets petiolulate or subsessile, of radical and lower cauline leaves ovate to suborbicular, 3-5-lobulate, of upper ones ovate-oblong or lanceolate, 0.8-2.5 x 0.2-1.2 cm, lobulate or serrate to almost entire, ciliate. Flowers crowded, white, 3-4 mm diam., in terminal racemes. Siliquae linear, 2-3.5 x *c*. 0.1 cm, glabrous, compressed.

Occ. : Common along water courses and in moist open places in forests up to *c*. 3100 m.

Fls. May-July ; *Frts.* July-Aug.

Satlanjan (3080 m) *B. M. Sharma* 19351; Aru (2400 m) *B. M. Sharma* 14680; Mundlan (2300 m) *P. S. Jamwal* 16341.

3. **C. hirsuta** Linn., Sp. Pl. 655, 1753; Busch in Kom., l.c. 159; Stewart, l.c. 304.

Annual, 10-25 cm tall, erect herb, with several, stiff, glabrous, sparingly leafy to almost leafless stems from a basal rosette of leaves. Leaves imparipinnately 5-11-foliolate; leaflets subentire to irregularly angled and toothed, of upper leaves smaller and elliptical; of the lower obovate to orbicular except the slightly larger, subreniform terminal one, 0.5-1 cm diam., hairy or glabrous. Flowers crowded, white, 3-4 mm diam., in terminal racemes; stamens 4. Siliquae erect, linear, 1.2-2.5 x 0.1-0.15 cm, glabrous, compressed; seeds ovoid, light brown, 1.2 x 0.8 mm, compressed, margined.

Occ. : In moist lawns and other open places in Pehalgam.

Fls. & Frts. April.

Pehalgam (2160 m) *B. M. Sharma* 14686.

8. **DRABA** Linn.

(Key adapted from Jafri, Fl. W. Pak. no. 55, 1973. Species marked with asterisk (*) not seen personally and, therefore, simply listed)

1. Flowering shoots aphyllous, scapose — **D. oreades**
1. Flowering shoots leafy, usually many-leaved (rarely 1-2-leaved):
    2. Annuals — **1. D. stenocarpa**
    2. perennials:
        3. Plants ± ciliately hairy with simple or branched hairs — **D. altaica**
        3. Plants tomentose with short, branched, white hairs (rarely mixed with simple ones):
            4. Homotrichous (with short and branched hairs) — **2. D. lanceolata**
            4. Heterotrichous (short branched hairs mixed with simple or forked long hairs) — **D. lasiophylla**

1. **D. stenocarpa** Hk.f. & Th., J. Linn. Soc. Bot. 5 : 153, 1861, Stewart, l.c. 311; Jafri in Nasir & Ali, Fl. W. Pak.

55 : 145, 1973. *D. linearis* Hk.f. & T. Anders., l.c. 144 (non Boiss. 1842).

Erect, up to 30 cm tall, annual herb, often branched from base. Leaves acute, like the stem covered with simple and furcate hairs, radical ones rosulate, obovate or linear-oblong, entire or subdentate, the cauline few, 1-3-toothed. Flowers yellowish, drying to white, *c.* 3.5 mm diam., laxly racemed. Siliquae linear oblong, erect, compressed, 0.8-2 cm long, on slender, ascending pedicels.

Fls. & Frts. Sept.

Amarnath (12500′) *K. M. Vaid*, DD. no. 118377.

2. **D. lanceolata** Royle, Ill. Bot. Himal. 1 : 72, 1834; Tolm. in Kom., l.c. 433; Stewart, l.c. 309. *D. incana* Hk.f. & T. Anders., l.c. 143 (non Linn.).

Perennial, leafy herb; shoots erect or semierect, hoary with short, whitish, branched hairs. Leaves acute, pubescent, mostly toothed, radical ones rosulate, lanceolate, the cauline ovate-, or oblong-lanceolate, sessile. Flowers 3-4 mm diam., laxly racemed. Fruits linear-oblong or elliptic lanceolate (6-14 x *c.* 1.5 mm), flattened, pubescent with branched hairs.

Fls. & Frts. July-Aug.

Lidder vy. near Sheshnag (12-13000′) *Duthie* 13327. Stated by Stewart (l.c.) to be common in Pehalgam. Astanmarg, Harnag etc. (11-15000′).

* **D. oreades** Schrenk syn. *D. alpina* Hk. f. & T. Anders., l.c. 142. Reported in Jafri, l.c. 131 from Mt. Kolahoi (*Stewart*) and Sangam vy. (*Duthie*).

* **D. altaica** Bunge syn. *D. fladnitzensis* Hk.f. & T. Anders., l.c. 143. Cited by Stewart (l.c. 308) from Harnag and Kolahoi.

* **D. lasiophylla** Royle - Cited by Stewart (l.c. 309) from Mt. Kolahoi; also reported in Jafri, l.c. 140 from Sheshnag (*Drummond*).

9. EROPHILA DC.

E. verna (Linn.) Besser, Enum. Pl. Volhyn. 71. 1822; Chevall., Fl. Gen. Env. Paris, 2 : 898, 1827; Walters in Fl. Eur. 1 : 312, 1964. *Draba verna* Linn., Sp. Pl. 642, 1753. *E. vulgaris* DC., Syst. 2:356, 1821; Hk.f. & T. Anders., l.c. 145.

Dwarf annual with solitary, slender, flowering stem, hairy below with short, branched hairs and glabrous above. Leaves all radical, rosulate, lanceolate, elliptic or oblong-ovate, entire to 1-2-toothed, more or less stellately hairy, especially so on the upper surface and towards the tips. Flowers white, *c.* 3 mm diam., in short terminal, initially dense racemes; sepals hairy; petals 2-fid. Siliculae oblong-elliptic, *c.* 8 mm long and about half as broad, glabrous.

Occ. : Common along roadsides and paths in Pehalgam.

Fls. April-May.

Pehalgam (2160 m) *B. M. Sharma* 14687; Pehalgam (2160 m) *P. S. Jamwal* 16327.

10. ARABIDOPSIS Heynh.

1. Bases of cauline leaves broad, sagittate-amplexicaul:

2. Racemes bracteate nearly to the apex
1. A. himalaica
2. Racemes ebracteate or at the most 2-3-bracteate only towards the base
2. A. mollissima

1. Bases of cauline leaves narrow, not sagittate-amplexicaul :
   3. Stem scabrid with short stellate pubescence along its entire length; flowers pink **3. A. stricta**
   3. Stem smooth above, only in lower part hairy with simple and 2-fid hairs; flowers white
   4. A. thaliana

1. **A. himalaica** (Edgew.) O.E. Schulz in Engler, Das Pflazenr. 86 (iv-105) : 283, 1924; Raizada, Indian For. 92 (5) : 301, 1966; Stewart, l.c. 298. *Arabis himalaica* Edgew., Trans. Linn. Soc. 20 : 31, 1846. *Sisymbrium himalaicum* Hk.f. & Th., J. Linn. Soc. 160, 1861; Hk.f. & T. Anders., l.c. 147. **Fig. 27**

Usually much branched, erect, 15-35 cm tall, stiff biennial or perennial herb, hispidly tomentose or hirsute with mixed simple and branched hairs. Leaves hairy like stem, radical short-petioled, obovate-lanceolate, the cauline sessile, oblong, up to 3.8 × 1 cm, sinuate-toothed, clasping the stem by their auricled bases. Flowers pink or white, *c.* 4 mm diam., in bracteate racemes; sepals hirsute. Siliquae terete, 1.5-3.5 cm long, erect or spreading, glabrous, with a foliose bract at the pedicel base.

Occ. : Common along banks in Chotta Khalen-Kolahoi tract and in open places along Chandanwari-Zojipal route.

Fls. & Frts. June-Aug.

Kolahoi (3350 m) *B. M. Sharma* 19353; Satlanjan (3100 m) *B. M. Sharma* 14682; Chandanwari (9500′) *H. D. Thapliyal* 26541; Nangneri nallah (2850 m) *P. S. Jamwal* 16313.

2. **A. mollissima** (C. A. Mey.) N. Busch, Fl. Sib. & Or. Extr. 1 : 136, 1913; Busch in Kom., l.c. 79. *Sisymbrium mollissimum* C. A. Mey., Ledeb. Fl. Alt. 3 : 140, 1831; Hk.f. & T. Anders., l.c. 147.

Mostly simple or with 2-3 stems from root; hardly distinct from the preceding except through key characters which too, however, frequently break down because of the occurrence of several intermediate forms. Flowering and fruiting periods and habitats same as those of *A. himalaica.*

Satlanjan (3100 m) *B. M. Sharma* 14683 & 14684; Pisu Ghati (3000 m) *P. S. Jamwal* 16317.

3. **A. stricta** (Camb.) N. Busch, Fl. Cauc. Crit. 3 (4) : 457, 1909; Schulz in Engler, Das Pflanzenr. 86 (iv-105) : 252; 1924. *Malcolmia stricta* Camb. in Jacq. Voy. Bot. Inde. 4 : 16, 1844. *Sisymbrium strictum* Hk.f. & Th. in J. Linn. Soc. Bot. 5 : 161, 1861; Hk.f. & T. Anders., l.c. 149.

Hoary, 30-60 cm tall, stiff, unbranched annual. Radical leaves perishing prior to flowering, the cauline sessile, linear-oblong, sinuate-toothed, 1.5-3.5 x 0.2-0.6 cm, densely stellate-hairy, gradually smaller upwards. Flowers pink, 3.5-4 mm diam., in short, dense, terminal racemes, the latter elongate in fruit; sepals and pedicels pubescent. Siliquae slender, glabrous, terete, 2.5-5 cm long, slightly curved, ascending.

Occ. : Among stones along the banks of streams.

Fls. June-July.

Liddarwat (2800 m) *B. M. Sharma* 19354; Nangneri nallah (2850 m) *P. S. Jamwal* 16312.

4. **A. thaliana** (Linn.) Heynh. in Holl & Heynh., Fl. Sachs. 1 : 538, 1842; Kitam., Fl. Afgh. 141. 1960; Stewert, l.c. 299. *Arabis thaliana* Linn., Sp. Pl. 665, 1753. *Sisymbrium thalianum* J. Gay & Monn., Annals Sci. nat. 1, ser. 7 : 399, 1826 in ad. not.; Hk.f. & T. Anders., l.c. 148.

Almost filiform to moderate, 5-30 cm tall annual; stems simple or branched, solitary or few from a basal rosette of leaves. Leaves hairy with simple, forked and stellate hairs, radical ones obovate-spathulate, 1-5 x 0.4-1.5 cm, toothed or entire, the cauline remote, oblong or obovate-oblong, sessile, narrowed below. Flowers white, 2.5-3 mm diam., laxly racemed; sepal-tips hairy. Siliquae glabrous, ascending to almost patent, on 0.5-1.2 cm long glabrous, pedicels; seeds ovoid, brownish, finely punctate.

Occ. : Common in lawns, waste places and on open dry slopes.

Fls. & *Frts.* April-June.

Aru Ghasan (2500 m) *B. M. Sharma* 19359; Pehalgam (2160 m) *B. M. Sharma* 14685; Sonsal nallah, Lidder vy. (12-13000′) *Duthie* 13345; Aru (2425 m) *P. S. Jamwal* 16335; Pehalgam (2160 m) *P. S. Jamwal* 16328.

11. **ERYSIMUM** Linn.

Hairs on the leaf surface mostly 2-partite; sepals 8-9 mm long; fruiting pedicels 0.7-1.5 cm long
1. E. melicentae

Hairs on the leaf surface mostly 3-4-partite, a few sometimes 2-partite; sepals 5-6 mm long, fruiting pedicels 4-5 mm long
2. E. hieraciifolium

1. **E. melicentae** Dunn, Kew Bull. 1920: 336, 1920; Jafri, Notes R. bot. Gdn. Edinb. 22 : 108, 1956; Stewart, l.c. 314. *E. odoratum* Hk.f. & T. Anders., l.c. 154 (non Ehrh. 1792). *E. parkeri* O. E. Schulz, Notizbl. Bot. Gart. Berl. 9 : 1083, 1927. **Fig. 28**

Erect, 45-70 cm tall, biennial or perennial herb; stem branched high above the ground or simple, rarely with 1-2 additional shoots from the base, appressedly hairy with 2-partite hairs. Leaves hairy, the radical ones petioled, rosulate, lanceolate 5-10 x 0.5-1 cm, sinuate-toothed, or entire, the cauline similar, sinuate-toothed, smaller and sessile or subsessile upwards. Flowers yellow or orange-yellow, 1-1.5 cm diam., congested, in terminal racemes, the latter elongate in fruit. Siliquae linear, 3-6 cm long, almost quadrangular, appressedly hairy; seed oblong, *c.* 2 mm long, smooth.

Occ. : Along banks and on humus laden rocks in torrents; common.

Fls. May-July ; *Frts.* July-Aug.

Liddarwat (2900 m) *B. M. Sharma* 14197; Lidder vy. above Jamian (9-10,000′) *Duthie* 13447; Chandanwari (9500′) *H. D; Thapliyal* 26532; Kolahoi, Lidder vy. *Inayat* 25497; Gagar pathri, Liddarwat (3000 m) *P. S. Jamwal* 14301.

2. **E. hieraciifolium** Linn. Cent. Pl. 1 : 18, 1755; Hk.f. & T. Anders., l.c. 153.

Erect, 30-60 cm tall, biennial herb; stem angled, mostly simple, appressedly hairy with 2-partite hairs, scabrous. Radical leaves petioled, usually perishing early, the cauline oblong, 2-2.5 cm long (in our specimens), sinuate toothed, sessile. Flowers orange yellow, 0.8-1 cm long, congested, in terminal racemes. Siliquae linear, erectopatent, 3-5 cm long, quadrangular; appressedly hairy.

Occ. : On banks and rocky cliffs about Aru.

Fls. & Frts. June-July.

Aru (2400 m) *B. M. Sharma* 14334; Aru (2400 m) *P.S. Jamwal* 16334.

12. **CHRISTOLEA** Camb.

C. stewartii (T. Anders.) Jafri in Notes Roy. Bot. Gard. Edinb. 22 (1) : 53, fig. 1, E. 1955 & Fl. W. Pak. 55 : 157, 1973. *Cherianthus stewartii* T. Anders. in Hk.f., l.c. 132. *Ermania stewartii* (T. Anders.) O.E. Schulz in Engl., Bot. Jahrb. 66 : 98, 1933; Stewart, l.c. 313.

Perennial, hoary herb with 5-12 cm long, semierect or spreading, pilose, leafy stems. Leaves pilose with simple and branched white hairs, radical ones rosulate, spathulate, 1-2.5×0.5-1 cm, at least some bluntly 3-5 (-10)-toothed at the apex, the cauline smaller, mostly linear, sessile. Flowers lilac, 4-5 mm diam., in bracteate, terminal racemes up to 6 cm in fruit. Siliquae glabrous secund, erectopatent, 1.5-3.5×0.25-0.3 cm, linear, compressed; stigma broad, sub-bilobed, nearly sessile.

Occ. : Along waysides in E. Lidder valley.

Fls. & Frts. July-Aug.

Pisu Ghati (300 m) *B. M. Sharma* 16675; Zojipal (3450 m) *B. M. Sharma* 16680.

13. **APHRAGMUS** Andrz.

A. oxycarpus (Hook. f. & Thoms.) Jafri, Notes Roy. Bot. Garden Edinb. 22 : 96, 1956. syn. *Braya oxycarpus* (Hook. f. & Thoms.) in J. Linn. Soc Bot. 5 : 169, 1861.

Reported in Nasir & Ali, Fl. W. Pak., l.c. 259 from Mt. Kolahoi (*Stewart*) and Liddar vy. (*Duthie)*.

14. **LIGNARIELLA** Baehni

L. obscura (Dunn) Jafri, Candollea 16 : 134. 1957. syn. *Draba obscura* Dunn.

Cited by Stewart (l.c. 298) from Upper Lidder vy.. Armium Glen to Harnag; also reported in Nasir & Ali, l.c. 260, from Astan marg (*Drummond*), Lidder vy. (*Duthie*) and above Lidderwat (*Duthie*).

15. **CAPSELLA** Medik.

C. bursa-pastoris (L) Medik., Pflanzeng. 85, 1792; Hk.f. & T. Anders., l.c. 159.

Fig. 29

Variable, erect, 4-45 cm tall annual herb, glabrescent or hairy with long simple and short stellate hairs; stems leafy to almost leafless, simple or branched, 1 or more from a basal rosette of leaves. Radical leaves usually petiolate pinnatifid or sinuate-toothed, up to 15×3 cm; cauline leaves smaller, stem-clasping, pinnatifid or toothed the upper almost entire. Flowers white. 2.5-3.5 mm diam., racemed. Siliculae flat obcordate-triangular, 5-9×4-7 mm; seeds oblong-ovoid, *c.* 1 mm long, reddish brown, punctate.

Occ. Common in cultivated fields, wastelands and along paths.

Fls. & Frts. May-Aug.

Liddarwat (2850 m) *B.M. Sharma* 19348; Aru (2400 m) *B.M. Sharma* 16601; Mahagunus Pass (13000′) *K. M. Vaid, D.D.* s.n.; Chandanwari (2700 m) *P.S. Jamwal* 16354.

16. MEGACARPAEA DC.

M. polyandra Benth. in Hk., Kew Journ. Bot. 7 : 356, t. 7, 1855; Hk.f. & T. Anders., l.c. 161; Blatter, Beaut. Fl. Kash. 1 : 44, Pl. 12, 1927; **Fig. 30**

Robust, *c.* 1 m tall, branched, perennial herb, thick and fleshy in the lower parts and pubescent with simple hairs. Basal leaves 25-45 cm long, pinnatisect into lanceolate, acuminate, serrate segments 10-20×1-5 cm; upper leaves gradually smaller, pinnatisect or, -fid, the upper most lanceolate or linear. Flowers cuplike, white, 0.8-1 cm diam., in dense, panicled, pubescent racemes; stamens 10-16, subequal. Siliculae didymous, much flattened; lobes broadly winged, suborbicular, 2.2-3 cm wide, one sometimes much smaller or abortive, 1-seeded; seeds reniform 1-1.2×0.9 cm, brown.

Occ. : Sparingly met with on slopes about Liddarwat and Kolahoi.

Fls. June-July; *Frts.* July-Aug.

Dudhsar (3775 m) *B. M. Sharma* 19344.

Fleshy roots are relished as pot herb or eaten raw; the leaves are prized as spinach.

17. THLASPI Linn.

1. Fruits orbicular to broadly obcordate, winged completely — **1. T. arvense**
1. Fruits oblong, oblong-elliptic or oblong-obovate, winged towards top or the wings suppressed or absent :
   2. Flowers 6-7 mm diam., fruits oblong-elliptic with entire, acute apex — **2. T. andersonii**
   2. Flowers 4-5 mm diam.; fruits oblong or oblong-obovate with slightly to distinctly emarginate apex :
      3. Fruits straight, winged towards the top at least — **3. T. montanum**
      3. Fruits often curved and twisted, wingless — **4. T. cochlearioides**

2. **T. arvense** Linn., Sp. Pl. 646, 1753; Hk.f. & T. Anders., l.c. 162.

Erect, glabrous, 15-50 cm tall, simple or sparingly branched annual with early perishing, oblanceolate, petiolate basal leaves. Stem leaves oblong-lanceolate, 2.5-10×0.5-2 cm, coarsely dentate to subentire, the upper with amplexicaul, sagittate bases. Flowers white, 3-4.5 mm diam., racemed. Siliculae broadly winged, 1-1.8 cm. diam., with deep, narrow or open apical notch. Seeds black, ovoid, 1.5-2 mm long, concentrically ribbed, 4-6 in each cell.

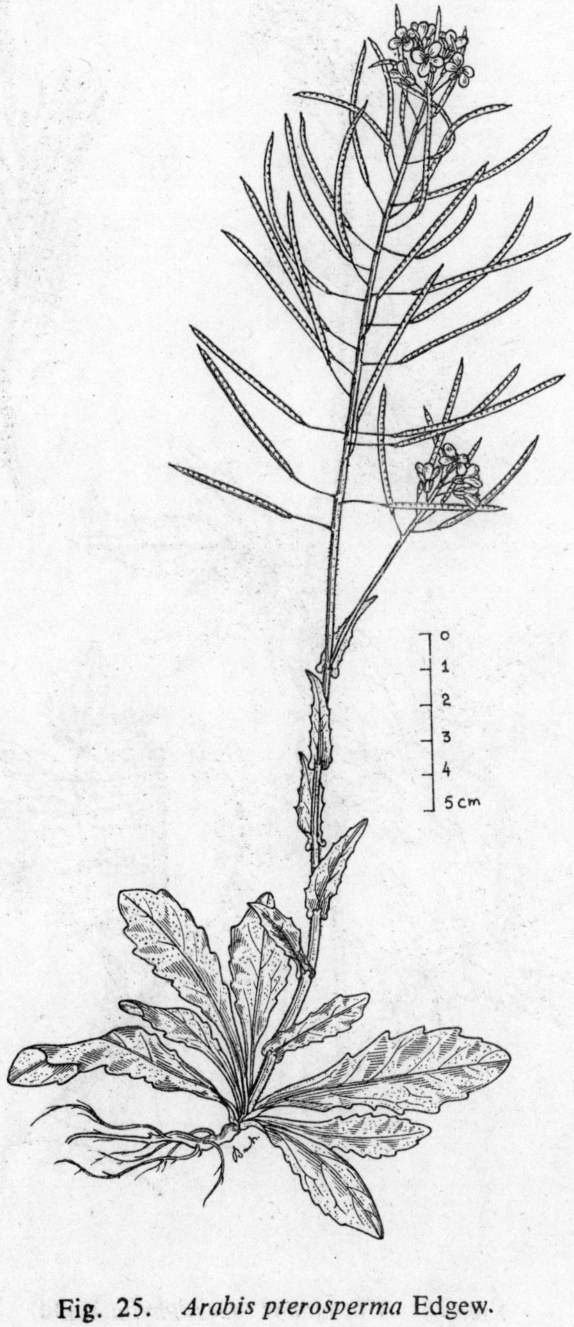

Fig. 25. *Arabis pterosperma* Edgew.

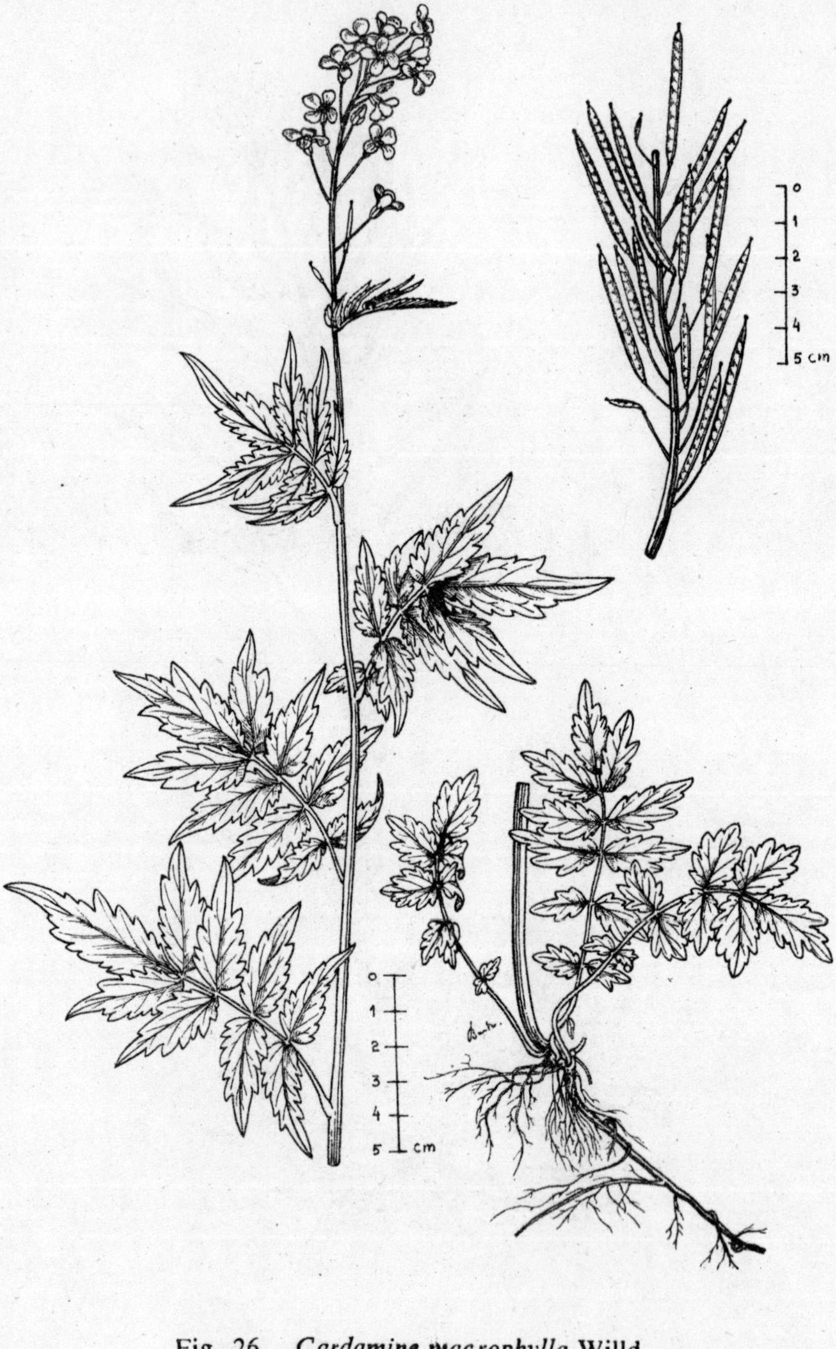

Fig. 26. *Cardamine macrophylla* Willd.

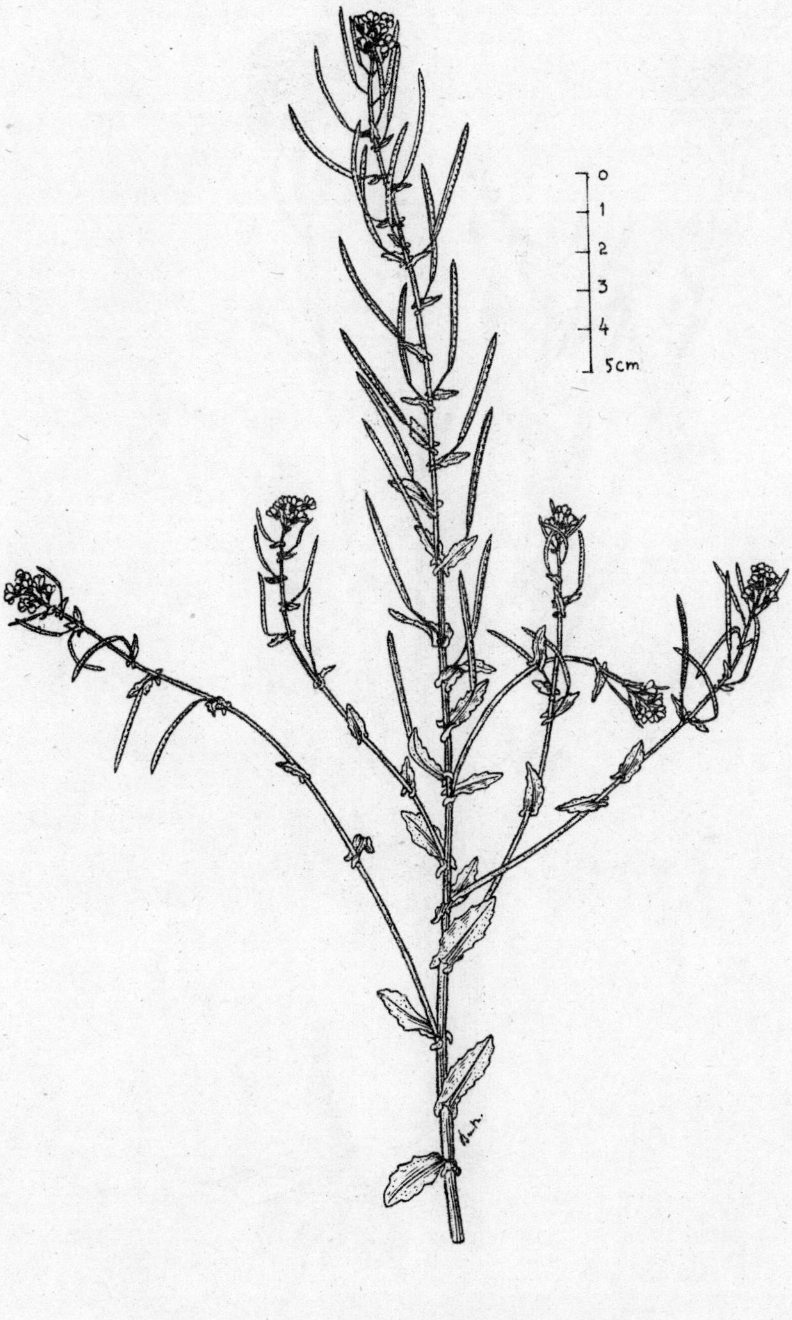

Fig. 27. *Arabidopsis himalaica* O.E. Schulz

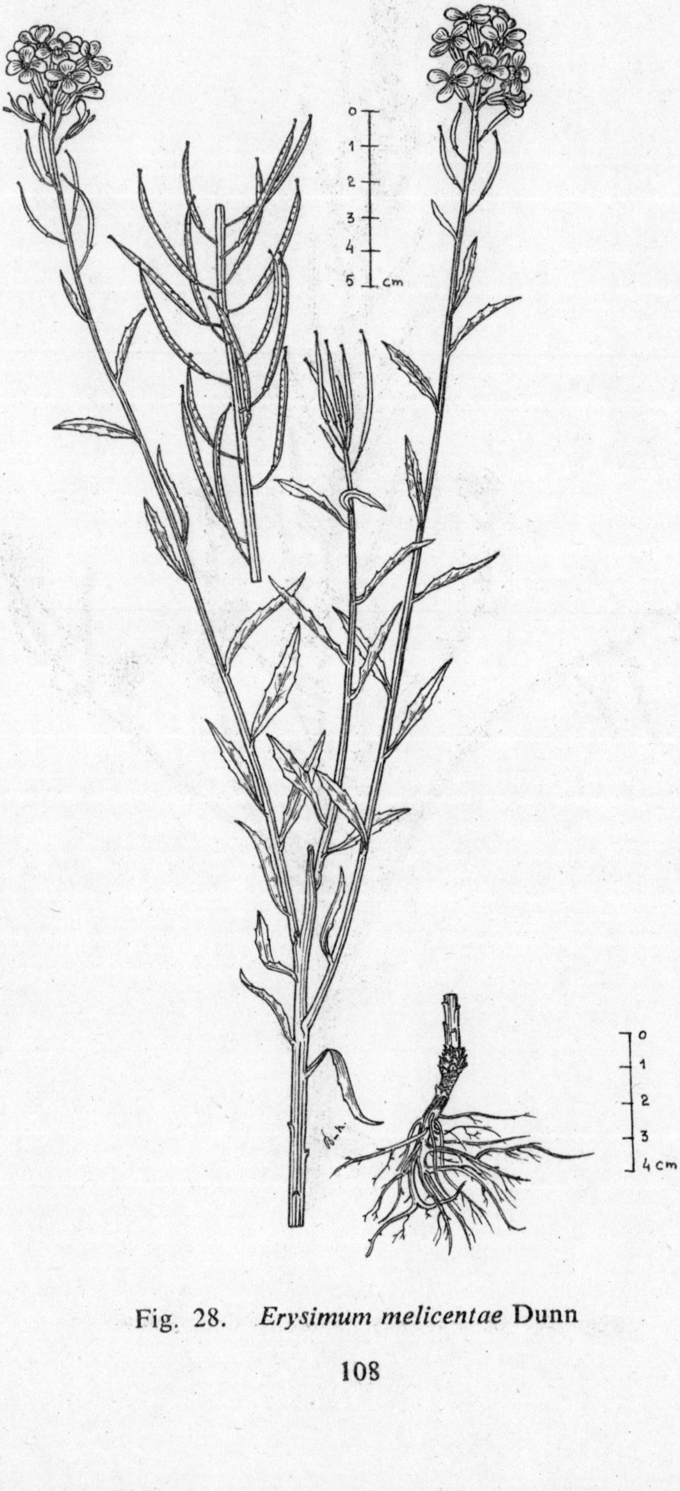

Fig. 28. *Erysimum melicentae* Dunn

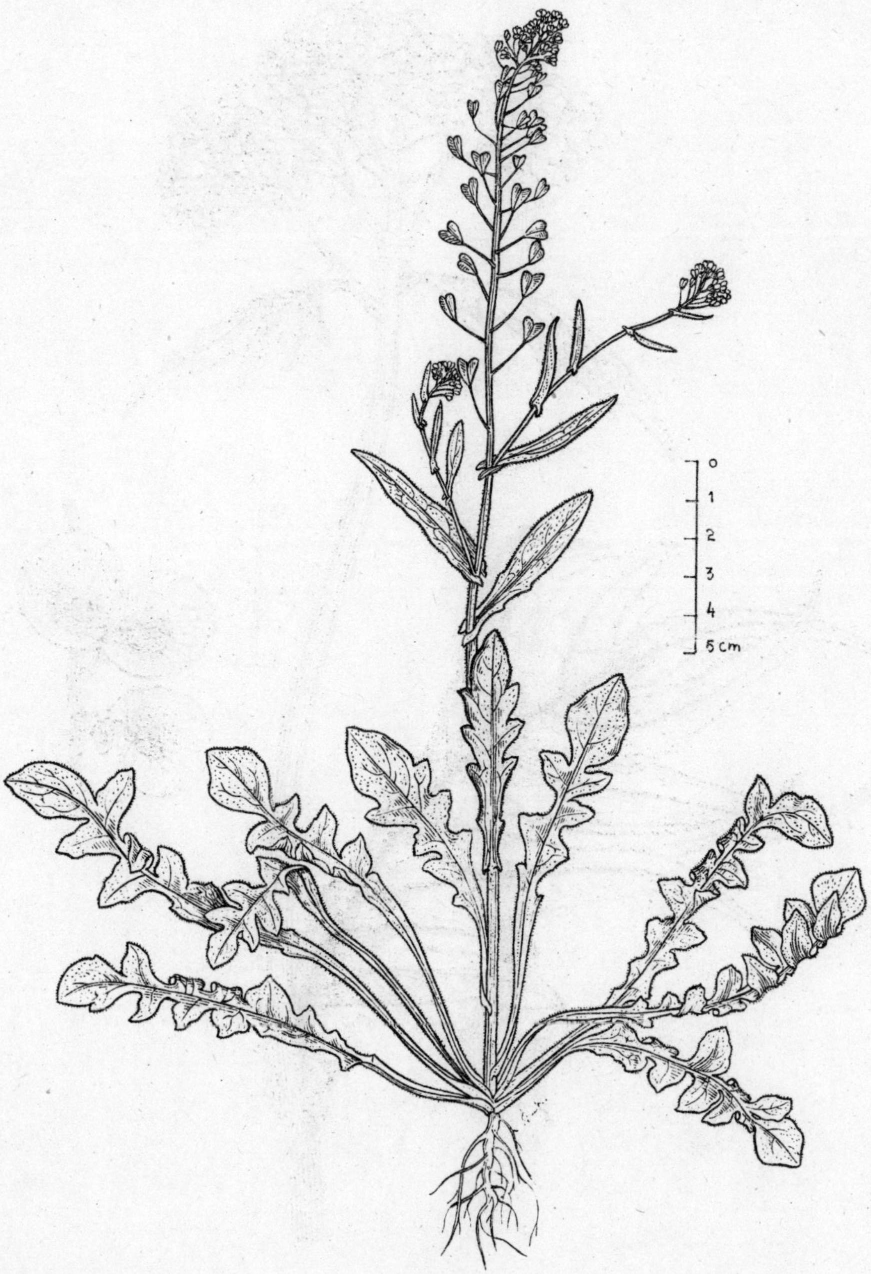

Fig. 29. *Capsella bursa-pastoris* Medik.

Fig. 30. *Megacarpaea polyandra* Benth.

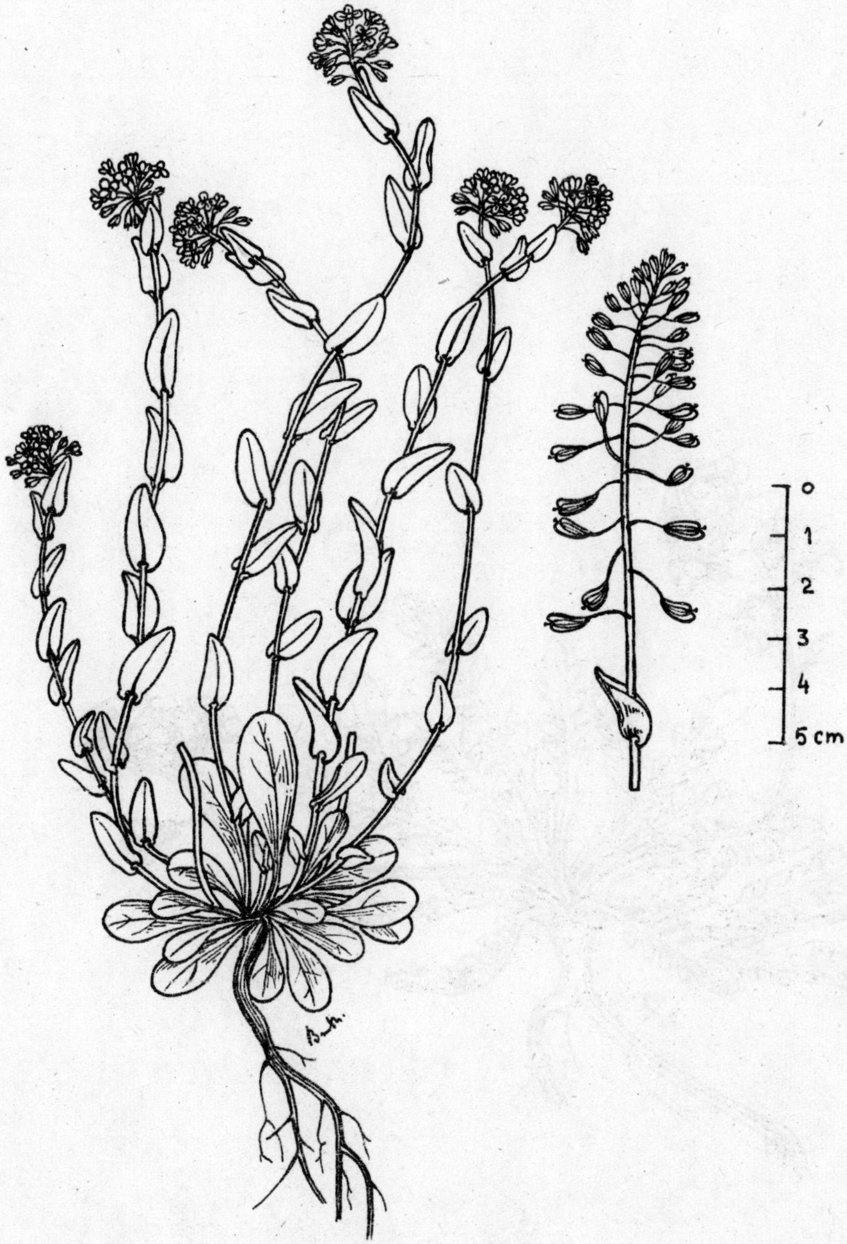

Fig. 31. *Thlaspi montanum* Linn.

Fig. 32. *Chorispora sabulosa* Camb.

Occ. : Common weed of cultivated fields.

Aru (2425 m) *B.M Sharma* 19345; Mundlan (2325 m) *B.M. Sharma* 14212; Aru (2400 m) *P.S. Jamwal* 14206.

2. T. **andersonii** (Hk.f. & Th.) O.E. Schulz in Anzeig. Akad. Wiss. Wien. Math.-Nat. 34. 98, 1826 in obs.; Stewart, l.c. 329. *Iberidella andersonii* Hk.f. & Th., J. Linn. Soc. Bot. 5 : 177 1861; Hk.f. & T. Anders., l.c. 163.

Glabrous, perennial herb with erect or ascending, 5-15 cm long flowering stems and weak, decumbent, non-flowering shoots. Radical leaves petiolate, spathulate 1-3×0.5-1.2 cm, denticulate, the cauline on flowering stems ovate-oblong, 0.6-1.4 × 0.5-0.8 cm, auricled, amplexicaul, those on the barren shoots spathulate to near orbicular and stalked. Flowers racemed, white or pinkish. Fruits 6-8×*c.* 2 mm, wingless or obscurely winged.

Occ. : Waste ground in Pehalgam and along the road to Aru.

Fls. April-July.

Pehalgam (2160 m) *B. M. Sharma* 16694 & 16695; near Mundlan (2300 m) *P. S. Jamwal* 16329 & 16330.

3 **T. montanum** Linn., Sp. Pl. 643, 1753 var. **montanum;** Holm. Mem. N. Y. Bot. Gard. 21(2) : 65-80, 1971; Bhaumik, J. Econ. Tax. Bot. 3(1) : 237-239, 1982. *T. cochleariforme* DC., Syst. Nat. 2 : 381, 1821 ssp. *griffithianum* (Boiss.) Jafri, Notes R. bot. Gdn Edinb. 22 : 118, 1956; For. Fl. Sgr. 169. *Carpoceros griffithianum* Boiss., Diagn. ser. 2 : 40, 1853. *T. griffithianum* (Boiss.) Boiss., Fl. Or. 1 : 329. 1867; Jafri in Nasir & Ali, l.c. 87. *T. alpestre* auct. non Linn.; Hk.f. & T. Anders., l.c. 162 pro parte.

Fig. 31

Glabrous, more or less glaucous, perennial herb with stiff, 10-30 cm tall, tufted stems from a basal rosette of spathulate or orbiculate, stalked leaves 1.2-5×0.5-1.5 cm. Cauline leaves oblong or ovate, 0.8-2 × 0.4-0.8 cm, entire or obscurely toothed, obtuse or subobtuse, amplexicaul, auricled. Flowers white, *c.* 5 mm diam., in racemes elongating up to 10 cm in fruit. Siliculae ± emarginate, 4-8 × 2.5-3 mm; wing narrow below, broadened upwards.

Occ. : On humus rich slopes and grassy ground near melting snow.

Fls. May-July.; *Frts.* July-Aug.

Seikwas (3350 m) *B. M. Sharma* 19346; Nangneri (2850 m) *B. M. Sharma* 19347; Lidder vy. above Kainmul (12000′) *Duthie* 13120; Mahagunus Pass (14700′) *K. M. Vaid* DD. no. 118319; Dudhsar (3400 m) *P. S. Jamwal* 14207.

4. **T. cochlearioides** Hk.f & Th., J. Linn. Soc. Bot. 5 : 177, 1861; Hk.f. & T. Anders., l.c. 162.

Small, glabrous, perennial herb with few, weak, decumbent stems; rootstock branched above and covered with withered leaf bases. Leaves more or less fleshy, entire or slightly toothed, the basal ones rosulate, spathulate or orbiculate, long-stalked, 1.5-3 × 0.4-0.8 cm, the cauline 2-5, small, elliptic-, or oblong-ovate, sessile. Flowers white, 4-5 mm diam., in short, corymbose racemes. Siliculae oblong (8-11 × 2.5 mm), emarginate, on spreading or ascending up to 3 mm long pedicels.

Occ. : Stated to be common in the morainal zone.

Fls. May-Aug.

Badzalkhod nallah, Lidder vy. (13-14000′) *Duthie* 13408 & 13409; Kinimola nala, Lidder vy. *Inayat* 25494.

18. CHORISPORA R. Br.

C. sabulosa Camb., Jacq. Voy. Bot. Inde. 15, t. 15, 1844; Hk.f. & T. Anders., l.c. 167. **Fig. 32**

Perennial herb from a thick, fleshy root branching at the top; flowering stems 5-20 cm long, semierect or ascending, glandular-hairy. Leaves all radical, 2-6 ×0.5-1 cm incl. petioles, oblong-lanceolate, pinnatifid or toothed to sometimes subentire. Flowers purple, 0.8-1 cm diam., in dense but subsequently elongate racemes; sepals 3-4 mm long, erect. Siliquae beaked, 1-1.5×0.2 cm, somewhat tuberculated, straight or curved, on 1-1.5 cm long, slender pedicels.

Occ. : In morainal ground about Kolahoi glacier snout; also in wet ground at Mahagunus.

Fls. June-July; *Frts.* July-Sept.

Kolahoi, near snout (3725 m) *B. M. Sharma* 14263; Sangam vy., Kashmir (13—14000′) *Duthie* 13287; Mahagunus Pass (14700′) *K. M. Vaid* DD. Herb. nos. 118298 & 118299.

XII VIOLACEAE

VIOLA Linn.

1. Flowers yellow — **1. V. biflora**
1. Flowers not yellow:
    2. Plants with evident, leafy, erect, ascending or trailing aerial stems bearing flowers, the latter sometimes also from the axils of rosette leaves:
        3. Stems trailing, rooting at the nodes; leaves mostly drawn into a long, pointed acumen — **2. V. pilosa**
        3. Stems erect, ascending or decumbent and then erect, never rooting at the nodes; leaves blunt or simply acute:
            4. Stems woody; leaves firm; stipules large, leaf-like, toothed or fimbriate mostly only to the lower half of their length — **3. V. jordani**
            4. Stems herbaceous, leaves membranous; stipules fimbriate along nearly their whole length:
                5. Stems 3-17 cm long; lateral petals bearded at the base — **4. V. sylvestris**
                5. Stems up to 1.5 cm long; lateral petals not bearded — **5. V. rupestris**
    2. Plants lacking leafy aerial stems but sometimes leaving stolons; flowers from the axils of rosette leaves alone:
        6. Stolons absent, leaves triangular; stigma terminal, 3-lobed — **6. V. betonicifolia**
        6. Stolons present; leaves orbicular or reniform; stigma bent down, not lobed — **7. V. odorata**

1. V. **biflora** Linn., Sp. pl. 936, 1753; Hk.f. et Th. in Hk.f., FBI. 1 : 182, 1872.

Perfectly glabrous to more or less pubescent herb; stems erect or ascending, 7-25 cm long from a slender, creeping rootstock. Leaves reniform, 2-8 cm wide, cordate, crenate; petioles 1-18 cm long; stipules ovate or oblong. Flowers 1.2 cm long, usually solitary on 1-4 cm long stalks; lower petal streaked with black, very shortly spurred; stigma 2-lobed.

Occ. : In damp, usually shaded and humus covered places in the forests; also under rocks where it usually forms dense, pretty bunches.

Fls. June-Aug.

Nangneri nallah (2850 m) *B. M. Sharma* 14132, Aru, Lidder Vy., *Inayat* DD s.n.; Dibyan, Lidder Vy., *Inayat* DD s.n.; Arnium nallah (2700 m) *P. S. Jamwal* 15657.

2. V. **pilosa** Bl., Cat. Gew. Buitnz. 57. 1823; Hara in Bull. Mus. Univ. Tokyo 8 : 83, 1975. *V. serpens* Wall. ex Ging. in DC., Prodr. 1 : 296, 1824. Hk.f. et Th. in Hk.f., l.c. 184.

Nearly stemless perennial herb, glabrous or with scattered white hairs and throwing leafy, up to 40 cm long, trailing, flowering shoots. Leaves long-petioled, broadly ovate to orbicular-reniform, 2.5-8 cm at their broadest width, deeply cordate, crenate-serrate, acuminate, the rosulate ones springing from a very short rootstock; stipules fringed or toothed. Flowers (bluish) solitary on axillary stalks. Capsule globoid, 5-6 mm, glabrous; seeds few, cream-white, ± 3 mm long.

Occ. : Common on wooded slopes on Aru-Chhumanaisar route; preferring damp places.

Fls. May-June; *Frts.* July-Aug.

Aru–Chhumanaisar route (2650 m) *B. M. Sharma* 14618; Lidder Vy. below Tanin (8-9000′) *Duthie* 13446, Armium Pathri vicinity (2850 m) *P. S. Jamwal* 19287.

3. V. **jordani** Hanry, Cat. Pl. Var. 159, 1853 var. **falconeri** (Hk.f. et Th.) W. Beck. in Beih. Bot Centralb. Abt. 2, 34: 301, 1917: Ban. et Pram. in Fasc. Fl. Ind. 12: 27, 1983. *V. falconeri* Hk. f. et. Th. in Hk.f., l.c. 185.

Perennial herb, stemless or with a very short stem; branches stout, erect or ascending, up to 25 cm high, smooth or scaberulous, very often flexuous. Leaves ovate-, or triangular-cordate, 2-5 × 1.5-3.5 cm, broadly crenate with rotund or pointed, obtuse apex. Stipules toothed, foliaceous; oblong or lanceolate, up to 2.5×0.7 cm, like the leaves densely dotted with black glands. Flowers pinkish, 1.8 cm across, solitary on axillary stalks; spur saccate. Capsule ovoid, 8-9 mm long, glabrous; seeds ovoid, 3 mm long, white.

Occ. : Rocky and gravelly roadsides along Mundlan Aru route.

Fls. May-June; *Frts.* July-Aug.

Aru (2400 m) *B. M. Sharma* 15603; Mundlan roadside (2350 m) *B. M. Sharma* 14639; roadside near Mundlan (2320 m) *P. S. Jamwal* 15663.

4. V. **sylvestris** Lam., Fl. Fr. 2: 680, 1778; Stewart, l.c. 493. *V. sylvatica* Fries ex Hartman in Bot. Notiser, 81, 1841. *V. canina* var. *sylvatica* (Fries) Hk.f. et Th., l.c. 185. **Fig. 33**

Variable, perennial herb with several glabrous, decumbent and subsequently erect, 3-17 cm long, axillary, leafy, flowering stems and a basal leafy rosette from a rather woody stock.

Leaves ovate to nearly orbicular or reniform, 1.5-4 cm long and nearly as broad, cordate, crenate-serrate, obtuse; stipules long-fimbriate. Flowers pale blue or pale violet to almost white, 1.6-1.8 cm long; spur stout, straight or curved; style head glabrous, rarely ± papillulose; stigma decurved, ending in an orifice.

Occ. : Common in woods and shaded habitats throughout from Pehalgam to Aru and near about places.

Fls. April-June.

Aru Ghasan (2500 m) *B. M. Sharma* 14218; Mundlan, roadside (2350 m) *P. S. Jamwal* 15641.

5. **V. rupestris** F. W. Schmidt in Neue. Abh. Boehm. Ges. 1:60, 1791; Stewart, l.c. 492.

Perennial, caespitose, ± glabrous herb with up to 1.5 cm long, axillary, ascending, leafy stems from a basal rosette of leaves. Leaves reniform or ovate-orbicular, 1 -1.8 × 0.8-2 cm, from cordate base, crenate, glandular punctate; petioles 1.3-3 cm long, stipules triangular lanceolate, deeply fimbriate. Flowers violet, 1.5-1.8 cm diam., on up to 4 cm long stalks. Sepals oblong-lanceolate, acute; petals obovate, 1.2×0.8 cm, the lower with a slightly recurved *c*. 4 mm long spur. Stigma beaked.

Occ. : A denizen of alpine meadows.

Fls. July.

Dudhsar (4200 m) *B.M. Sharma* 14037; near Shesh Nag, Lidder vy. (12-13000′) *Duthie* 13334.

6. **V. betonicifolia** J. Sm. in Rees. Cyclop. 37: Viola n. 7. 1819 ssp. **betonicifolia**; Ban. et Pram., l.c. 15. *V. betonicifolia* J. Sm. ssp. *napaulensis* W. Beck. in Engl. Bot. Jahrb. 54 Beibl. 120: 167, 1917. *V. patrinii* sensu Hk.f. et Th., l.c. 183 (pro parte), non DC. **Fig. 34**

Stemless herb. Leaves all radical triangular ovate to triangular-oblong, 2.5–6 × 1.5–4 cm, crenate, obtuse, from a distinctly cordate to nearly truncate base, on long, pubescent petioles usually winged upwards; stipules adnate more than half way, fimbriate along their entire length. Flowers violet, solitary on 2.5-12 cm long, axillary stalks. Capsule glabrous, many seeded; valves 0.7-1.4 cm long, acute,Seeds brown, ovoid, 2 mm long.

Occ. : Common in open, grassy places and shady nooks under stones.

Fls. May-July; *Frts.* July-Aug.

Gagar Pathri near Satlanjan (2900 m) *B. M. Sharma* 14081; Aru-Liddarwat route (2750 m) *B.M. Sharma* 14661.

7. **V. odorata** Linn., Sp. Pl. 934, 1753; Hk.f. et Th., l.c. 184.

Small, perennial herb throwing scaly, subterranean, rooting stolons from a woody stock. Leaves pubescent or glabrous, rosulate, broad-ovate to orbicular-reniform, up to 2.5 cm along either length, obtuse, crenate, from cordate base; stipules lanceolate, toothed or subentire. Flowers fragrant, bluish or bluish-purple to deep violet, occasionally white, ± 1.5 cm long, with a short, blunt, nearly straight spur; sepals obtuse.

Occ. : In woods and open places; common.

Fls. April-July.

Rewil meadow land (± 3300 m) *B.M. Sharma* 19373; Roadside below Aru (2375 m) *P.S. Jamwal* 15642.

XIII. POLYGALACEAE

1. POLYGALA Linn.

P. sibirica Linn., Sp. Pl. 702, 1753. var. **monopetala** (Camb.) Chodat in Mem. Soc. Phys. Hist. Nat. Geneve, 31(2) : 348, 1893; Bennett in Hk.f., FBI. l.c. 205. Stewart, l.c. 443; For. Fl. Sgr. 150. *P. monopetala* Camb. in Jacq., Voy. Bot. Inde. t. 27, 1844. **Fig. 35**

Curly pubescent perennial herb with several, slender, 10-20 cm long, ascending stems from a woody rootstock. Leaves linear, elliptic or lanceolate, 0.5-2.5× 0.1-0.5 cm, sharply pointed. Flowers 5-7 mm long, bluish purple, in lax, 5-12-flowered racemes, wing sepals up curved, obliquely oblong or obovate, ciliate, exceeding the corolla; keel with prominently fimbriate crest. Capsule suborbicular, 4-5 mm in diam., distinctly margined; seeds black with a 3-lobed strophiole, 3 mm long, densely white-hairy.

Occ. : On drier slopes along the road between Mundlan and Aru; rather uncommon.

Fls. June-July; *Frts.* July-Aug.

Aru (2350 m) *B. M. Sharma* 14378; Liddar Valley (6-7000′) *Duthie* 13079; Mundlan-Aru road (2300 m) *P.S. Jamwal* 15629.

XIV. CARYOPHYLLACEAE

1. Sepals combined into a distinctly gamosepalous calyx; petals clawed:
  2. Calyx furnished at base with closely appressed, imbricate bracts ... **1. Dianthus**
  2. Calyx without bracts:
    3. Calyx 5-veined; styles 2 :
      4. Seeds boat-shaped, peltate, with facial hilum ... **2. Petrorhagia**
      4. Seeds reniform, with lateral hilum ... **3 Gypsophila**
    3. Calyx 10-many-veined; styles 3-5 :
      5. Fruit fleshy, baccate ... **5. Cucubalus**
      5. Fruit a capsule :
        6. Styles 3; capsule teeth twice the number of styles ... **4. Silene**
        6. Styles 5; capsule opening by 4-5 teeth, the latter usually splitting into twice their number ... **6. Lychnis**
1. Sepals free, or only at the very base coalescent; petals subsessile or 0 :
  7. Stipules absent :
    8. Leaves linear–subulate :

9. Styles as many as there are sepals **13. Sagina**
9. Styles fewer than there are sepals **11. Minuartia**

8. Leaves linear or not, never subulate :
10. Petals absent **9. Stellaria**
10. Petals present :
11. Petals shortly or deeply 2-lobed :
12. Capsule dehiscing by as many shortly 2-fid teeth as there are styles : **8. Myosoton**
12. Capsule dehiscing by twice as many teeth as there are styles :
13. Plants bearing small napiform subterranean tubers **10. Pseudostellaria**
13. Plants without tubers :
14. Petals lobed at the most to about one-third of their length; capsule cylindric or conic **7. Cerastium**
14. Petals lobed generally to at least half of their length; capsule globose, ovoid or oblong **9. Stellaria**
11. Petals entire **12. Arenaria**
7. Stipules present, scarious **14. Spergularia**

1. DIANTHUS Linn.

1. **D. jacquemontii** Edgew. et Hk.f. in Hk.f., FBI. 1 : 214, 1874.

Glabrous, up to 50 cm high perennial; stems slender, stiff, nearly erect, many from a tufted woody stock. Leaves sessile, coriaceous, linear, 2-4.5×0.1-0.25 cm, serrulate, acute, the radical longer. Flower white, 2-2.5 cm across, solitary, terminal. Bracts, ovate broadly scarious, abruptly long-cuspidate. Calyx 2-3 cm long, finely striate; teeth rigid; petal limb toothed.

Occ. : On rocky slopes and drier gravelly places, more frequent in E. Liddar Valley.

Fls. July-Aug.

Mundlan-Aru road (2300 m) *B. M. Sharma* 14294 and 14610.

2. PETRORHAGIA Link

P. alpina (Habl.) P.W. Ball et Heywood, Bull. Brit. Mus. (Bot.) 3 : 145, 1964; P.W. Ball in Tutin et al., Fl. Eur. 1 : 187, 1964; Stewart, l.c. 249. *Gypsophila alpina* Habl. in J.G. Gmel., It 4 : 178. *Tunica stricta* (Bunge) Fisch. et Mey. Ind. Sem. Hort. petrop. 4 : 50. 1837; Edgew. et Hk.f. in Hk.f. l.c., 216. **Fig. 36**

Erect, 10-30 cm tall, glabrous annual; stems slender, 1-few from the base, simple or branched. Leaves linear or spathulate, 0.6-3×up to 0.3 cm; margins scaberulous. Flowers white, 4 mm long, on slender solitary or laxly clustered pedicels. Calyx turbinate, with 5 green veins; petals undivided. Capsule oblong, 5 mm long, 4-toothed; seeds wrinkled.

Occ. : Gravelly waysides and edges; not common.

Fls. & Frts. July-Aug.

Seikwas route, one and a half km from Liddarwat (2990 m) *B. M. Sharma* 14016; Hamwas (3200 m) *B. M. Sharma* 14637.

3. GYPSOPHILA Linn.

G. cerastioides D. Don. Prodr. Fl. Nep. 213, 1825; Edgew. et Hk.f. in Hk.f., l.c. 217. **Fig. 37**

Hoary pubescent, much branched perennial herb; stems many from root, 6-25 cm long, decumbent. Leaves obovate or spathulate, 0.5-5×0.3-1.8 cm, the lower distinctly stalked. Flowers white or pale pink, streaked with purple, 0.6-1.3 cm across, in terminal and axillary bracteate cymes. Calyx bell-shaped; petals slightly notched; capsule 4-valved to about the middle.

Occ. : Common and often gregarious on shaded gravelly banks; frequent also in moister places in alpine pastures.

Fls. July-Aug.

Liddarwat (2975 m) *B. M. Sharma* 19368; Seikwas (3400 m) *B. M. Sharma* 14612; Chanderwari-Zojpal (9300′) *R. L. Badhwar* RL. Herb. no. 1162.

4. SILENE Linn.

1. Annual. Calyx 30-nerved; petals entire or slightly emarginate **1. S. conoidea**
1. Perennials. Calyx 10–, or 20-nerved; petals 2–partite or deeply 2–fid :
    2. Calyx 20-nerved **2. S. vulgaris**
    2. Calyx 10-nerved :
        3. Petal-claw auricled on either side **3. S. tenuis**
        3. Petal-claw without auricles **4. S. kunawurensis**

1. **S. conoidea** Linn., Sp. Pl. 418, 1753; Edgew. et Hk.f. in Hk.f., l.c. 218.

Erect, glandular-viscid, 15-40 cm tall, simple or branched herb. Leaves oblong or lanceolate, sessile, 3.5-10×0.3-1.2 cm, the lower spathulate. Flowers pink, 1.5-2 cm across, in terminal panicles. Calyx tubular, glandular-pubescent, 2-2.5 cm long, conic-ovoid in fruit. Capsule crustaceous, long-drawn upwards; seeds brown, reniform, 1.2-2.5 mm, tubercled.

Occ. : A field weed at lower altitudes. *Fls.* May-July; *Frts.* June-July.

Aru (2400 m) *B. M. Sharma* 14390 & 14607.

2. **S. vulgaris** (Moench) Garcke, Fl. Nord. Mittel Deutschl. ed. 9 : 64, 1869; Coode et Cullen in Davis, Fl. Turkey 2 : 210, 1967; Stewart, l.c. 252. For. Fl. Sgr. 172. *Behen vulgaris* Moench, Meth. 709, 1794. *S. venosa* (Gilile) Aschers., Fl. Brandenb. 86, 1964; Raizada, Indian For. 92(5) : 323, 1966. *Cucubalus venosus* Gilile, Fl. Lithuan 2 : 165, 1782. *S. cucubalus* Wibel, Prim. Fl. Werth. 241, 1799. Kitam., Fl. Afgh. 116, 1960. *S. inflata* Sm., Fl. Brit. 467, 1800; Edgew. et Hk.f. in Hk.f., l.c. 218. **Fig. 38**

Branched, glabrous herb; stems 0 3-1 m high, erect or ascending from a decumbent, woody base. Leaves ovate, oblong or elliptic to lanceolate, 3-7×0.8-2.5 cm acute. Flowers white, 1.3-2 cm across, drooping, in loose terminal panicles. Calyx inflated, reticulately veined, 1.5-2 cm long, at times flushed with purple on one side. Capsule globose, 6-7 mm long; seeds reniform, grey or chestnut, 1.2-1.8 mm, tubercled.

Occ. : Common among scrubs, generally in shaded situations in the forests.

Fls. June-Aug. *Frts.* Aug.-Sept.

Aru (2400 m) *B. M. Sharma* 14700; Nangneri (2900 m) *B. M. Sharma* 19369; Amarnath Cave (12500′) *K.M. Vaid* DD. Herb. no. 117545.

3. **S. tenuis** Willd., Enum. Hort. Berol. 474, 1809; Edgew. et Hk.f. in Hk.f., l.c. 219.

Glabrous herb; stems 15-30 cm long, simple, slender, erect or ascending, many from a tufted, woody base. Leaves chiefly radical, linear or lanceolate, 1.5-5 × 0.1-0.3 cm; margins ciliate. Flowers dirty yellowish-brown, crowded in terminal racemes; calyx membranous, tubular-campanulate 0.7-1.2 cm long. Capsule oblong, 7 mm long.

Occ. : Collected only once from an alpine grassy slope.

Fls. July.

Below Tarsar (3750 m) *B. M. Sharma* 14676.

4. **S. kunawurensis** Benth. in Royle, Ill. Bot. Himal. 79, 1835; Edgew. et Hk.f. in Hk.f. l.c. 220.

Cited by Stewart (l.c. 251) from Upper Liddar Vy. Not seen by us.

5. CUCUBALUS Linn.

C. baccifer Linn., Sp. Pl. 414, 1753; Edgew. et Hk.f. in Hk.f., l.c. 222.

Fig. 39

Straggling or semiscandent, pubescent perennial; stems weak, divaricately branched, up to 1 m long, often purple. Leaves ovate - lanceolate, 2.5-8 × 0.8-3.5 cm, cuneate into a short petiole. Flowers greenish-white or, -yellow, 0.8-1.5 cm across solitary in forks of branches or in lax, bracteate clusters of up to 3; bracts leaf-like. Calyx broadly bell shaped; teeth revolute in fruit. Petals deeply 2-fid, bearing a pair of coronal scales at the base of the limb. Fruit globose, 0.8-1.2 cm diam., black. Seeds reniform, 1.5 mm, chestnut, shining.

Occ. : Very common in forest shade, especially in Liddarwat and near about places.

Fls. May-July; *Frts.* July-Sept.

Liddarwat (2975 m) *B. M. Sharma* 19367; Lidder Vy. (6-7000′) *Duthie* 13066; Pisu Ghati, on way to Amarnath (11000′) *K.M. Vaid* DD. Herb. no 117543; Armiun Glenn (2750 m) *P. S. Jamwal* 15622; Pehalgam vicinity (7500′) *R. L. Badhwar*, RL. Herb. nos. 1211 & 1212.

6. LYCHNIS Linn.

1. Stem mostly 1-flowered; seeds winged, neither granulated nor tubercled **1. L. apetala**
1. Stem few-, to many-flowered; seeds reniform, granulated, tubercled or echinate :
   2. Calyx campanulate, 1.2-2 cm long; petals much longer **2. L. indica**
   2. Calyx globose, 1 cm or less in length; petals scarcely longer **3. L. nutans**

1. **L. apetala** Linn., Sp. Pl. 1 : 437, 1753; Edgew. et Hk.f. in Hk.f., l.c. 222. Stewart, l.c. 246. *Melandrium apetalum* (Linn.) Fenzl in Ledeb., Fl. Ross. 1 : 326, 1842. *Silene wahlbergella* Chowdhuri, Notes R. Bot. Gard. Edinb. 22 : 237, 1957; For. Fl. Sgr. 173.

Fig. 33. *Viola sylvestris* Lam.

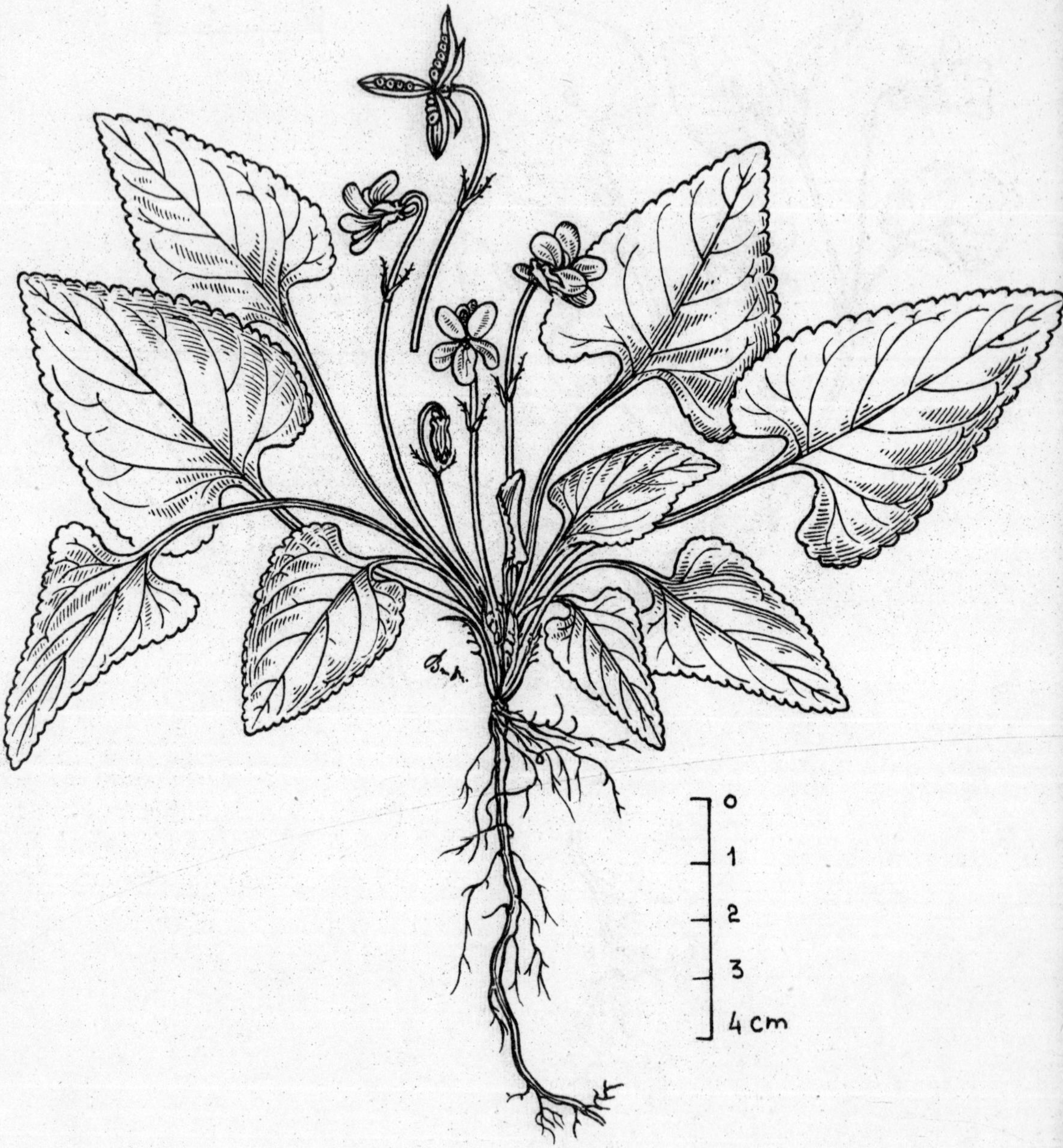

Fig. 34. *Viola betonicifolia* J. Sm.

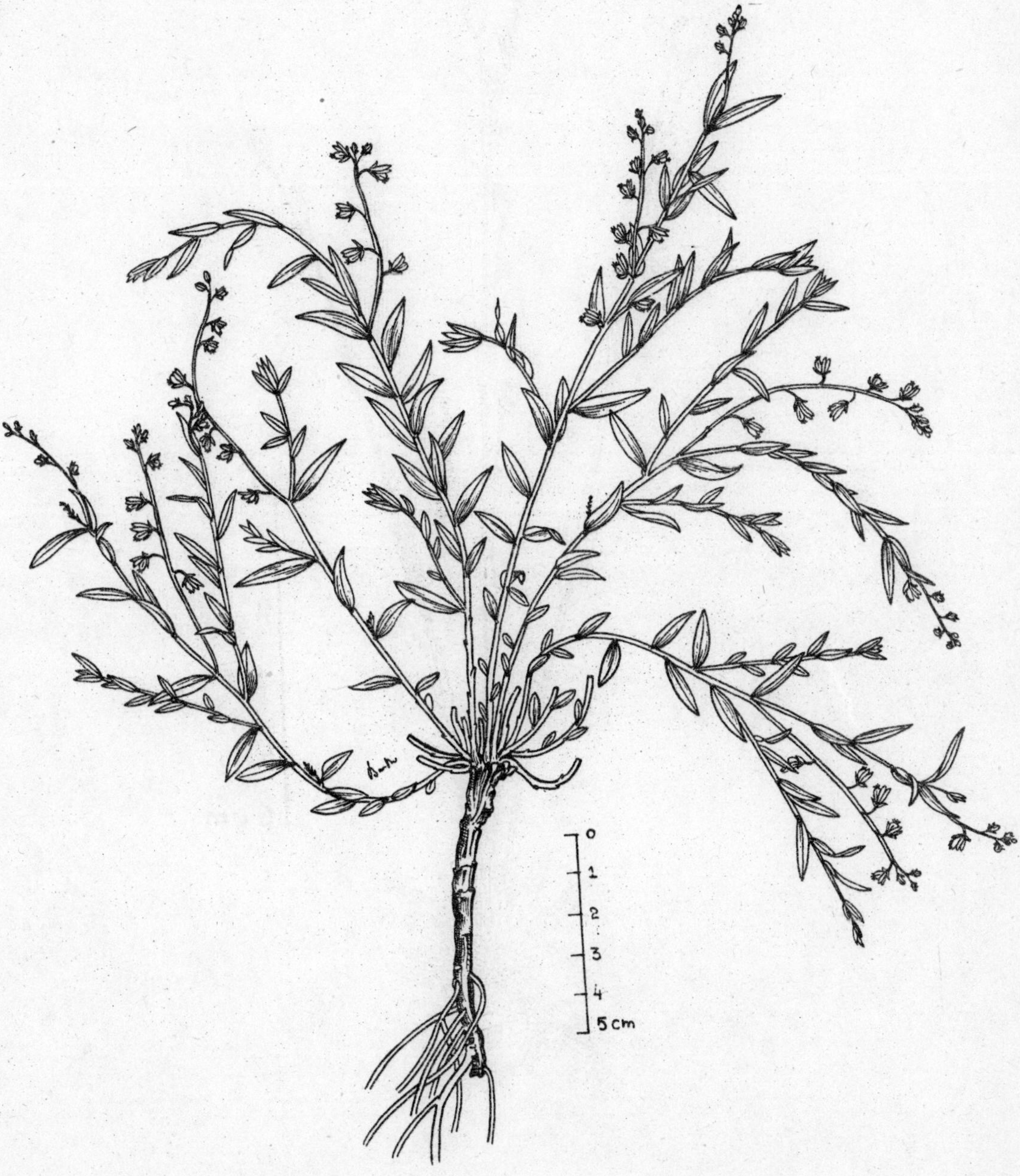

Fig. 35. *Polygala sibirica* Linn. var. *monopetala* Chodat

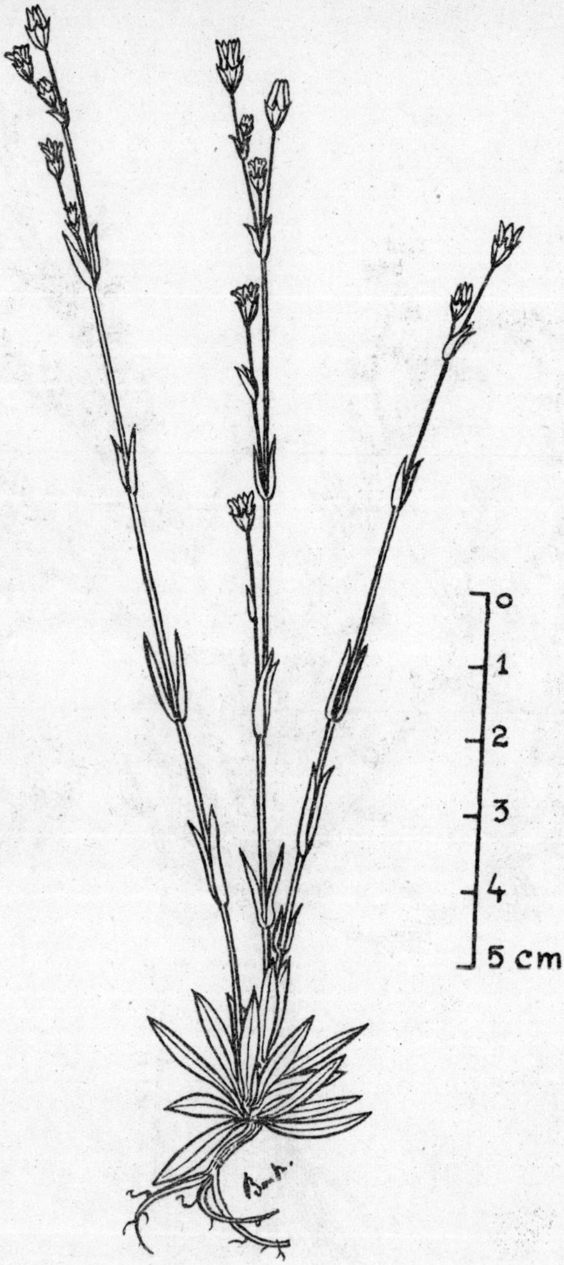

Fig. 36. *Petrorhagia alpina* Ball et Heyw.

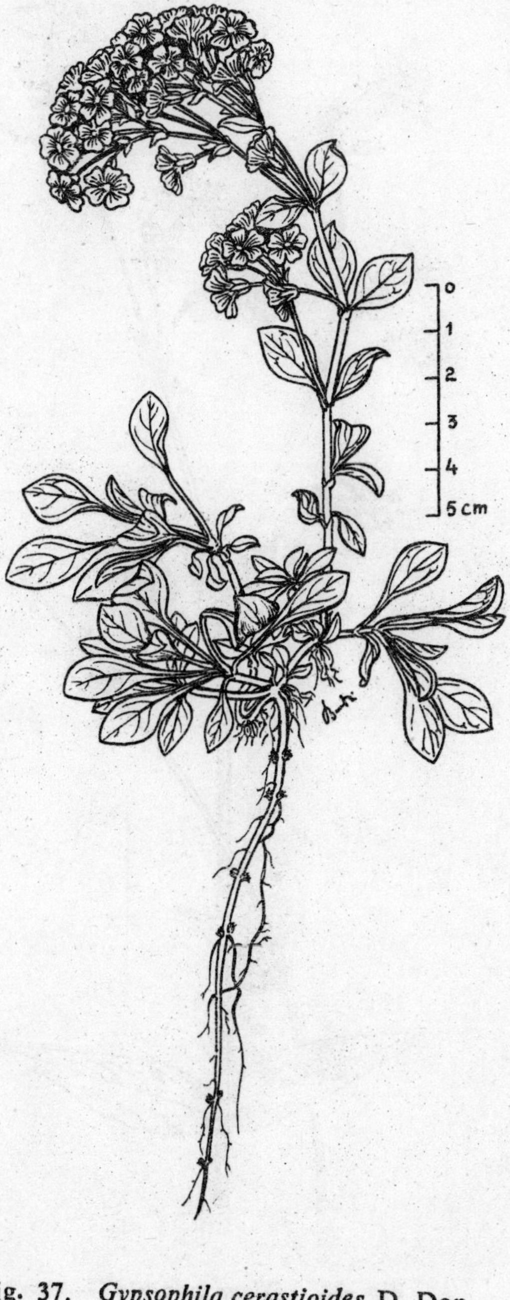

Fig. 37. *Gypsophila cerastioides* D. Don

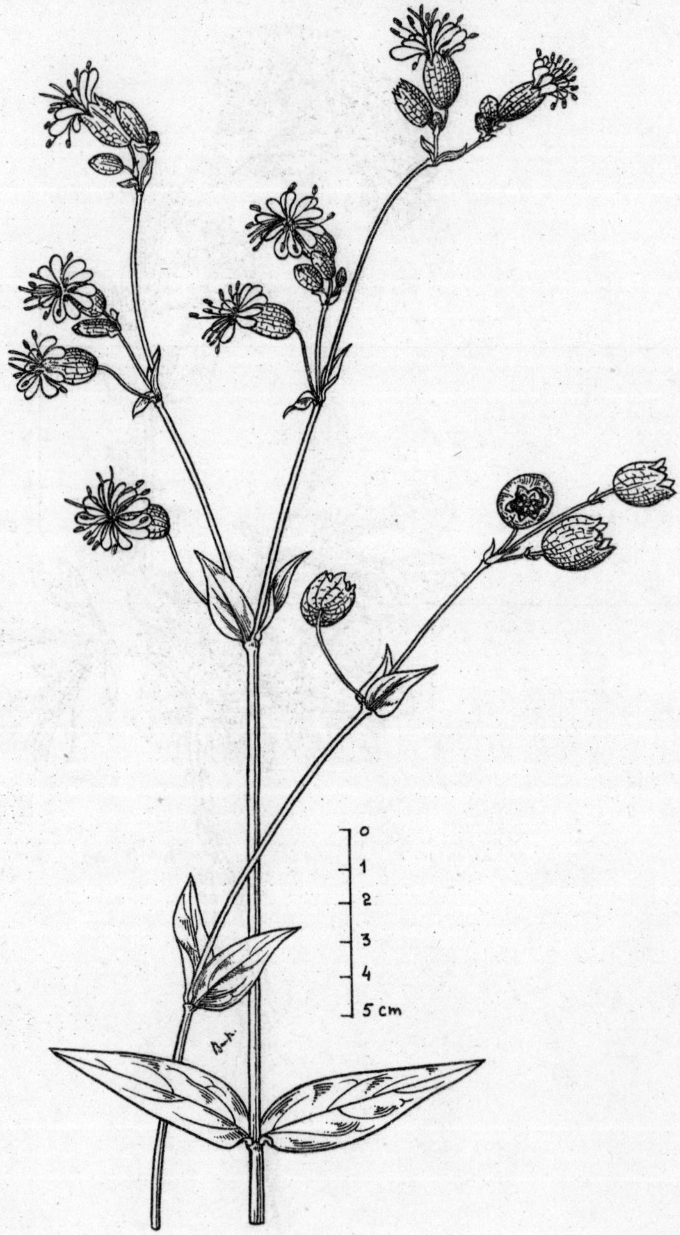

Fig. 38. *Silene vulgaris* Garcke

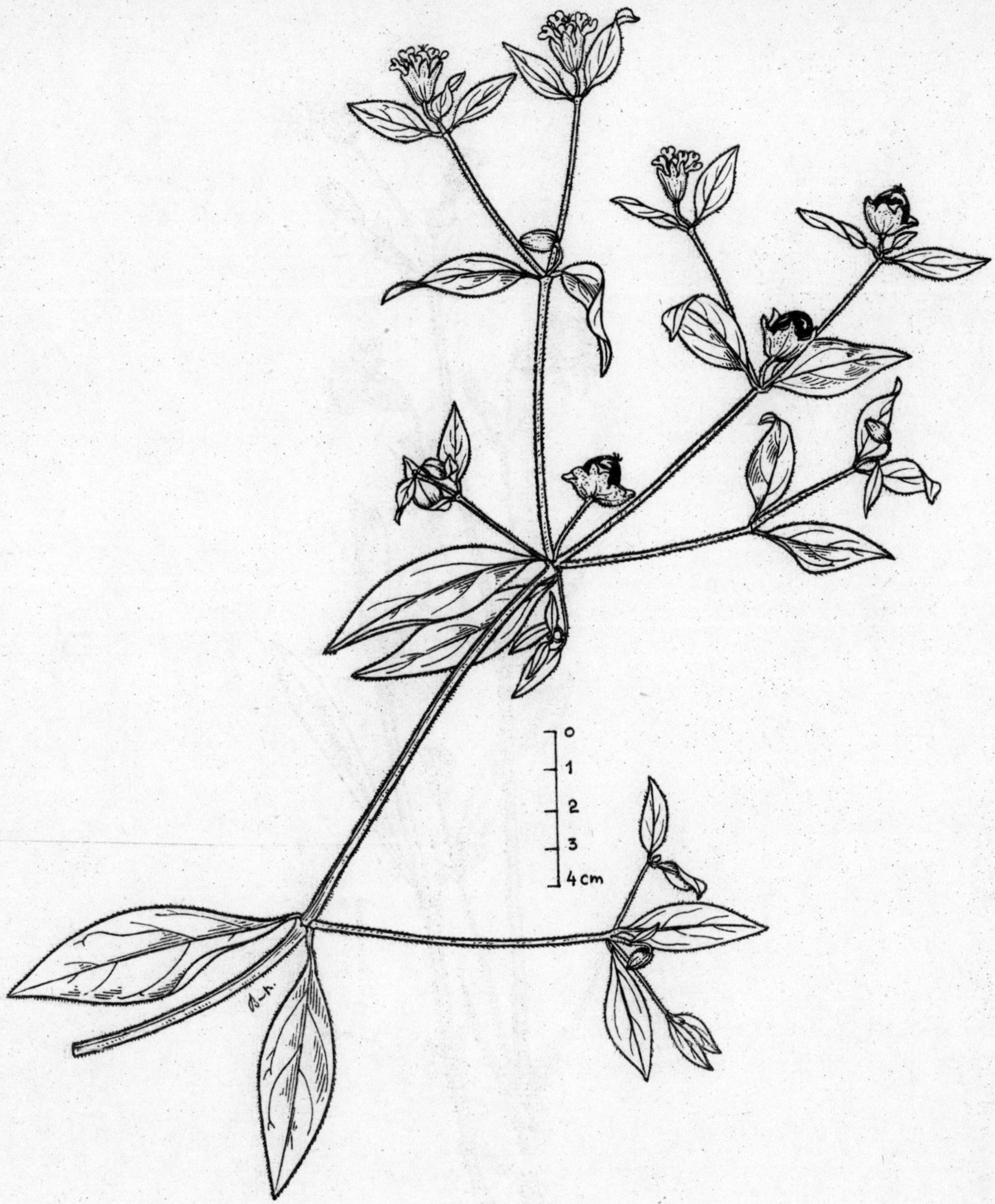

Fig. 39. *Cucubalus baccifer* Linn.

Fig. 40. *Lychnis indica* Benth. var. *fimbriata* Edgew. et Hk.f.

Erect, herbaceous perennial; stem simple, up to 30 cm tall, glandular-pubescent, downwards more or less glabrous. Leaves 2.5-6×0.2-0.7 cm, cauline 1 to 2 pairs, linear-lanceolate, the radical ones generally spathulate, acute or obtuse. Flowers solitary (—3), light pinkish purple, nodding; calyx whitish between the purple-brown nerves, inflated, 1-1.5 cm long, contracted at mouth.

Occ. : Open places in stony nallahs; also on alpine, grassy slopes. Not common.

Fls. July-Aug.

Nangneri (2950 m) *B. M. Sharma* 14120; Yemher Pass, Kashmir (13200′) *M.V. Laurie* 16800.

2. **L. indica** (Roxb.) Benth. in Royle, Ill. Bot. Himal. 81, 1834 var. **fimbriata** Edgew. et Hk.f. in Hk.f., l.c. 225; Stewart, l.c. 246. *L. fimbriata* Benth. in Royle, l.c. 1834. *Melandrium indicum* var. *fimbriatum* Rohrb., Linnaea 36: 235. *Silene cashmeriana* (Benth. in Royle) Majumdar, J. Indian bot. Soc. 42 (4):649, 1963. **Fig. 40**

Erect or ascending, glandular-viscid especially upward, 0.6-1 m tall, perennial herb. Leaves lanceolate, 3.5-9×0.4-1.2 cm, acute, the lower shortly stalked and often oblanceolate. Flowers erect or nodding, in panicled; glandular-sticky cymes, 2-2.5 cm across, purple or cream-white, often both colours on the same plant; petal limb lacerate, with the segments 2-lobed. Capsule ovoid.

Occ. : Common in the forest clearings about Liddarwat, also among open scrubs in areas about Aru.

Fls. July-Sept.

Nangneri (2875 m) *B.M. Sharma* 19371; Kainmul, Lidder Vy. (10-11000′) *Duthie* s.n. DD. (under *L. cashmeriana* Royle); Shishnag, *K.S. Ahluwalia* RL. Herb. nos. 1241, 1243, 1246, 1247 & 1249 (all under *L. cashmeriana* Royle); Aru (2400) m) *P.S. Jamwal* 15624.)

3. **L. nutans** Benth. in Royle, Ill. Bot. Himal. 80, 1835; Edgew. et Hk.f. in Hk.f., l.c. 225. *Silene thomsonii* Majumdar, l.c 650.

Diffuse, dichotomously branched perennial herb, 40-80 cm high, viscid-hairy especially upwards. Leaves ovate, 4-8×1.7-3.7 cm, acute or acuminate, the lower lanceolate, elliptic or oblanceolate, up to 6×1.5 cm. Flowers nodding, purple, in glandular-hairy panicles. Calyx inflated; petal limb shortly fringed. Capsule ovoid, 1.5 cm long, shining, crustaceous; seeds reniform, 1.5-2 mm, rough, strongly echinate dorsally.

Occ. : In habitats similar to those for the preceding species.

Fls. July-Aug.; *Frts.* Aug.-Sept.

Nangneri (2875 m) *B.M. Sharma* 19370; Liddarwat (2950 m) *B.M. Sharma* 14608.

7. CERASTIUM Linn.

1. Stems usually with a line of hairs running down the internodes; styles 3; capsule opening by 6 teeth **1.** C. cerastoides
1. Stems glabrous or pubescent all round; styles 5; capsule opening by 10 teeth:
   2. Glabrous; bracts herbaceous; petals for exceeding the sepals; sometimes nearly twice their length; capsule

straight; tips of capsule-teeth revolute **2. C. dahuricum**

2. Pubescent; bracts, at least the upper, with scarious margins; petals hardly longer than sepals; capsule curved; tips of capsule-teeth straight **3. C. fontanum**

1. **C. cerastoides** (Linn.) Britton, Mem. Torrey club, 5: 150, 1894; Gartner in Fedde, Repert. (Beih,) 113, 1939; Majumdar, J. Indian bot. Soc. 44(1) : 140, 1965; Stewart, l.c. 240; For. Fl. Sgr. 170. *Stellaria cerastoides* Linn., Sp. Pl. 604, 1753. *C. trigynum* Vill., Hist. Pl. Dauph. 3 : 645, 1789; Edgew. et Hk.f. in Hk.f., l.c. 227.

Laxly tufted slender perennial with 8-20 cm long, decumbent or ascending flowering branches from spreading and rooting, much branched stems. Leaves linear to oblong, mostly curved, 1-2.5 × 0.1-0.3 cm, obtuse. Flowers white, 1.5-2 cm across, solitary or up to 3 in a cyme; petals deeply 2-fid; pedicels glandular. Capsule cylindric, nearly twice the length of the calyx.

Occ : Common and often gregarious in alpine pasture lands, especially in limestone tracts.

Kolahoi (3300 m) *B.M. Sharma* 14311; Tarsar (3750 m) *B.M. Sharma* 14017; Seikwas (3400 m) *P. S. Jamwal* 15627.

2. **C. dahuricum** Fisch. ex Spreng., Pl. Min. Cog. Pugill. 2 : 65, 1815; Edgew. et Hk.f. in Hk.f., l.c. **Fig. 41**

Stout, straggling or semiscandent perennial; stems shining branched, 0.5-2 m long. Leaves ovate or oblong, 3-10 × 1.5-3 cm, pointed or blunt, auricled or slightly connate at base. Flowers white, 2.5 cm across, in large,repeatedly forked, spreading cymes. Capsule 1.5-2 cm long, striate, shining; seeds dark brown, 1.5 mm, acutely tubercled.

Occ. : Among *Viburnum* bushes in Liddarwat and its environs, also among boulders and junipers in Kolahoi.

Fls. July-Aug.; *Frts.* Aug.-Sept.

Kolahoi (3700 m) *B.M. Sharma* 19227; Lidder Vy. near Sheshnag (12-13000') *Duthie* s.n. DD.

3. **C. fontanum** Baumg., Enum. Stirp. Trans. 1 : 425, 1816 ssp. **triviale** (Link) Jalas, Arch. Soc. Zool.-Bot. Fenn. Vanamo, 18(1) : 63, 1963 and in Tutin et al. Fl. Eur. 1 : 142, 1964; Stewart, l.c. 240; For. Fl. Sgr. 170. *C. holosteoides* Fries, Nov. Fl. Suec. ed. 2 : 126, 1928. *C. caespitosum* Gilib., Fl. Lituan. 2 : 159, 1781. *C. vulgatum* Linn. (1762) **var.** *triviale* (Link) Edgew. et Hk.f. in Hk. f., l.c. 228.

Glandular hairy herb with decumbent and then ascending laxly caespitose, 5-30 cm long, usually unbranched flowering stems. Leaves sessile, lanceolate to oblong-ovate, 0.8-2 ×0.2-0.8 cm acute, the lower spathulate. Flowers white 0.8-1 cm across, in few-flowered, rather open spreading cymes; petals slightly longer than scarious-margined sepals. Capsule (10-12 mm) about twice the length of the sepals; seeds tubercled.

Occ. : Common along waysides and shaded sloping ground.

Fls. May-July; *Frts.* June-Aug.

Kolahoi (3700 m) *B.M. Sharma* 14392; Aru (2400 m) *B.M. Sharma* 14615.

8. MYOSOTON Moench

M. aquaticum (Linn.) Moench, Meth. 225, 1794; Clapham in Tutin et al. Fl. Eur. 1 : 146, 1964; Stewart, l.c. 249; For. Fl. Sgr. 171. *Cerastium aquaticum* Linn., Sp. Pl. 439, 1753. *Malachium aquaticum* (Linn.) Fries., Fl. Hall. 77, 1817. *Stellaria aquatica* (Linn.) Scop., Fl. Carn. ed. 2, 1 : 319, 1772; Edgew. et Hk.f. in Hk.f., l.c. 229. **Fig. 42**

Much branched, straggling perennial; stems viscidly pubescent, 0.25-1 m long, decumbent or spreading. Leaves ovate, 2-7 × 1-3.5 cm, from a cordate or truncate base, the lower petiolate. Flowers white, 1.3-1.5 cm across, in leafy, terminal and axillary cymes; petals 2-fid almost to base, longer than viscid, obtuse, accrescent sepals Capsules exserted, 1 cm long, on patent pedicels; seeds subreniform, brown, ± 1 mm, tuberculate.

Occ. : Common among Viburnums and other bushes; also in shady places along streams.

Fls. June-July; *Frts.* Late July-Aug. Liddarwat (2950 m) *B.M. Sharma* 14389; Aru (2450 m) *B.M. Sharma* 14616; Lidder Vy., near Tanin (9-10000′) *Duthie* 13448; On way to Chandanwari, *L.D. Kapoor*, RL. Herb. no. 1300; Chandanwari, *P.S. Jamwal*, RL. Herb. nos. 10009, 10011 & 10013.

9. STELLARIA Linn.

1. Leaves, at least lower, distinctly petiolate **2. S. media**
1. Leaves all sessile :
   2. Leaves 6-20 cm long; ovary 3-celled **1. S. monosperma**
   2. Leaves never exceeding 2.5 cm length; ovary 1-celled :
      3. Stamens 5 **4. S. subumbellata**
      3. Stamens 8-10 :
         4. Perfectly glabrous; pedicels deflexed in fruit at least; seeds many. Leaves 1.2-2.5 cm long **3. S. alsine**
         4. Stems, at least upwards, and pedicels pubescent, the latter not deflexed; seeds 2-8. Leaves at the most 8 m long **5. S. cherleriae**

S. monosperma Buch.-Ham. ex D. Don, Prodr. Fl. Nep. 215, 1825; Mizushima, J. Jap. Bot. 32 : 248, 1957, Stewart, l.c. 254; Majumdar, Bull. bot. Surv. India, 15 (1&2) : 43, 1973. *S. crispata* Wall. ex Edgew. et Hk f. in Hk.f. l.c. 229. **Fig. 43**

Straggling or decumbent herb; stem shining, 0.6-1.3 m long, hairy at nodes, often with a line of fine hairs running between the latter. Leaves oblong or lanceolate from usually cordate base, 1-2.5 cm at broadest width, acuminate; margin flat or crispate, with a faint intramarginal nerve. Flowers small, white, in glandular pubescent cymes; sepals viscid, longer or shorter than petals. Seeds 1-2, large, granulated

Occ. : Common along banks and on humus covered ground under coniferous forests throughout the Armiun glenn-Aru-Liddarwat area.

Fls. July-Sept.

Liddarwat (2730 m) *B.M. Sharma* 14182; Girwar . nallah (2800 m) *P.S. Jamwal* 15623.

2. **S. media** (Linn.) Vill., Hist. Pl. Dauph. 3 : 615, 1789; Edgew. et Hk.f. in Hk.f., l.c. 230. *Alsine media* Linn., Sp. Pl. 272, 1753. *S. media* Cyr., Ess. Pl. Char. 36, 1784.

Diffuse, much branched annual herb; stems decumbent, or semierect, 10-50 cm long, with a line of hairs between the nodes. Leaves ovate or elliptic, up to 2.5 cm long, acute, the upper usually sessile. Flowers white, 6-8 mm across, axillary and in terminal cymes; sepals viscid, obtuse, longer than the petals. Capsule ovoid, 5-6 mm long, exserted; seeds brown, acutely tubercled.

Occ. : Very common weed of paths, disturbed slops and in shady places among bushes.

Fls. April-Sept.

Satlanjan (3050 m) *B. M. Sharma* 14203; Aru (2500 m) *P.S. Jamwal* 15625.

3. **S. alsine** Grimm, Nova Acta, Acad. Leop. -Carol. 3, App : 313, 1767; Chater et Heywood in Tutin et al; Fl. Eur. 1 : 135, 1964; Stewart, l.c. 253. *S. uliginosa* Murray, Prodr. Gott. 55, 1770; Edgew. et Hk f., l.c. 233.

Much branched, leafy annual; stems weak, quadrangular, 15-30 cm long, decumbent or ascending from a creeping base. Leaves lanceolate, 1-5 mm broad, acuminate. Flowers greenish, 6-7.5 mm across, in terminal and axillary cymes; pedicels slender. Sepals glabrous, with white shining margins, acute, 3-nerved; petals minute or 0. Capsule ovoid, equalling or hardly exceeding the sepals.

Occ. : In dense patches on wet ground near stones.

Fls. July-Sept.

Kolahoi (3700 m) *B. M. Sharma* 19372; Kolahoi Vy. near Lidderwat (11-12000′) *Duthie* 13514; Lidder Vy. above Kainmul, *Duthie* 13100.

4. **S. subumbellata** Edgew. in Hk.f., FBI. 1 : 233, 1874; Stewart, l.c. 255.

Slender, glabrous herb; stem ascending. Leaves linear or elliptic, acute. Flowers small, greenish, apetalous, solitary or in axillary cymes; sepals acute, 3-nerved, scarious-margined; pedicels capillary deflexed in fruit. Capsule cylindric, twice as long as the sepals.

We have not beenable to collect this, probably as a result of oversight. In Dehradun Herbarium there is one sheet ticketed "Sangam Vy. above Lidderwat (13-14000′) *Duthie*." In Stewart, l.c., the species has been cited from Pehalgam as well.

5. **S. cherleriae** (Fisch.) Williams, Bull. Herb. Boiss. ser. 2, 7 : 830, 1907; Stewart, l.c. 254. *Arenaria cherleriae* Fisch. ex DC., Prodr. 1 : 409, 1824. *S. decumbens* Edgew. in Trans. Linn. Soc. 20 : 35, 1846; Edgew. et Hk.f. in Hk.f. l.c. 234.

Shining, bright-green or yellowish perennial herb, forming densely intricated cushions or mats. Leaves ovate or oblong, 3-6×1-2.5 mm, acicular-pointed, often closely imbricated; margins usually ciliate. Flowers white, 6-7 mm across, solitary axillary; sepals 4-5, coriaceous, 3-nerved, scarious margined; petals narrow, short, deeply 2-parted.

Occ. : Common along banks throughout in morainal zone; often forming extensive, uninterrupted mats near melting snow.

Fls. July-Sept.

Kolahoi (3730 m) *B.M. Sharma* 14068; Kolahoi Vy. near Lidderwat, *Duthie* 14136; Budzalkhod nallah, Lidder Vy. (13-14000′) *Duthie* 13403.

10. PSEUDOSTELLARIA Pax

P. himalaica (Franch.) Pax in Engler & Prantl, Pflanzenfam. ed. 2, 18 C. 318, 1934; Stewart, l.c. 249. *Stellaria davidi* (Franch.) Hemsl. ex Forb. var. *himalaica* Franch., Pl. Delav. 99, 1889. *P. bulbosa* (Nakai) Ohwi in Acta Phytotax. et Geobot. 4 : 34, 1935. *P. cashmiriana* Schaeftelein in Phyton (Ann. Rei. Bot.) 7 : 195, 1957; Majumdar, l.c. 140; For. Fl. Sgr. 172. *S. bulbosa* Edgew. et. Hk.f. in Hk.f., l.c. 231 (non Wulf.).

Erect, slender perennial from a creeping rootstock; stem simple with a line of hairs between the nodes. Leaves ovate or lanceolate, in 2-6 pairs, up to 5×1.2 cm, sessile or contracted into a short petiole. Chasmogamic flower solitary; 1 cm across, white, or erect, subterminal pedicel.

Occ. : Humus laden ground under coniferous trees.

Fls. June.

Aru (2450 m) *B.M. Sharma* 14201

Our specimen is apparently similar to *P. heterophylla* (Miq.) Pax f. *nepalensis* Majumdar in the upper two pairs of leaves being disposed in a pseudowhorl, but differs in having only 2 styles and the petals scarcely 2-lobed.

11. MINUARTIA Linn.

M. lineata Bornm. in Beih. Bot. Centralb. 27, Abt. 1 : 318, 1910 f. **kashmirica** (Edgew.) R.R. Stewart, Annot. Cat. Vasc. Pl. W. Pak. et Kash. 248, 1972. *M. kashmirica* Mattfeld in Engl. Jahrb. 57, Beibl. 126 : 32, 1921; Majumdar, l.c. 140. *Arenaria kashmirica* Edgew. in Hk.f., l.c. 236. **Fig. 44**

Glandular-hairy perennial herb with several slender, 8-16 cm long, erect flowering branches from a densely tufted, creeping base. Leaves narrow, 0.5-1.8 cm long, rather stiff, 3-nerved. Flowers white, 0.8-1 cm across, on long-peduncled, subumbellate cymes, rarely solitary; petals like capsules longer than the sepals. Capsule ovoid; seeds pale brown, dorsally granulate.

Occ. : Common alpine of cliffs, open rocky ground and high meadows.

Fls. July-Sept.

Saswarian (3450 m) *B.M. Sharma* 14170; Chandanwari, *H.D. Thapliyal* 26545.

M. lineata f. *foliosa* R.R. Stewart shares the distribution and habitats of the above taxon and is distinguished from the latter chiefly by its glabrous branches—a character hardly reliable (see Stewart l.c.).

Sheshnag, *K.S. Ahluwalia* RL. Herb. nos. 1318 & 13770; Lidder Vy. above Kainmul (11-12000′) *Duthie* 13101 & 13142.

12. ARENARIA Linn.

1. Styles 3; seeds rough :
   2. Leaf margins ciliate for their entire length; sepals dorsally hairy all over but for the scarious margins **1. A. serpyllifolia**
   2. Leaf margins ciliate towards base alone; sepals glabrous except sometimes for a fine dorsal line of hairs on the nerve **2. A neelgherrensis**
1. Styles 5; seeds smooth **3. A. orbiculata**

1. **A. serpyllifolia** Linn., Sp. Pl. 423, 1753; Edgew. et Hk.f. in Hk.f., l.c. 239.

Diffuse, rather slender, pubescent annual; stems pubescent all round, 5-25 cm long, decumbent or semierect, many from the root. Leaves ovate or elliptic, 0.5-1.3×0.2-0.8 cm, acute or acuminate. Flowers white 5-7 mm across, axillary and in terminal cymes; sepals nearly twice the length of the petals, 3-nerved, the inner with a scarious margin. Capsule ovoid, about the length of the sepals; seeds black.

Occ. : Common on bare ground along the paths; also in shade of bushes and as a field weed in Aru.

Fls. May-June; *Frts.* June-Aug.

Satlanjan (3060 m) *B.M. Sharma* 14400; Chota Khalen (2900 m) *P. S. Jamwal* 15626.

2. **A. neelgherrensis** Wt. & Arn., Prodr. Fl. Ind. Or. 43, 1834; Edgew. et Hk.f., in Hk.f., l.c. 239.

Much branched annual herb; stems many from the root, procumbent, rather stiff, up to 40 cm long, on the same plant pubescent all round or on one side or with two lines of pubescence running between the nodes. Leaves glabrous, distant, elliptic or obovate, 0.6-1.5× 0.2-0.6 cm, shortly stalked or subsessile, apiculate. Flowers white, 6-7 mm across, solitary-axillary and in few-flowered terminal cymes; sepals longer than petals, 1-nerved, all with both margins scarious. Capsule ovoid, exerted.

Occ. : Mostly among the scrubs; common in Aru-Liddarwat area.

Fls. May-July; *Frts.* July-Aug.

Liddarwat (2750 m) *B.M. Sharma* 14003; Aru (2425 m) *B.M. Sharma* 14606.

3. **A. orbiculata** Royle ex Edgew. et Hk.f. in Hk.f., FBI. 1 : 240, 1874.

Decumbent; branches very slender. Leaves orbicular or elliptic, 3-4 mm in diam., capsule 10-valved.

Stewart, l.c. 239, cites this from Harnag, Upper Lidder Valley. From our area there is no specimen at Dehradun Herbarium, nor have we been able to collect any.

13. SAGINA Linn.

S. **saginoides** (Linn.) Karsten, Deutsche. Fl. Pharm.-med. Bot. 539, 1882; Mizushima in J. Jap. Bot. 35 : 194, 1960; Butcher, New Ill. Brit. Fl. 1 : 447. 1961; Stewart, l.c. 250; For. Fl. Sgr. 172. *Spergula saginoides* Linn., Sp. Pl. 441, 1753. *Sagina procumbens* Edgew. et Hk.f. in Hk.f., l.c. 242 (non Linn.)

Glabrous, tufted, small perennial; stems several, slender 2-8 cm long, prostrate and decumbent. Leaves connate at base into a white sheath, mucronate, 0.4-1 cm long, the radical nearly twice as long. Flowers white, 2 mm long, 5-merous, solitary on slender, axillary and terminal pedicels; petals about equalling obtuse, white-margined sepals. Capsule exserted, 3 mm long, 5-valved almost to base; seeds minute, brown, finely granulate.

Occ. : Common in moister places throughout most of the area of work.

Fls. & Frts. June-Aug.

Kolahoi (3350 m) *B.M. Sharma* 14219; Liddarwat (2950 m) *B.M. Sharma* 14605; Budzalkhod Nallah, Lidder Vy. (12-13000′) *Duthie* 13413.

14. **SPERGULARIA** J. & C. Presl

S. rubra (Linn.) J. & C. Presl, Fl. Cechica 94, 1819; Kitam., Fl. Afgh. 117, 1960; Stewart, l.c. 253. *Arenaria rubra* Linn., Sp. Pl. 423, 1753. *Spergula rubra* Edgew. et Hk.f. in Hk.f., l. c. 244.

Low growing, diffuse herb; stems many, 5-12 cm long, spreading or decumbent from a woody root, glandular hairy upwards at least. Leaves narrowly linear, 0.5-1.2 cm long, fasciculate, mucronate; stipules silvery. Flowers pink, 5-6 mm across; petals shorter than obtuse, white-margined sepals. Capsule ovoid, about equalling the sepals; seeds rough, minute, brown.

Occ. : Throughout along roadsides and paths from Pehalgam to Kolahoi-n-kot; preferring calciferrous habitats.

Fls. & Frts. May-July.

Kolahoi (3700 m) *B.M. Sharma* 14384; Sheshnag, *L.D. Kapoor* 1460 (under *Arenaria festucoides* Benth.)

XV. TAMARICACEAE

MYRICARIA Desv.

M. germanica Desv. in Annls Sci. nat. ser. 1, 4 : 349, 1825; Thiselton Dyer in Hk. f , l. c. 250.

Glabrous, up to 2 m tall, festigiate shrub; branches striate, purplish-brown, glaucous when young. Leaves sessile, somewhat glaucous, linear-lanceolate, 1-6 mm long, rather close-set. Flowers pink, 6-7 mm long, in 5-45 cm long spike-like racemes terminating mostly the main branches; bracts broadly scarious-margined, 3-4-times the length of the pedicels; stamens 10, connate into a tube for about half their length, alternately long and short. Capsule 7-10 mm long; seeds numerous, brown, crowned by a stipitate coma.

Occ. : Common and gregarious on gravelly islands in stream beds at Pehalgam, rare elsewhere in the upper reaches.

Fls. & Frts. July-Sept.

2 km below Aru (2350 m) *B. M. Sharma* 14269; Pehalgam (7000′) *K. M. Vaid*, DD. no. 7304; near Pehalgam (2300 m) *P. S. Jamwal* 15633.

XVI. HYPERICACEAE

HYPERICUM Linn.

Margins of sepals devoid of any gland-tipped teeth **1. H. perforatum**
Margins of sepals furnished with short, gland-tipped teeth **2. H. napaulense**

1. **H. perforatum** Linn., Sp. Pl. 785, 1753; Thiselton Dyer in Hk. f., l. c. 255. **Fig. 45**

Perennial, 30-60 cm high, glabrous herb; stems erect, 2-angled, few to many from a woody rootstock, branched upwards. Leaves oblong or ovate, pellucid punctate, 0.5-2.5×0.15-1 cm, sprinkled with black dots. Flowers yellow, 2-2.8 cm across, in terminal corymbs; sepals lanceolate, acute, like the petals black-dotted on margins; stamens numerous, 3 adelphous, with black-dotted anthers. Capsule conical, 8 mm long, 3-valved.

Occ. : Common in open grassy places in Aru, also on the roadside slopes between Mundlan and Aru.

Fls. May-Aug.; *Frts.* July-Sept.

Aru (2375 m) *B.M. Sharma* 14386; the weak-stemmed specimen with oblong-linear leaves collected from drain-side at Aru (2425 m) *P. S. Jamwal* 19365 is var. *debile* Royle.

2. **H. napaulense** Choisy in DC., Prodr. 1 : 552, 1824; Thiselton Dyer in Hk.f., l.c. 255.

Stewart, l. c. 486, has cited this from Aru and Pehalgam. At Dehradun Herbarium there is a solitary sheet from the area; Lidderwat (9000′) *M. V. Laurie* 16756. However, we have not been able to spot it in the field.

XVII. MALVACEAE

1. Epicalyx of 6 or more segments, united at the base — 1. **Althaea**
1. Epicalyx of 3 segments :
   2. Segments united at the base — 2. **Lavatera**
   2. Segments ununited — 3. **Malva**

1. ALTHAEA Linn.

A. **rosea** (Linn.) Cav. Diss. 2 : 91, t. 29, f. 3, 1790; Masters in Hk.f., l.c. 319. *Alcea rosea* Linn., Sp. Pl. 687, 1753; For. Fl. Sgr. 150.

Occurring as an escape from cultivation; flowers large; pinkish-white. Rare.

Fls. July-Aug.

Aru (2400 m) *B. M. Sharma* 14275.

2. LAVATERA Linn.

L. **kashmiriana** Camb. in Jacq. Voy. Bot. Inde, 4 : 29, t. 32, 1844; Masters in Hk. f., l.c. 319. **Fig. 46**

Branched, 1-1.5 m tall, stellately downy perennial herb. Leaves crenate, lower orbicular-cordate with 5 rotund lobes and petiole about equalling the blade, the upper with pointed 3 to 5 lobes and shorter petioles. Flowers purplish-pink, with darker veins, 6–9 cm across, solitary, axillary. Calyx accrescent; lobes lanceolate, longer than broadly-ovate, acute, epicalyx-segments. Carpels reniform, smooth and black when ripe.

Occ. : Sporadic, on humus rich soil in forest clearings; more comon on Pehalgam-Chandanwari route.

Fls. July-Aug.; *Frts.* Aug.-Sept.

Fig. 41. *Cerastium dahuricum* Fisch. ex Spreng.

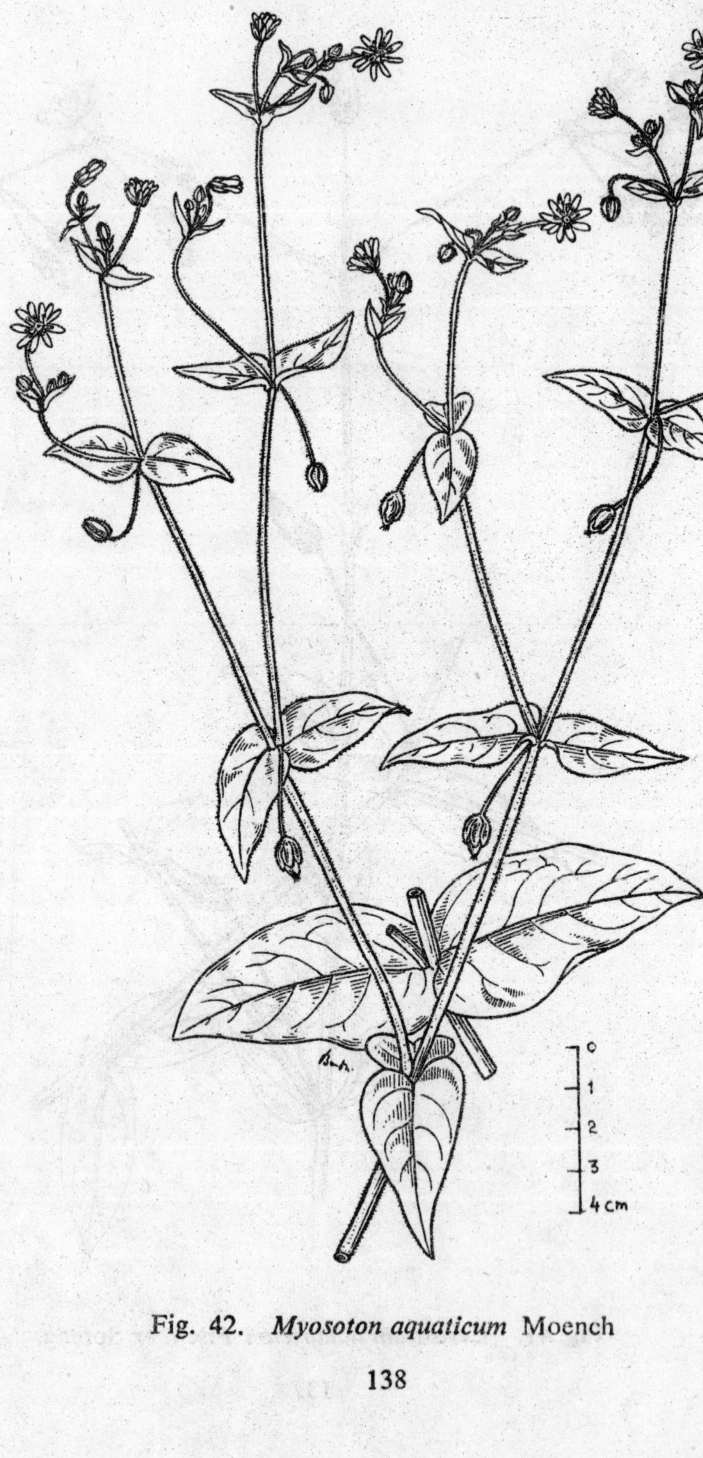

Fig. 42. *Myosoton aquaticum* Moench

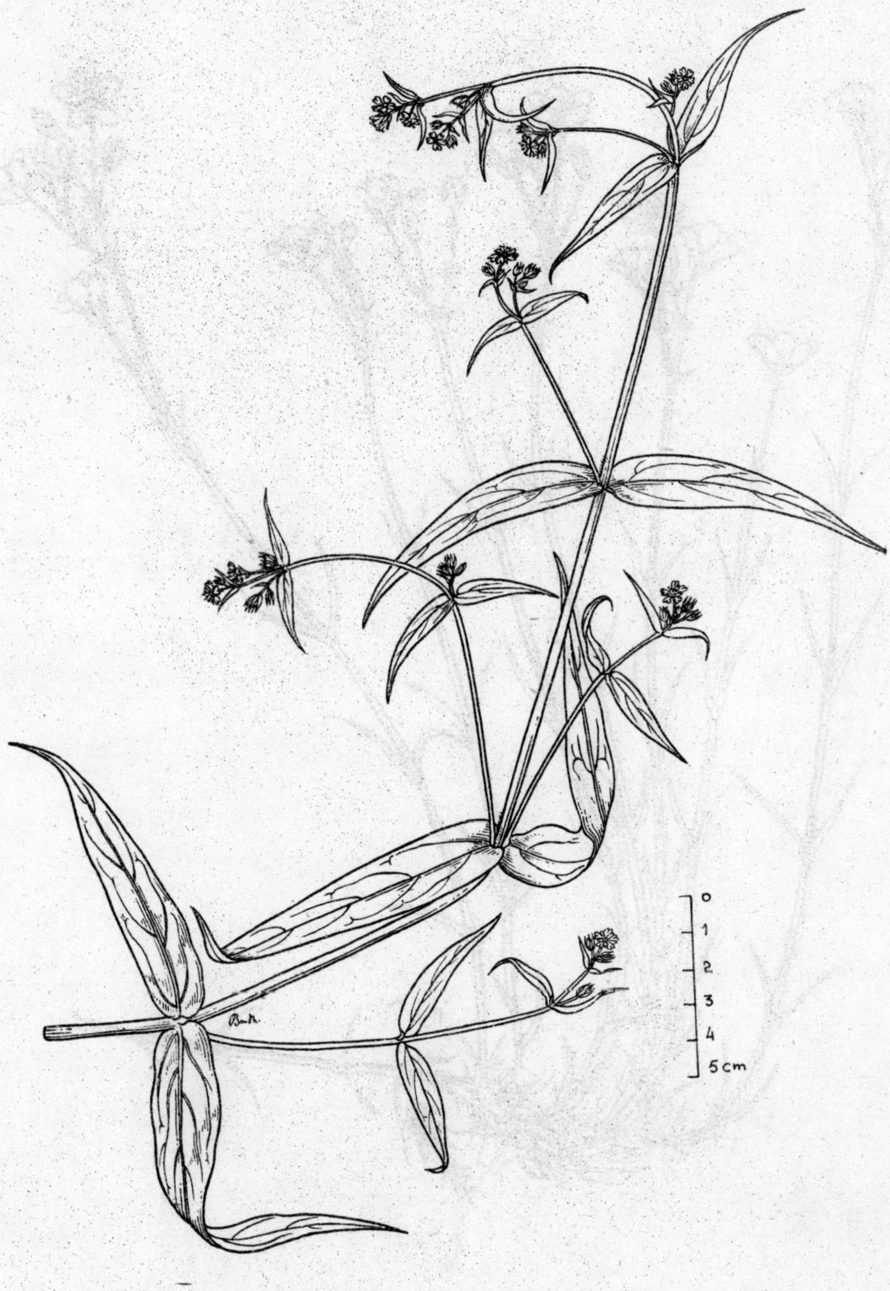

Fig. 43. *Stellaria monosperma* Buch.-Ham. ex. D. Don

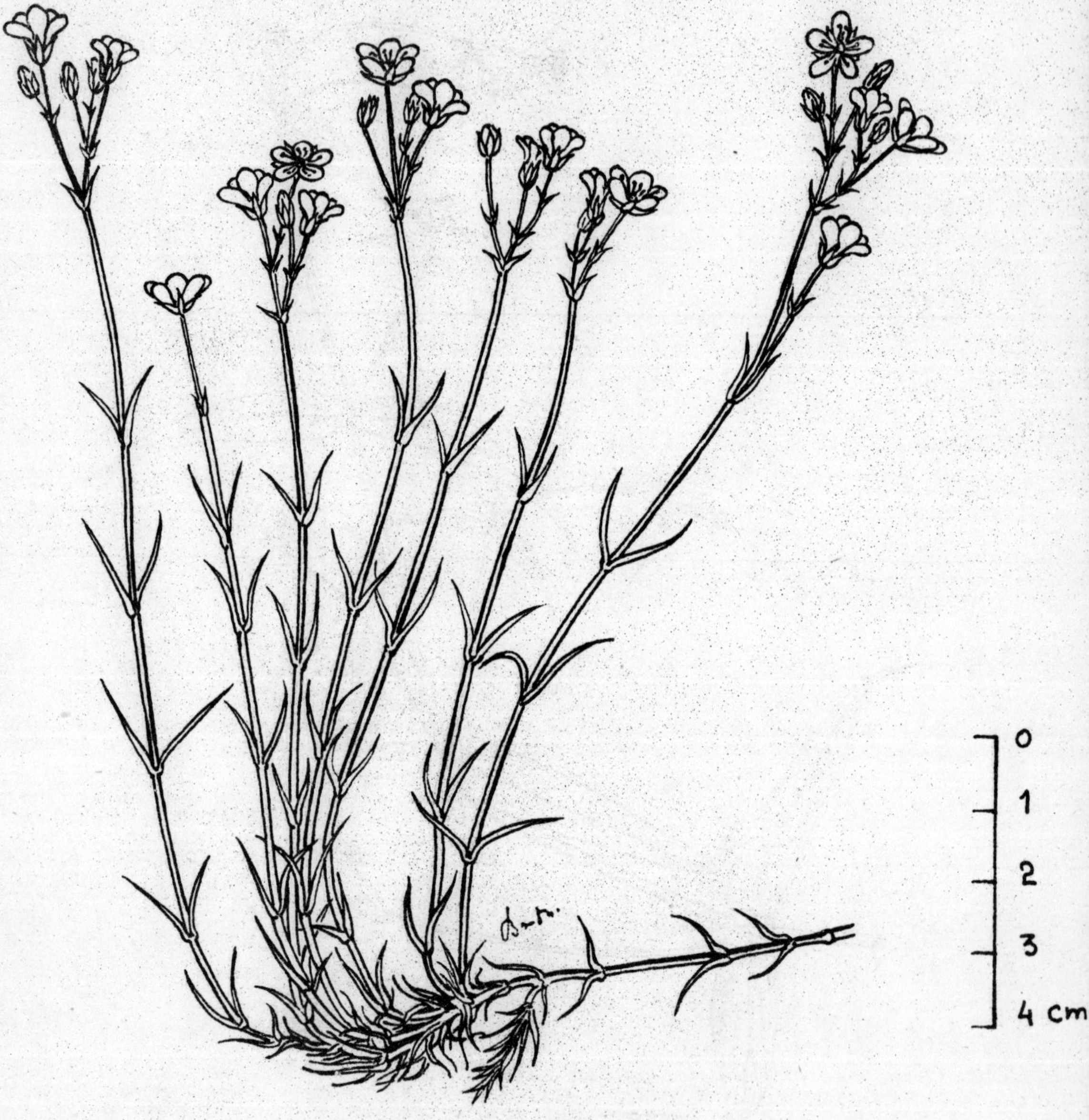

Fig. 44. *Minuartia lineata* Bornm.f. *kashmirica* R.R. Stewart

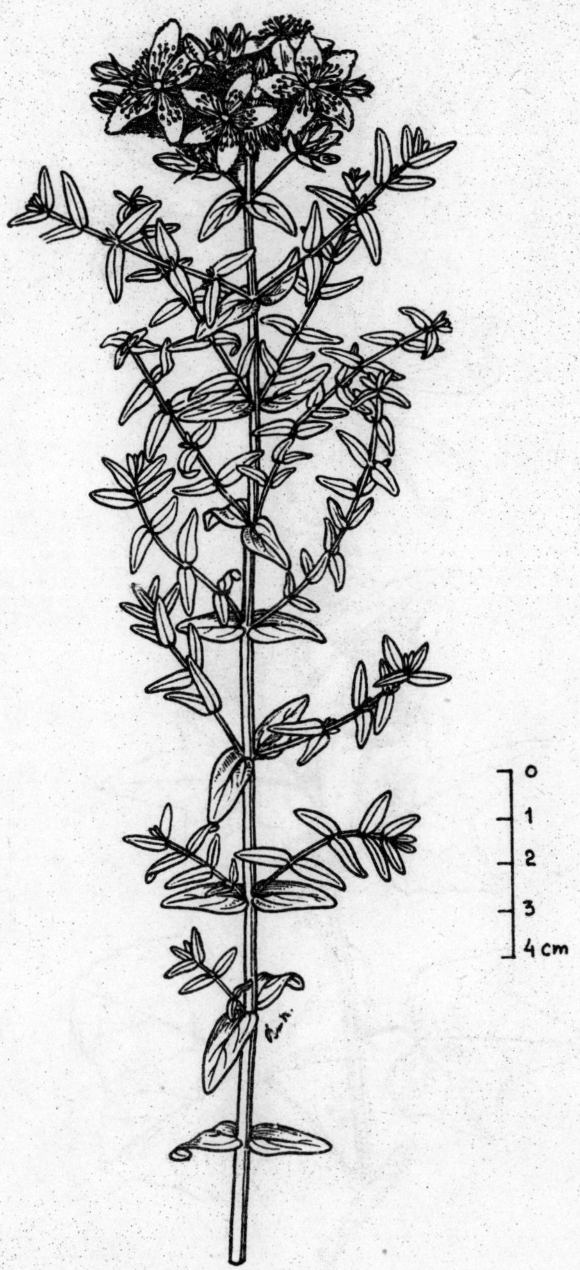

Fig. 45. *Hypericum perforatum* Linn.

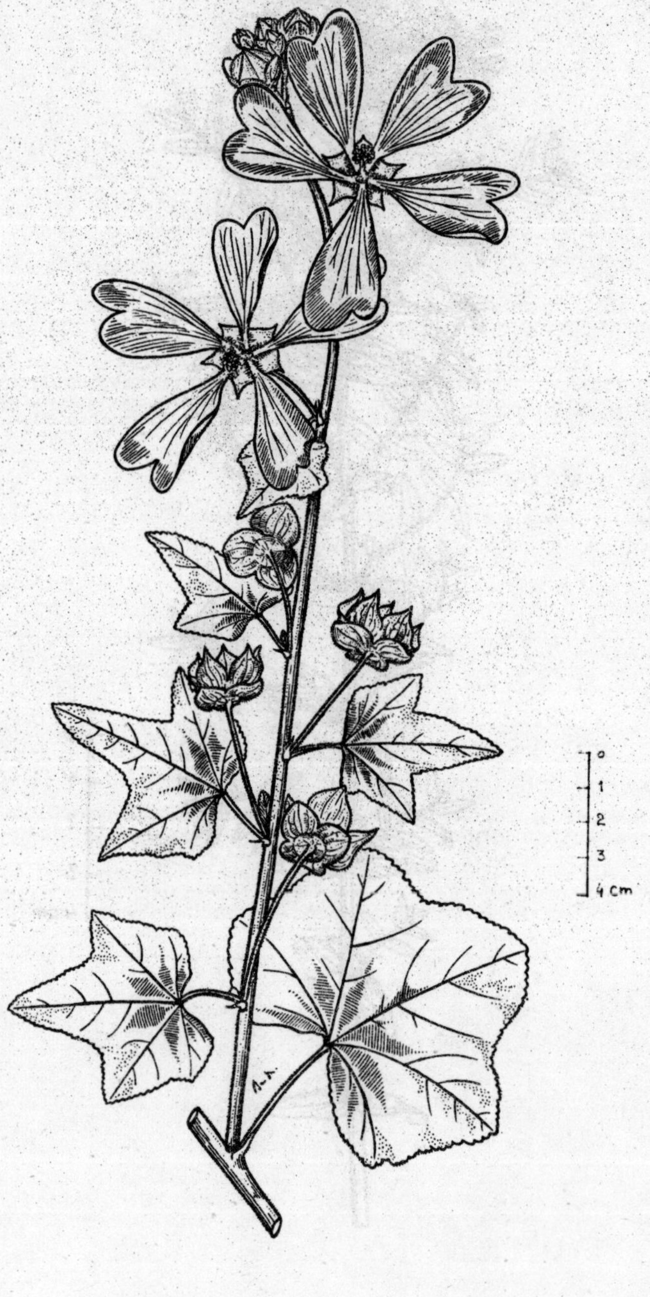

Fig. 46. *Lavatera kashmiriana* Camb.

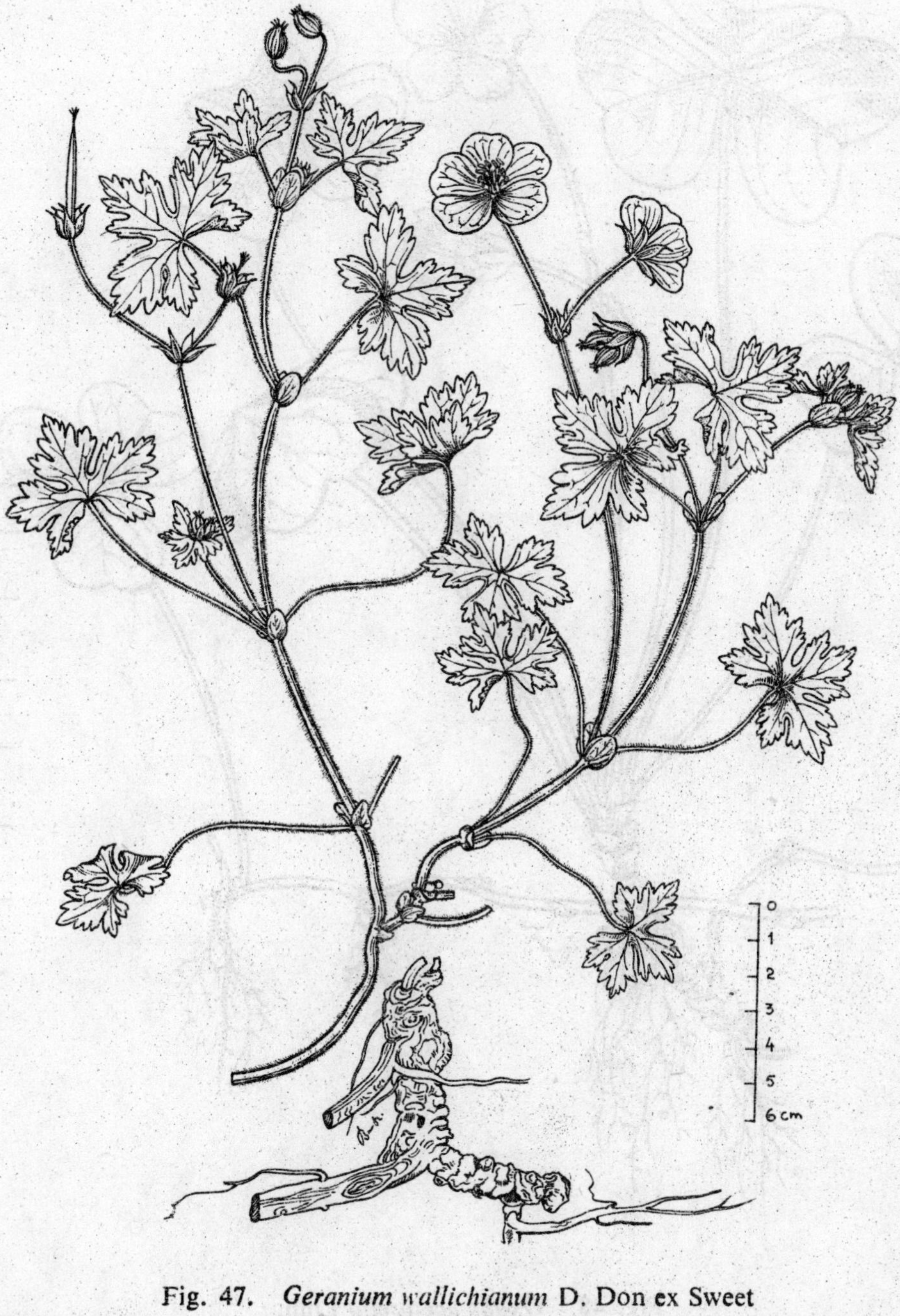

Fig. 47. *Geranium wallichianum* D. Don ex Sweet

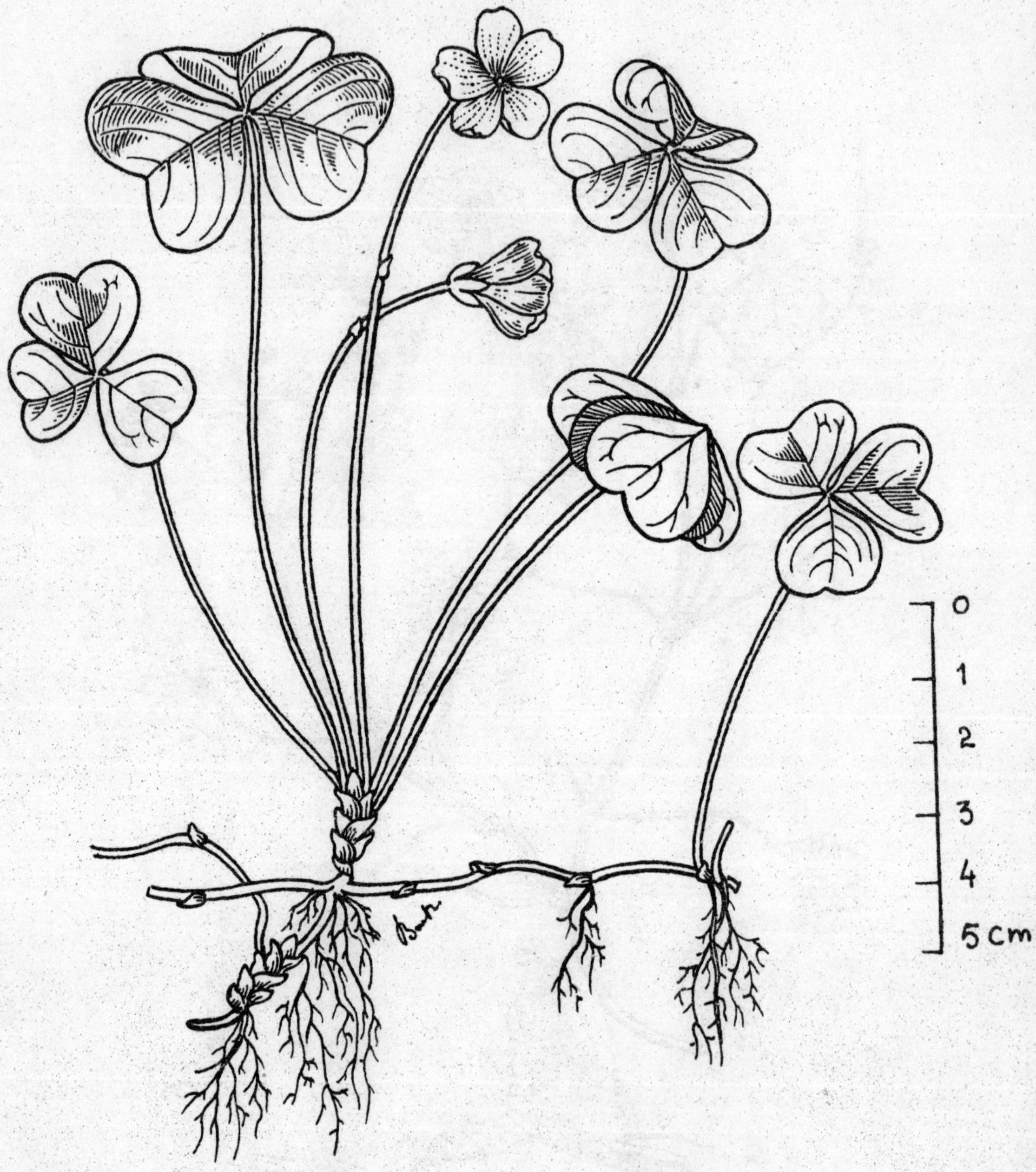

Fig. 48. *Oxalis acetosella* Linn.

Aru (2300 m) *B.M. Sharma* 14293; Roadside below Chandanwari (2600 m) *P. S. Jamwal* 15671.

3. **MALVA** Linn.

Flowers in nearly sessile clusters; petals less than twice as long as the sepals; calyx strongly accrescent **1. M. verticillata**
Flowers long-stalked; petals at least twice as long as the sepals; calyx not accrescent **2. M. neglecta**

1. **M. verticillata** Linn., Sp. Pl. 689, 1753; Masters in Hk. f., l.c. 320.

Remotely stellate-hairy, 0.6-1 m tall annual herb; branches downy. Leaves long-stalked, reniform to suborbicular, 5-15 cm across, from a cordate base, 3-6-lobed; lobes crenate, orbicular, those of the upper leaves deltoid. Flowers pink, 1.2-1.4 cm across, in dense axillary clusters; petal-claw with few, weak hairs. Ripe carpels brown, 2-2.5 mm, strongly reticulate laterally, weakly transversely ridged dorsally.

Occ. : In cultivated fields and near habitations.

Fls. July-Aug.; *Frts.* Aug.-Sept.

Aru (2400 m) *B.M. Sharma* 19285; On way to Chumanisar, near Aru (2450 m) *P. S. Jamwal* 15628.

2. **M. neglecta** Wallr., Syll. Pl. Nov. Ratisb. 1 : 140, 1824; Iljin in Fl. URSS. 15 : 56, 1949; Kitam., Fl. Afgh. 271, 1960; Stewart, l.c. 481; For Fl. Sgr. 151. *M. rotundifolia* auct. plur., non Linn.; Masters in Hk.f., l.c. 320.

Stellately pubescent, perennial herb with 8-35 cm long, procumbent or decumbent stems radiating from the rootstock. Leaves orbicular, 1-6 cm across, shallowly 5-7-lobed; lobes rotund, crenate; petiole up to 17 cm long. Flowers pink, pale lilac or whitish, 1.5-2.5 cm across, in 3-6-membered axillary fascicles; pedicels unequal, 1-5 cm long in fruit. Ripe carpels pale brown, 2 mm, smooth, dorsally stellately downy.

Occ. : Common along either side of the route from Liddarwat to Kolahoi, more often in stony ground.

Fls. & Frts. June-Sept.

Satlanjan (3100 m) *B.M. Sharma* 14078 & 19362; Lidder Valley (8-9000′) *Duthie* 13451.

XVIII. GERANIACEAE

Leaves palmately (in *G. robertianum* ternately) divided or lobed; stamens 10, all anther-bearing **1. Geranium**
Leaves pinnately divided; stamens 10. only 5 anther-bearing **2. Erodium**

1. GERANIUM Linn.

1. Flowers 2.5 cm and above in diameter :
   2. Stipules bifid or bipartite; filaments densely villous with long hairs **1. G. grevilleanum**
   2, Stipules undivided; filaments ciliate :
      3. Stipules small, subulate-lanceolate, acute **2. G. pratense**

3. Stipules 1-2.5 cm long, broadly oblong-ovate to near-orbicular, obtuse — 3. **G. wallichianum**

1. Flowers 2 cm and less in diameter :

4. Leaves very deeply 3-5-partite, their primary divisions finely 2-pinnately dissected; stipules ovate; petals about twice the length of the sepals — 4. **G. robertianum**

4. Leaves palmately lobed, their primary divisions simply toothed or lobed; stipules subulate-lanceolate; petals about as long or scarcely longer than sepals — 5. **G. nepalense**

1. **G. grevilleanum** Wall., Pl. As. Rar. 3,4. t. 209, 1832; Edgew. et Hk.f. in Hk.f., l.c. 430. *G. eriostemon* D. Don, Prodr. Fl. Nep. 208, 1825.

Stout, pubescent, semierect or procumbent, perennial herb, glandular-hairy upwards. Leaves long-petioled, pentagonal, 10-18 cm wide, with 5-7 rhomboid, acute or acuminate, toothed lobes, the upper smaller, usually 3-lobed, sessile; stipules paired or 2-lobed. Flowers rose purple, 3-4.5 cm across, on long, slender, generally 2-, or 3-flowered peduncles. Fruit 3 cm long; carpels hairy; seeds dark-brown, ovoid, 3-4 cm long.

Occ. : On humus covered soil in forest shade.

Fls. July-Aug.; *Frts.* Aug. Sept.

Chhumanai nallah bank, *B.M. Sharma* 14091; Lidder Vy. above Pehalgam (8-9000′) *Duthie* 14142; Lidder Vy. Kainmul nallah (8-9000′) *Duthie* 13164; Aru, Lidder Vy., *Inayat* 25546; 3 km from Aru, on way to Chhumanaisar, *P.S. Jamwal* 15631.

2. **G. pratense** Linn., Sp. Pl. 681, 1753; Edgew. et Hk.f. in Hk.f., l.c. 429.

Rather stout, semierect or ascending, patently glandular-hairy perennial herb; stems 35-75 cm long, many from a thick rootstock. Leaves orbiculate, 5-10 cm wide, palmately deep-cut into 5-7(-9), pinnatifid lobes; segments acute. Flowers bright blue-purple, 3.5-4.5 cm across, paired on long, stout peduncles; fruiting pedicels bent but ultimately straight. Fruit 4 cm long; carpels hirsute; seeds ovoid, brown, 4 mm long, netted.

Occ. : Common in high altitude pasture lands about Seikwas and Rewil.

Fls. June-Aug.; *Frts.* July-Aug.

Seikwas (3350 m) *B.M. Sharma* 19363; Sheshnag (11000′) *D. F. O., M. P. Utilization Div.*, DD no. 52860.

3. **G. wallichianum** D. Don ex Sweet, Geran. 1, t. 90, 1821; Edgew. et Hk.f., in Hk. f., l.c. 430. **Fig. 47**

Much branched, hairy perennial herb with erect or ascending, 30-75 cm long, rather robust stems from a thick rootstock. Leaves orbicular, 3.5-10.5 cm wide, palmately divided into 3-5, cuneate, irregularly toothed lobes. Flowers pinkish-, or bluish-purple with dark-red veins, 3-4.5 cm across, paired. Fruit 3-3.5 cm long; carpels hairy; seeds smooth.

Occ. : Common usually in forest shade in areas about Aru and Liddarwat; occurring also in open, along streams and among junipers.

Fls. June-Aug.

Seikwas (3350 m) *B. M. Sharma* 19364; Liddarwat stream bank (2730 m) *P.S. Jamwal* 15635.

4. G. **robertianum** Linn , Sp. Pl. 681, 1753; Edgew. et Hk.f. in Hk.f., l.c. 432.

Strongly scented, 30-60 cm high, softly viscid-hairy, much branched herb; stems erect or ascending, rather succulent and brittle, very often red. Leaves triangular, 3-8 cm wide, 3-5-partite to the midrib into deeply 2-pinnatisect lobes; ultimate segments mucronate or apiculate. Flowers streaked, bright pink, 1.2-1.5 cm across, paired on long peduncles. Fruit 1.8-2.5 cm long; carpels keeled, transversely ridged upwards.

Occ. : Shaded places along roadsides between Mundlan and Aru; also collected from a damp curvature near Liddarwat bridge.

Fls. & Frts. June-July.

Liddarwat, *B.M. Sharma* 14376; Lidder Vy. (6-7000′) *Duthie* 13078; Thanan, Lidder Vy., *Inayat* 25545; Lidderwat, *K M. Vaid* DD. No. 118085; Roadside near Mundlan (2325 m) *P. S. Jamwal* 15631.

5. **G. nepalense** Sweet, Geran. 1, t. 12, 1820; Edgew. et Hk.f. in Hk.f., l.c. 430.

Much branched, diffuse, perennial herb, with 15-45 cm long, rather slender, softly hairy stems. Leaves pentagonal, 2.5-7 cm wide, palmately divided into 3-5, irregularly toothed and segmented lobes. Flowers pink, fading to white, streaked, 0.8-1.5 cm across, solitary or very often twin on long peduncles; fruiting pedicels reflexed. Fruit 1.5-2 cm long; seeds shining, dark chestnut, 2.5 mm long, finely netted.

Occ. : Very common along waysides and shaded slopes throughout between Pehalgam and Liddarwat, especially abundant in nitrophilous soils.

Fls. June-Sept.; *Frts.* July-Sept.

Liddarwat (2700 m) *B. M. Sharma* 14188. Aru-Chhumanisar route (2600 m) *P.S. Jamwal* 15636.

2. ERODIUM L′ Herit

E. **cicutarium** (Linn.) L′ Herit in Aiton, Hort. Kew. ed. 1, 2 : 414, 1789; Edgew. et Hk.f. in Hk.f., l.c. 434. *Geranium cicutarium* Linn., Sp. Pl. 680, 1753.

Variable, more or less viscid, hairy annual with diffuse or spreading, up to 40 cm long, slender or stoutish stems, swollen at joints. Leaves radical and opposite, oblong, 2-12 cm long, pinnately divided into pinnatifid lobes; segments acute. Flowers small, pink, often purple-dotted, in 2-many-flowered, umbellate clusters on erect, axillary peduncles; petals unequal. Fruit 2-4 cm long; carpels hispid, pitted at top, with beaks silky on inside; seeds brown, 2.5-3 mm long, smooth.

Occ. : Waysides between Pehalgam and Aru, also on disturbed roadside slopes in the area.

Fls. May-July; *Frts.* June-Aug.

Aru (2400 m) *B.M. Sharma* 14262- Mundlan-Aru road (2300 m) *P.S. Jamwal* 15653.

XIX. Oxalidaceae

1. OXALIS Linn.

Stems much branched, rooting at nodes; flowers in small subumbellate clusters; capsule cylindric tomentose **1. O. corniculata**

Stemless; flowers solitary; capsule ovoid, glabrous **2. O. acetosella**

1. **O. corniculata** Linn., Sp. Pl. 435, 1753; Edgew. et. Hk.f. in Hk.f., l.c. 436, 1874.

Pubescent, creeping or ascending, diffuse perennial herb; branches slender, 6-30 cm long. Leaves long-stalked, alternate, 3-foliolate; leaflets nearly sessile, 0.5-2 cm wide, broadly obcordate, emarginate, with rounded lobes. Flowers yellow, 0.8-1.5 cm across, on slender, axillary peduncles. Capsule beaked, 1.2-2 cm long; seeds brown, transversely rugose.

Occ. : Almost everywhere along waysides, preferably in shady habitats, along Pehalgam-Satlanjan route.

Fls. May-July; *Frts.* June-Aug.

Satlanjan (3000m) *B.M. Sharma* 14217; Aru (2425 m) *P.S. Jamwal* 15630.

2. **O. acetosella** Linn., Sp. Pl. 433, 1753; Edgew. et Hk.f. in Hk.f., l.c. 436. **Fig. 48**

Perennial herb with creeping, scaly rootstock. Leaves 3-foliolate, all radical, on 2.5-10 cm long petioles; leaflets as in the preceding species, 1-2.5 cm wide, Flowers campanulate, 1-1.5 cm long, on 2-bracteate, leaf-exceeding radical stalks.

Occ. : Common and sometimes gregarious in shady, damp places about Aru.

Fls. May-June

Near Aru (2475 m) *B.M. Sharma* 19329; Kinimola nala Lidder Vy., *Inayat* 25550.

XX. Balsaminaceae

1. IMPATIENS Linn.

1. Flowers pink, purple or crimson :
   2. Leaves opposite or in 3-membered verticils; sepals sac abruptly contracted to spur **1. I. glandulifera**
   2. Leaves opposite or alternate; sepal sac gradually tapering to spur :
      3. Leaves crenate; flowers 2-3 cm long **2. I. bicolor**
      3. Leaves mostly serrate, flowers below 2 cm in length **3. I. thomsonii**
1. Flower colour not as described above :
   4. Flowers white, below 8 mm in length, spurless or minutely spurred **4. I. brachycentra**
   4. Flowers yellow, streaked with red inside, about 2.5 cm long including the spur **5. I. edgeworthii**

1. **I. glandulifera** Royle, Ill. Bot. Himal. 151, t. 28, f. 2, 1834; descr. 1835; Sprague, Kew Bull. 386, 1933; Moore in Tutin et al., Fl. Eur. 2 : 240, 1968; Stewart, l.c. 465; For. Fl. Sgr. 204. *I. roylei* Walpers, Rep. Bot. Syst. 1 : 475, 1842; Hk.f. et Th. in J. Linn. Soc. Bot. 4 : 127, 1859; Hk.f., FBI. 1 : 468, 1875. **Fig. 49.**

Glabrous, 0.25-1.2 m tall annual; stem thick, succulent, sometimes red. Leaves lanceolate or elliptic from a cuneate base, 3.5-15×1.5-6 cm, serrate, glandular on petioles and often so on lower serratures. Flowers purplish pink, freckled with brown towards the base of the sepal sac, 2.5-4 cm long, excluding the cylindric, recurved, rather thick, 2.5-6 mm long spur, in terminal umbellate clusters or racemes. Capsule clavate, 1.7-2.5 cm long, 5-8 mm at its broadest width; seeds grey, wrinkled, broadly obovoid, 3.5-4 mm long.

Occ. : Common in damp places, in fields and along the streams throughout to 3200 m.

Fls. June-Sept.; *Frts.* Aug.-Sept.

Aru (2400 m) *B.M. Sharma* 14337; Satlanjan (3100 m) *B.M. Sharma* 14156; Lidder Vy. above Tanin (10-11000′) *Duthie* 13308; Aru, Lidder Vy., *Inayat* DD s.n.; Aichnar Nallah, Lidder Vy., *Inayat* DD s.n.; Aru, Lidder Vy., *Inayat* DD s.n.; Chhuth Panjal Nallah, Lidder Vy., *Inayat* DD s.n.; Kolahoi, Lidder Vy., *Inayat* DD s.n.; Lidderwat, Lidder Vy., *Inayat* DD s.n.; Nafran Nallah, Lidder Vy., *Inayat* DD s.n.; Armiun Glenn (2750 m) *P.S. Jamwal* 15638.

I. gigantea Edgew., cited by Stewart (l.c., 465) from Lidder Vy., has been stated by him to be close to *I. glandulifera* but distinguishable by its alternate, crenate leaves and narrowly clavate fruits. We have not come across this in the area.

2. **I. bicolor** Royle, Ill. Bot. Himal. 151, t. 28, 1835; Raizada, Indian For. 92 (5) : 312, 1966; Stewart, l.c. 464; For. Fl. Sgr. 203. *I. amphorata* Edgew. var. *umbrosa* (Edgew.) Hk.f., l.c. 476. *I. umbrosa* Edgew. in Trans. Linn. Soc. 20 : 39, 1846.

Glabrous, slender or stout, 30-90 cm tall annual. Leaves alternate, ovate or lanceolate, 6-15×2.5-6 cm, acuminate; petioles sometimes glandular; stipular glands sessile. Flowers purplish pink, in racemes or umbellate clusters, often paniculate; pedicels remotely and minutely glandular. Sepal sac broadly funnel-shaped, drawn into a slender incurved spur of variable length. Standard keeled with crested tip; wings deeply 2-lobed. Capsule 2-3 cm long, subtorulose; seeds obovoid-oblong, 4 mm long, compressed, wrinkled.

Occ. : Very common on shady and damp humus rich slopes along water courses where it becomes sometimes almost gregarious.

Fls. July-Aug.; *Frts.* Aug-Sept.

Aru (2400 m) *B.M. Sharma* 14617; Gha Shangan (2950 m) *B.M. Sharma* 15609; Chandanwari-Pehalgam (8000′) *R. L. Badhwar* RL Herb. nos. 1672 & 1673 (both under *I. amphorata* Edgew).

3. **I. thomsonii** Hk.f., J. Linn. Soc. Bot. 4 : 128, 1860; Hk.f., l.c. 469, 1875.

Glabrous, 12-60 cm tall, slender or robust annual. Leaves petioled, opposite or the upper alternate or whorled, lanceolate, 2.5-12 (5-15)×1.2-4 cm, coarsely serrate, acuminate; stipules pulvinate. Flowers pink or pale pink, 1-1.8 cm long, in umbellate clusters; sepal sac funnel-shaped, freckled with brown or yellow, tapering into a slender, curved spur. Capsule 2 cm long, narrowly clavate.

Occ : Common in situations similar to those for the preceding species.

Fls. & Frts. July-Sept.

Gha Shangan (2950 m) *B.M. Sharma* 14299; Liddarwat (2900 m) *B.M. Sharma* 14167; Lidder Vy. above Kainmul (11-12000′) *Duthie* 13157; Kolahoi, Lidder Vy., *Inayat* DD s.n.; Lidderwat, Lidder Vy. *Inayat* DD s.n.; Nafran Nallah, Lidder Vy , *Inayat* DD s.n.

4. **I. brachycentra** Kar. et Kir. in Bull. Soc. Nat. Moscow, 15 : 179, 1842; Hk.f., l.c. 481.

Glabrous, 12-50 cm tall, slender, simple ar branched annual. Leaves alternate, ovate-lanceolate, 2-8 × 1.2-3 cm excluding petioles, with usually purplish, gland-tipped teeth. Flowers 5-8 mm long, in racemes or subumbellate clusters on slender, up to 7 cm long, axillary and subterminal peduncles; sepal sac boat-like, almost spurless or with a minute conical spur. Capsule narrowly clavate, 1-1.4 cm long.

Occ. : Very common in damp, wooded or open places throughout up to about 3200 m.

Fls. June-Aug.; *Frts.* July-Sept.

Aru (2350 m) *B.M. Sharma* 14353; Aru, Lidder Vy., *Inayat* DD s.n.; Chutt Panjal Nallah, Lidder Vy., *Inayat* DD s.n.; Lidderwat, Lidder Vy., *Inayat* DD s n.; Chandanwari, *L. D. Kapoor* R. L. Herb. nos. 1682 & 1683; Liddarwat (2750 m) *P.S. Jamwal* 15639.

5. **I. edgeworthii** Hk.f., FBI. 1 : 476, 1875 et Rec. bot. Surv. India, 4 : 9, 1904. *I. chrysantha* Hk.f., Bot. Mag. t. 7786.

Glabrous, 0.3-1.2 m tall, branched annual. Leaves alternate, elliptic or oblong-elliptic, 6-20 × 3-8 cm, long-petioled, acuminate, crenate-serrate, with teeth tipped by a bristle-like gland. Flowers 2.2-2.5 cm long, on whorled or panicled pedicels arising from long peduncles springing mostly from the axils of the upper, often crowded leaves. Sepal midrib prominently thickened; basal lobe of the wings protruded. Capsule linear, glabrous.

Occ. : About water courses in forest openings; restricted only to a few localities.

Fls. July-Sept.

Mundlan-Aru road (2350 m) *B.M. Sharma* 15608; Armiun nallah (2760 m) *B.M. Sharma* 14609; Aichnar nallah, Lidder Vy., *Inayat* DD s.n.; Aru, Lidder Vy.; *Inayat* DD s.n.; Nafran nallah, Lidder Vy,, *Inayat* DD s.n.; below Aru (2350 m) *P. S Jamwal* 15640.

XXI. Rutaceae

1. SKIMMIA Thunb.

S. **laureola** Sieb. et Zucc. ex Walp., Rep. 5 : 405, 1845-46; Hk.f., FBI. 1 : 499, 1875. **Fig. 50**

Erect, strongly aromatic, 0.6-2 m high, gregarious ever-green shrub with whitish bark. Leaves thick, softly coriaceous, oblong-lanceolate or oblanceolate, 6-15 × 1.5-4.5 cm; crowded towards extremities, orange-smelling when crushed; lateral nerves obscure. Flowers 5-merous, white or greenish-yellow, 0.8-1.2 cm across, polygamous, in erect, 3-4 cm long, dense panicles; calyx lobes imbricate; stamens 5, alternate with and equalling the petals, 0 in female flowers. Drupe fleshy, ovoid, 0.6-1.5 cm long, red, with 2 (sometimes 1 or 3), 1-seeded stones.

Occ. : In pure patches on either side of Nangneri nallah base; also mixed with other shrubs under firs on the slopes near Liddarwat Rest House.

Fls. May-July; *Frts.* July-Sept.

Nangneri (2800 m) *B.M. Sharma* 19289; Liddarwat (2950 m) *P.S. Jamwal* 15637.

XXII. Celastraceae

1. EUONYMUS Linn.

Style short but distinct; capsule deeply 3-4-lobed, not winged **1. E. hamiltonianus**
Style minute or obsolete; capsule with 4 long, tapering wings **2. E. fimbriatus**

1. **E. hamiltonianus** Wall. in Roxb., Fl. Ind. (ed. 1) 2 : 403, 1824; Lawson in Hk.f., l.c. 612. **Fig. 51**

Deciduous shrub or a small tree with rough, grey, corky bark; twigs terete, smooth. Leaves membranous or at length rather coriaceous, elliptic, oblong or broadly ovate-elliptic, 5-13×2-6 cm; finely serrate, mostly acuminate; stipules filiform. Flowers white, 0.7-1 cm across, in 2-, or 3-chotomous, axillary or intra-axillary cymes; petals oblong, entire, with reflexed margins. Capsule turbinate, 0.7-1 cm across the lobes.

Occ. : In forest openings up to about 2500 m. Scarcely distributed.

Fls. June-July; *Frts.* July-Sept.

Aru (2425 m) *B.M. Sharma* 14251; Lidder Vy. (6-7000′) *Duthie* 13067; Aru, Batkot (7000′) *L. D. Kapoor* 1092; Pehalgam (7200′) *R. L. Badhwar* 3800; Aru (2400 m) *P.S. Jamwal* 15655.

2. **E. fimbriatus** Wall. in Roxb., Fl. Ind. (ed. 1) 2 : 408, 1824; Lawson, in Hk.f., l.c. 611. *E. lacerus* Buch.-Ham. in D. Don, Prodr, Fl. Nep. 191. 1825.

Deciduous shrub or small tree, distinguishable from the preceding species through the characters of the key.

From Lidder valley there is a solitary specimen stored at FRI Herbarium Dehradun and ticketed as "Lidder Vy. above Tanin (9-10,000′) *Duthie* 13324".

XXIII. Aceraceae

1. ACER Linn.

Leaves pale and glaucous beneath; lobes pointed, not drawn into tail-like tips **1. A. caesium**
Leaves green beneath, lobes drawn into linear, tail-like tips **2. A. acuminatum**

1. **A. caesium** Wall. ex Brandis, For Fl. 111, t. 21, 1874; Hiern in Hk.f., l.c. 695. **Fig. 52**

Deciduous. 20-22 m tall tree with straight. rough, dark-grey trunk, often scattered with knotty protuberances; shoots glabrous. Leaves palmately 5-lobed, 12-20 cm across their longest breadth. cordate, serrate, red when young; lobes triangular, the outer 2 small to almost obscure, acuminate. Flowers yellow, in terminal, corymbose panicles, contemporaneous with young leaves. Fruit of 2 conjoined samaras having unevenly gibbous nuts and 3-5.5 cm long, divergent wings.

Occ. : Fairly common throughout in the valley from Pehalgam to Bara Khalen, more often forming isolated patches in open places.

Fls. April-May; *Frts.* June-Sept.

Saswarian nallah base (2900 m) *B.M. Sharma* 14161; Lidder Vy. above Tanin (10-11000′) *Duthie* 13309; Liddarwat (2750 m) *P.S. Jamwal* 15694.

2. A. **acuminatum** Wall. ex D. Don, Prodr. Fl. Nep. 249, 1825; Stewart, l.c. 461; Fl. Bash. Himal. 60, 1977. *A. caudatum* Hiern in Hk.f., l.c. 695 (non Wall. 1831)

Stewart (l.c.) has cited this trom Lidder Vy., Pehalgam to Aru. From this area there is no sheet in Dehradun Herbarium, nor have we been able to collect one.

XXIV. Hippocastanaceae

1. AESCULUS Linn.

A. indica (Wall. ex Camb.) Hk.f., Curtis Bot. Mag. 85, t. 5117, 1859; Heirn in Hk.f., l.c. 675. *Pavia indica* Wall. ex Camb. in Jacq. Voy. Bot. Inde, 31, t. 35, 1844.

Deciduous, small or medium tree with glabrous, white-lenticellate young shoots; buds scaly. Leaves opposite, digitately 5-9-foliolate, on 8-20 cm long petioles; leaflets oblong-lanceolate or oblong-oblanceolate, 10-30 × 3.5-10 cm, serrate, acuminate, from acute base, the middle ones the largest. Flowers a mixed white, pink and yellow, zygomorphic, 2-2.5 cm long, in short pubescent cymes arranged in terminal, erect, large, pyramidal panicles. Calyx shortly 5-lobed, split longitudinally in open flowers; petals 4; of unequal width; stamens 7, longer than petals. Capsule ovoid, brown 2.5-5 cm long.

Occ. : Scarce. Only a few trees, in the nallah below Aru.

Fls. May-June.

Aru (2375 m) *B.M. Sharma* 14373; Below Aru (2350 m) *P. S. Jamwal* 15656.

XXV. Fabaceae

1. Rachis of leaves transformed at the apex into a simple or branched tendril or drawn out into a bristle :
    2. Staminal tube truncate at the top; style bearded lengthwise on the inner side only, otherwise glabrous **12. Lathyrus**
    2. Staminal tube very oblique at the top; style with a bunch of hairs near the apex or all round downy near the tip **11. Vicia**
1. Rachis of leaves neither ending in a tendril nor drawn out into a bristle :
    3. Leaves 3-foliolate :
        4. Stipules free, often very small and caducous **9. Lespedeza**
        4. Stipules adnate to the petiole, persistent :
            5. Terminal leaflet stalked; corolla free from the staminal tube, the petals falling after flowering :

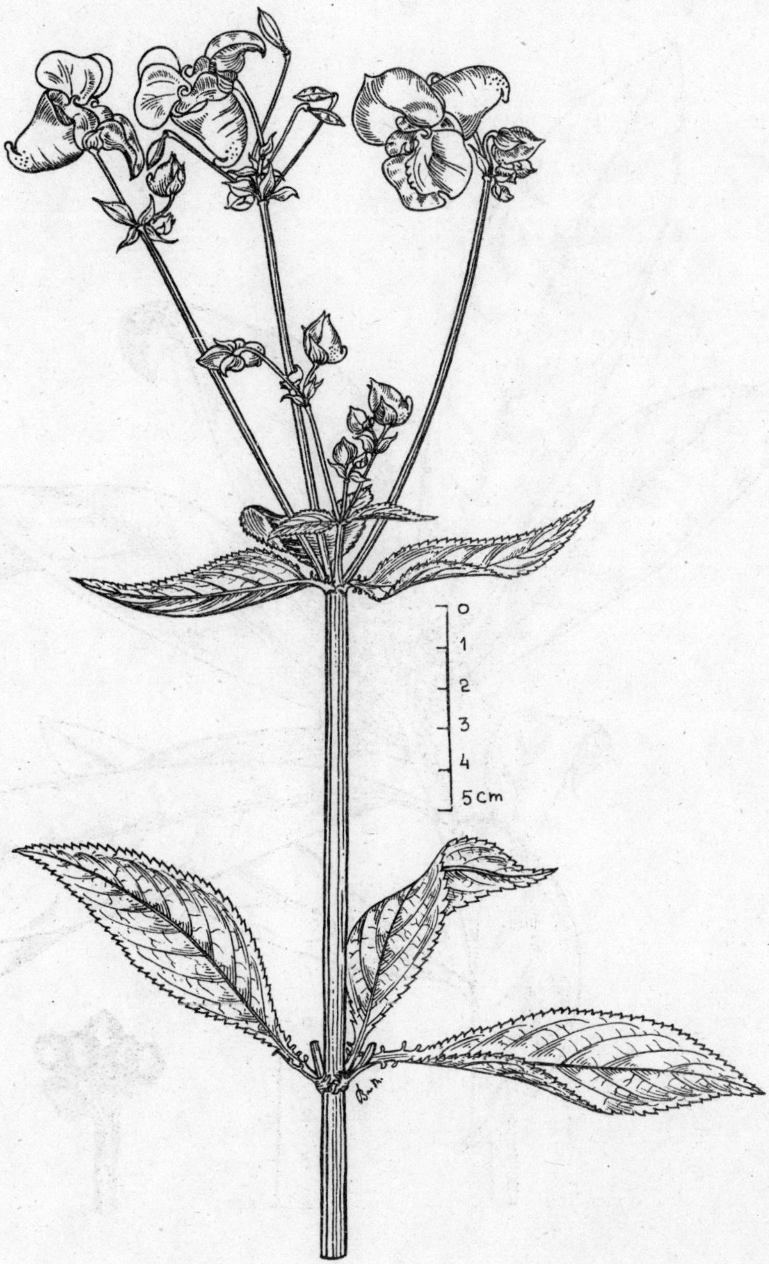

Fig. 49. *Impatiens glandulifera* Royle

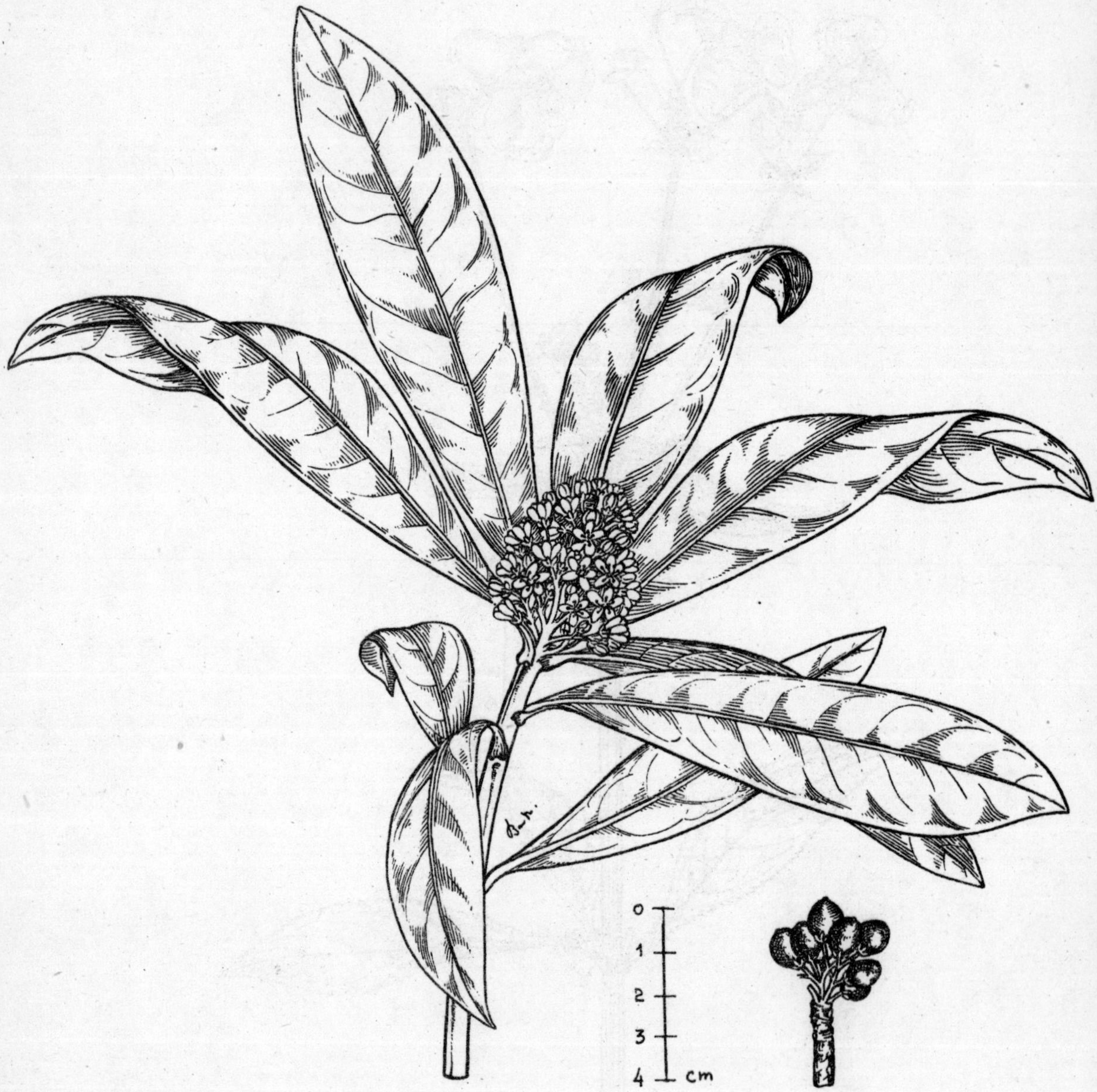

Fig. 50. *Skimmia laureola* Sieb. et Zucc. ex Walp.

Fig. 51. *Euonymus hamiltonianus* Wall. ex Roxb.

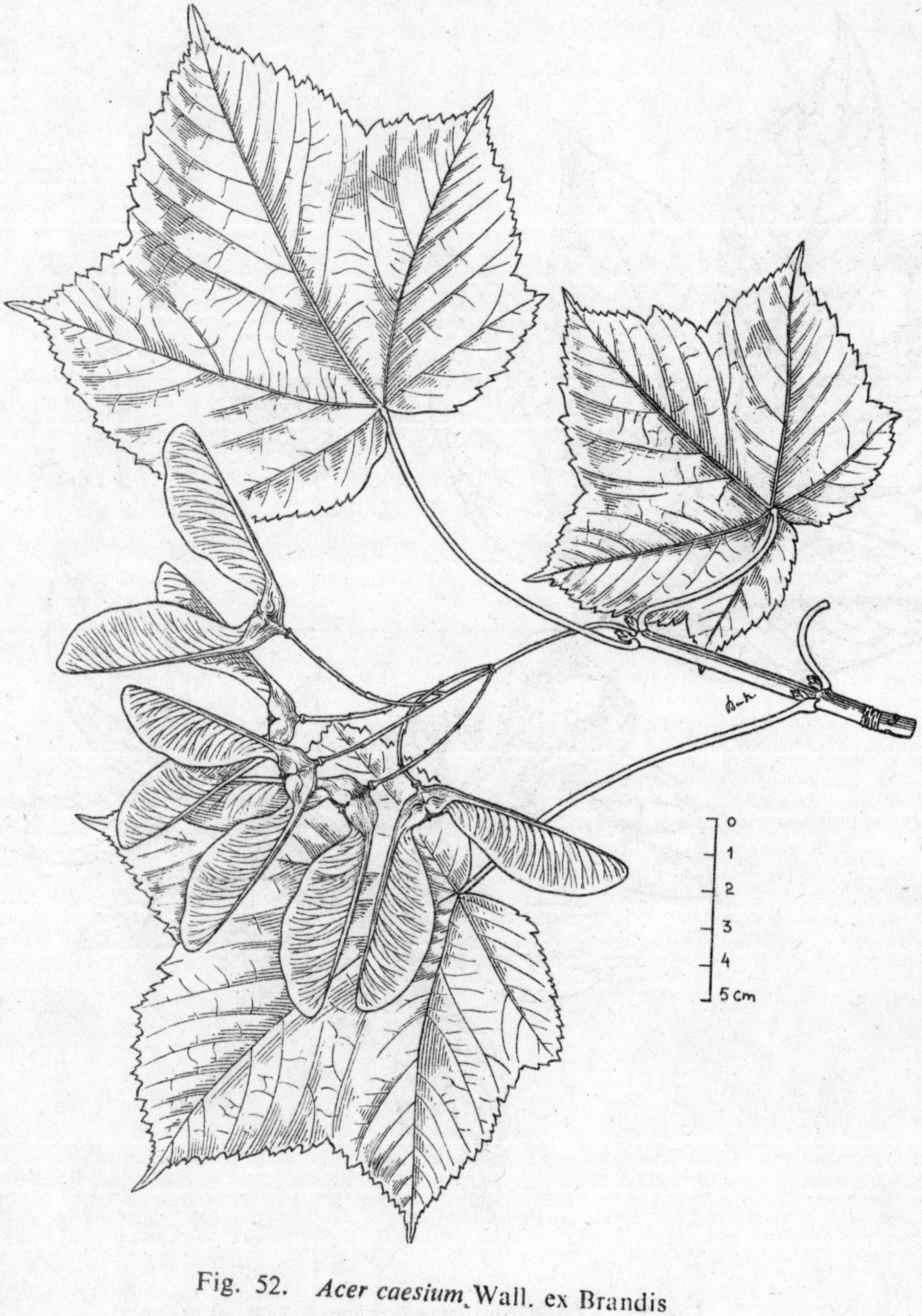

Fig. 52. *Acer caesium* Wall. ex Brandis

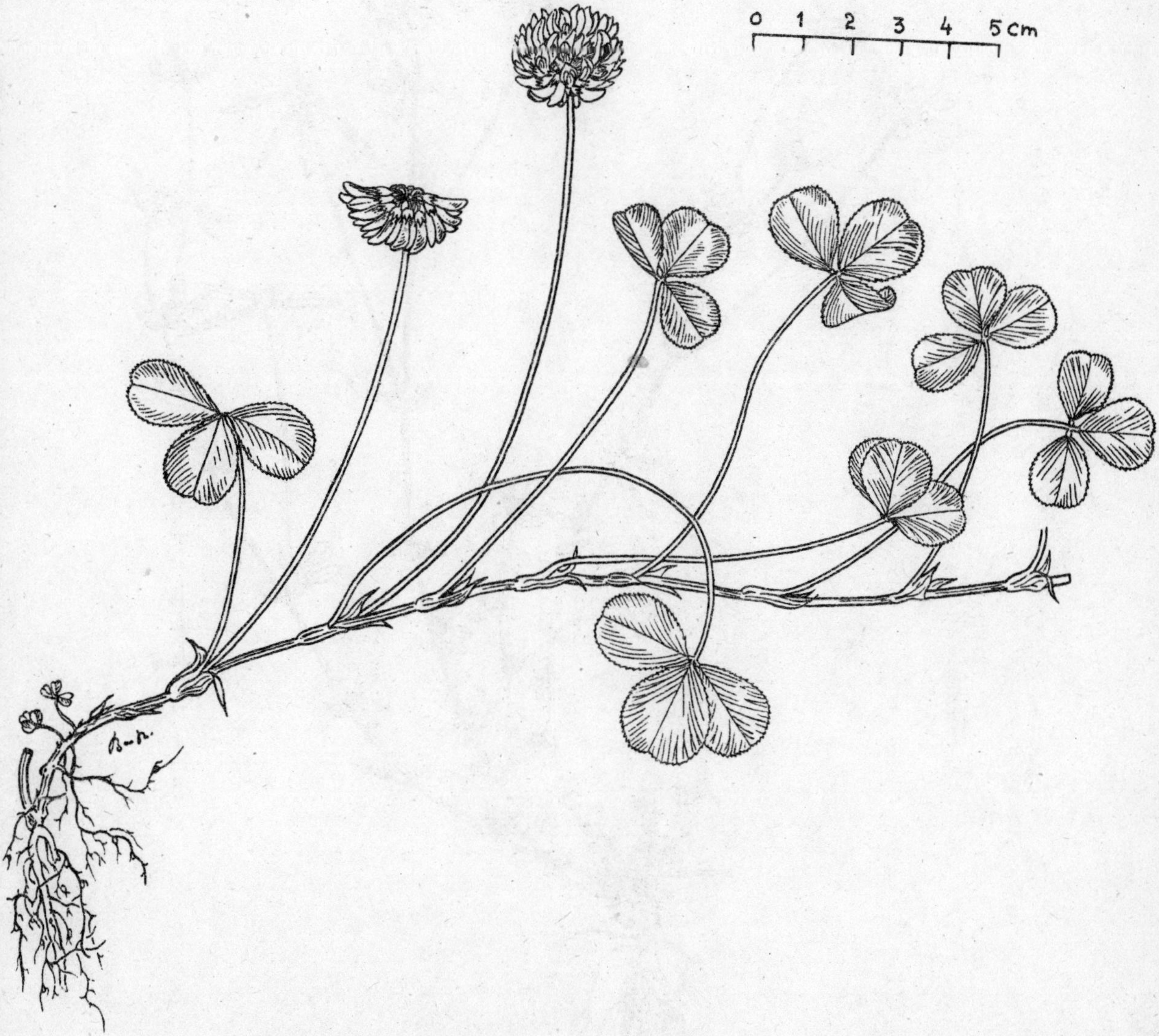

Fig. 53. *Trifolium repens* Linn.

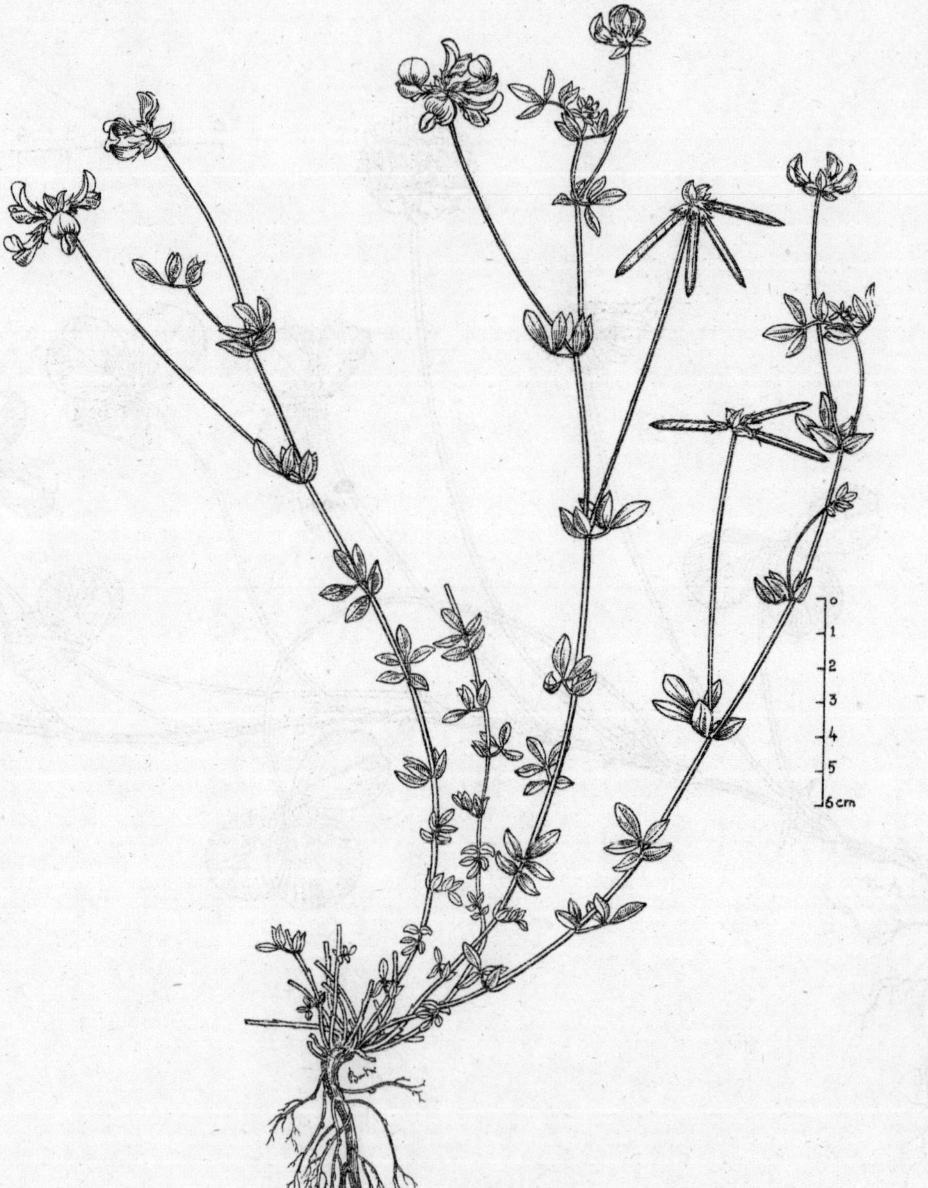

Fig. 54. *Lotus corniculatus* Linn.

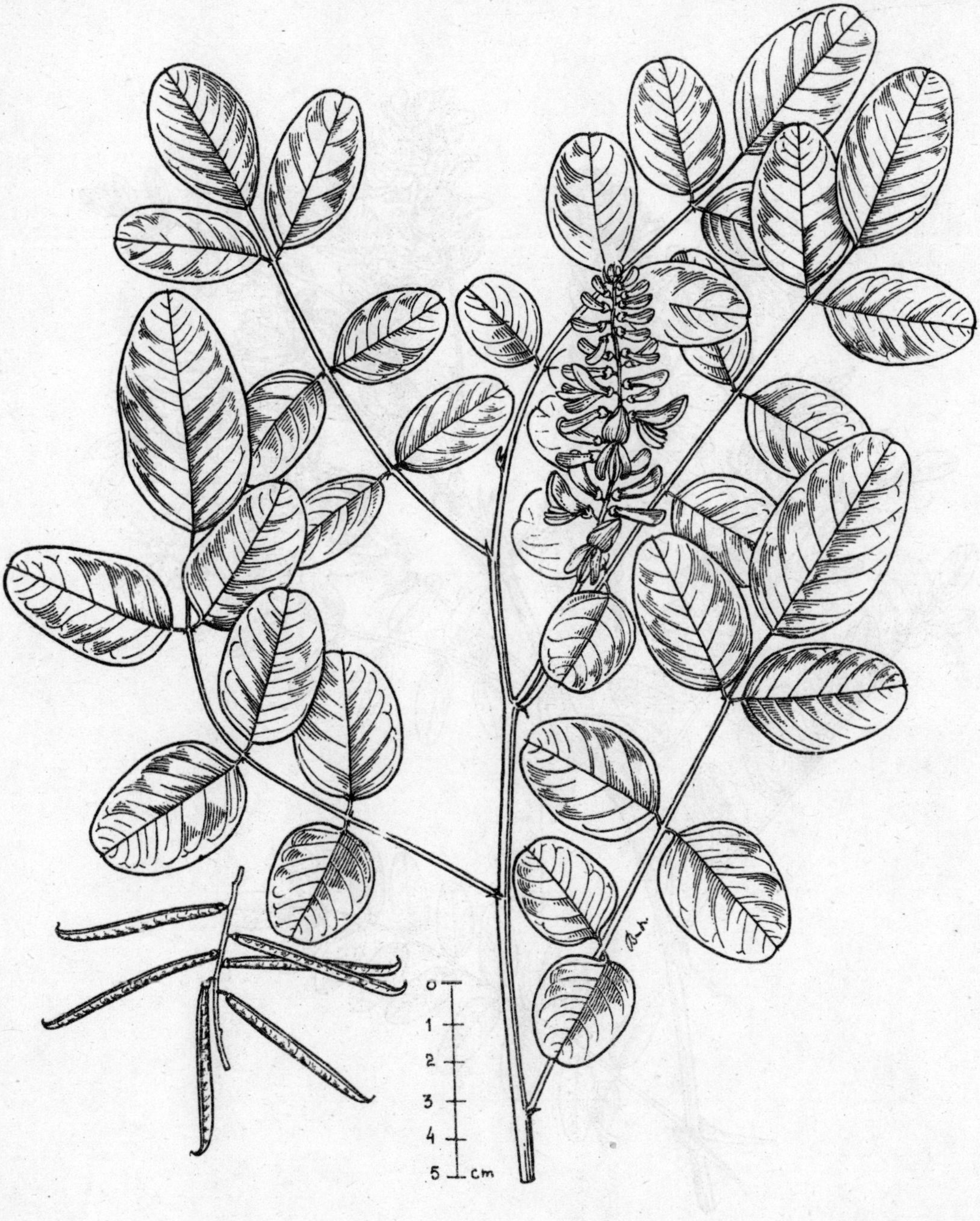

Fig. 55. *Indigofera hebepetala* Benth. var. *hebepetala* Ali

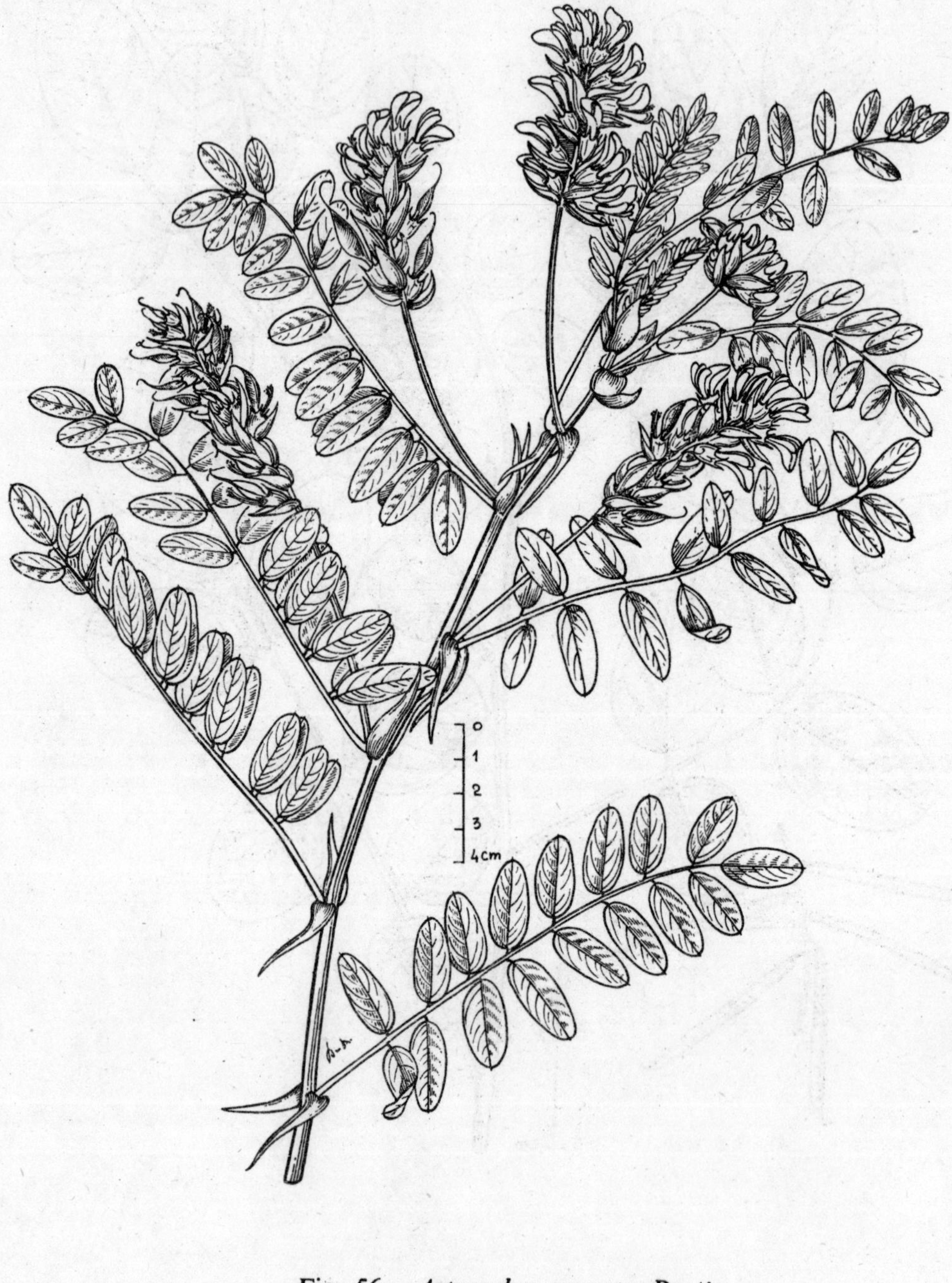

Fig. 56. *Astragalus concertus* Benth.

6. Pod spirally coiled or strongly falcate **4. Medicago**
6. Pod neither spiral nor strongly falcate, rarely only slightly recurved :
7. Pod short, subglobose or ovoid, 1-2-seeded **3. Melilotus**
7. Pod elongated, straight or nearly so, generally 4-many-seeded **2. Trigonella**
5. Terminal leaflet sessile; corolla adnate to the staminal tube, marcascent **1. Trifolium**
3. Leaves 5-multi-foliolate :
8. Lowermost pair of leaflets resembling stipules; free upper part of alternate filaments dilated or expanded up wards **5. Lotus**
8. Lowermost pair of leaflets similar to others, not stipule-like; free upper part of all filaments filiform :
9. Pod distinctly 1-3-jointed; joints indehiscent **10. Hedysarum**
9. Pod not jointed, dehiscent :
10. Anthers tipped with a gland; fruit septate between the seeds **6. Indigofera**
10. Anthers not tipped with a gland; anther loculii separated at the apex; fruit lengthwise septate or imperfectly so by the intrusion of seed-bearing suture, often inflated or turgid :
11. Keel-petal obtuse **7. Astragalus**
11. Keel-petal mucronate or acuminate or with recurved apex **8. Oxytropis**

1. TRIFOLIUM Linn.

Calyx with narrowly triangular-lanceolate, glabrous, subequal teeth separated by acute sinuses; pod 3-4-seeded **1. T. repens**
Calyx with bristle-like, ciliate teeth separated by broad sinuses, the lowest one tooth much longer than the rest; pod 1-seeded **2. T. pratense**

1. **T. repens** Linn., Sp. Pl. 767, 1753; Baker in Hk.f., l.c. 2 : 86, 1876.

Fig. 53

Glabrous, trailing perennial herb, rooting from nodes. Leaves long-petioled; leaflets obcordate-cuneate or broadly obovate and retuse at apex, 0.6-2×0.5-1.8 cm, denticulate; stipules oblong, membranous, drawn into a subulate apex. Flowers scented, 1-1.2 cm long, white or pink-tinged, forming 1.5-2.7 cm wide, globose heades on naked, leaf-exceeding peduncles; calyx glabrous. Pod linear, minute, compressed.

Occ. : Common along waysides and on alpine meadows throughout the area.

Fls. May-Aug.

Liddarwat (3000 m) *B.M. Sharma* 19374; Rewil nallah (3150 m) *P.S. Jamwal* 15670.

2. **T. pratense** Linn., Sp. Pl. 768, 1753; Baker in Hk.f., l.c. 86.

Decumbent or erect, hairy, perennial herb; stems up to 70 cm long. Leaflets ovate or broadly elliptic to obovate or obovate-oblong, 1.2-4×0.8-2.3 cm, at times nearly orbicular, finely denticulate or entire, with round or emarginate apex; stipules large, ovate, conspicuously veined, upwards triangular and suddenly contracted into a long, setaceous, ciliate point. Flowers pink-, or red-purple, in an ovoid or globose, 2-3.5 cm wide, sessile or shortly peduncled, usually involucrate head; calyx tube appressedly hairy. Pod ovate, very small.

Occ. : Common in fields and along waysides throughout in Aru-Pehalgam-Chandanwari area; both cultivated and wild.

Fls. June-Sept.

Liddarwat (2700 m) *B.M. Sharma* 14193; Pahalgam (7200′) *K M. Vaid*, DD no. 117562; near Chandanwari, *L.D. Kapoor*, RL. Herb. no. 2146; Aru (2450 m) *P.S. Jamwal* 15698.

2. TRIGONELLA Linn.

T. cachemiriana Camb. in Jacq. Voy. Bot. Inde. 36, 1843; Ali in Nasir & Ali, Fl. W. Pak. 100 : 299, 1977; Stewart, l.c. 423. *T. medicaginoidea* (Sirj.) Vass., Not. Syst. Herb. Inst. Bot-Kom. Acad. Sc. URSS. 14 : 231, 1951. *T. emodi* Benth. var. *medicaginoidea* (*medicagneae*) Sirj., Publ. Fac. Sci. Univ. Masaryk 102 : 20, 1928.

Perennial, much branched herb with 30-50 cm long, pubescent to tomentose, prostrate or diffuse stems. Leaf-stalks 4-8 mm long; stipules foliaceous and dentate or the upper linear and almost entire. Leaflets obovate, 0.5-2.2×0.4-1.7 cm, serrate, glabrous above pubescent beneath, mucronate. Flowers yellow, 1-1.2 cm long, in axillary, peduncled, umbellate inflorescence; keel and vexillum equal, 1-1.2 cm long; wing 0.8-1.1 cm long; calyx 4-5 mm long, pubescent to tomentose. Pod flat, 1.1-2.4×0.4-0.5 cm, glabrous, prominently transverse-veined; stipe 1-2 mm long.

Occ. : Open alpine and subalpine grassy slopes in both the valleys.

Fls. June-Sept.; *Frts.* Aug-Sept.

Slope below Tarsar (3600 m) *B.M. Sharma* 14282; Pisu Ghati : (3000 m) *B.M. Sharma* 14674; roadside below Aru (2380 m) *P.S. Jamwal* 16371; above Chandanwari (2940 m) *P.S. Jamwal* 16352.

3. MELILOTUS Miller

M. alba Medic., Vorles. Churpf. Phys.-Ok. Ges. 2 : 382, 1787; Baker in Hk.f., l.c. 89.

Branched, erect, fragrant herb, up to 1.5 m tall. Leaflets obovate to narrow oblong-obovate, 1-3×0.3-1.6 cm, toothed; stipules setaceous. Flowers white, 4-5 mm long, in slender axillary racemes, the latter in fruit 8-16 cm or longer even; wings and keel nearly equal, shorter than the standard. Pod reticulate, ovoid, 3.5-5 mm long, with persistent style; seeds 1-2, smooth.

Occ. : In depressions near the bridges in Pehalgam.

Fls. & Frts. July-Sept.

Pehalgam (2150 m) *B.M. Sharma* 14699; Lidder Vy. above Pehalgam (8000′) *Duthie* 13463; Pehalgam-Gurkum route (2300 m) *P.S. Jamwal* 15676.

4. MEDICAGO Linn.

M. lupulina Linn., Sp. Pl. 779, 1753; Baker in Hk.f., l.c. 90.

Diffuse or trailing, pubescent or downy herb; stems slender, 10-60 cm long, many from root. Leaflets obovate to near orbicular, 0.5-2×0.4-1.4 cm, denticulate in the upper part, rotund or emarginate; stipules toothed, ovate or lanceolate, acuminate. Flowers yellow, 1.5-2 mm long, in 4-6 mm wide, globose or ovoid heads on leaf-exceeding peduncles. Pod once coiled, glabrous or pubescent, 2-2.5 mm in diam., reticulate, black when ripe; seed solitary.

Occ. : Fields, open places, pastures and undar the shade of Viburnums and other bushes. Common.

Fls. & Frts. May-Sept.

Kolahoi (3300 m) *B.M. Sharma* 14205; Pehalgam (7200′) *K.M. Vaid* DD. no 118087; Aru (2400 m) *P.S. Jamwal* 15684.

5. LOTUS Linn.

L. corniculatus Linn., Sp. Pl. 775, 1753; Baker in Hk.f., l.c. 86. **Fig. 54**

Glabrous or subglabrous, herbaceous perennial with decumbent or ascending, 8-32 cm long, branching stems from a woody rootstock. Leaves gradually larger upwards; leaflets 5, rather fleshy, ovate, obovate or oblong, 0.4-1.5×0.15-0.8 cm, obtuse, acute or acuminate. Flowers yellow, 1-1.5 cm long mostly up to 6 in capitate, bracteate clusters on 1.5-10 cm long peduncles; calyx glabrous, pubescent or ciliate, with subequal, narrowly lanceolate teeth from a triangular base. Pod glabrous, 1.5-3 cm long, many-seeded.

Occ. : Common in open grassy places on hillsides.

Fls. June-Aug.

Liddarwat nallah (3500 m) *B. M. Sharma* 14183; Lidder Vy above Kainmul (10-11000′) *Duthie* 13160; above Chandanwari (2800 m) *P. S. Jamwal* 15680.

6. INDIGOFERA Linn.

1. Leaflets 2-7.5 cm long; bracts boat shaped, quite enclosing the flower buds **1. I. hebepetala**
1. Leaflets below 1 cm long; bracts minute :
    2. Ovary glabrous; calyx teeth longer than the tube **2. I. himalayense**
    2. Ovary apressedly argento-pubescent; calyx teeth as long as the tube **3. I. heterantha**

1. **I. hebepetala** Benth. ex Baker in Hk.f., l.c. 101 var. **hebepetala** Ali in Bot. Notiser, 111 : 572, 1958; Stewart, l.c. 407; For. Fl. Sgr. 139. **Fig. 55**

Deciduous, sparingly branched, 1.5-3 m tall shrub; branches erect with glabrous or sparsely pubescent extremities. Leaves 9-25 cm long; leaflets mostly 5-11, ovate or ovate-oblong, 1-4.5 cm broad, appressedly pubescent on either face, pale beneath. Flowers purplish-red. 1-1.3 cm long, in 10-20 cm long racemes; bracts long-cuspidate; calyx *c.* 2.5 mm long, with short, rounded on deltoid teeth. Pod glabrous, 3.5-5.5 cm long, 8-10-seeded.

Occ. : Shaded, roadside places between Mundlan and Aru. Scarcely distributed.

Fls. May-June; *Frts.* July-Aug.

Saswarian nallah base (2800 m) *B.M. Sharma* 9302; Mundlan-Aru roadside (2300 m) *P.S. Jamwal* 15697.

1. **I. himalayense** Ali in Bot. Notiser, 111 : 574, 1958; Stewart, l.c. 408; For. Fl. Sgr. 138.

Much branched, 0.6-1.2 m high, deciduous shrub with erect, greyish, smooth twigs; young shoots appressedly argento-pubescent. Leaves odd-pinnate; leaflets 11-21, obovate, elliptic or oblanceolate, 4-8 mm long, obtuse apiculate, appressedly white-pubescent on both sides. Flowers deep bluish purple, 1-1.2 cm long, in shortly peduncled to nearly sessile, 2.5-3 cm long, axillary racemes; standard *c*. 1×0.7 cm, hairy without. Calyx campanulate, 2.5-3 mm long, pubescent; teeth linear from a deltoid base. Pod cylindric, 1.2-1.5 cm long, glabrous.

Occ.: In forest openings and around borders of fields about Aru.

Fls. June-July

Aru (2400 m) *B.M. Sharma* 14360 & 14698; Aru (2400 m) *P.S. Jamwal* 15677.

3. **I. heterantha** Wall. ex Brandis, For. Fl. N. W. & C. Ind. 135, 1874; Ali, l.c. 566; Stewart, l.c. 407; For. Fl. Sgr. 138. *I. gerardiana* Wall. ex Baker var. *heterantha* Baker in Hk.f., l.c. 100.

Much resembling the previous but in the leaflets up to 31, more conspicuously canescent, calyx teeth equalling the tube, standard narrower and ovary appressedly argento-pubescent.

Fls. June-Aug.

Chandanwari (8-9000′) *K.M. Vaid* DD, no. 118087. Quite likely to be met with in the W. Liddar Vy. as well.

7. ASTRAGALUS Linn.

1. Leaf rachis terminating into a leaflet :
    2. Ovary and fruit glabrous :
        3. Stipules connate, 1.2-3 cm long and 4 mm or more broad **1. A. concertus**
        3. Stipules free, below 8×2 mm :
            4. Leaflets 13-17; calyx teeth deltoid, below 1 mm long; fruit 6-10-seeded **2. A. chlorostachys**
            4. Leaflets 17-25; calyx teeth not deltoid, c. 1.5-3 mm long; fruit 10-12-seeded **3. A. bakeri**
    2. Ovary and fruit pubescent :
        5. Fruit unilocular **4. A. himalayanus**
        5. Fruit bilocular :
            6. Calyx 7-9 mm long; vexillum 1.5-1.8 cm long **5. A. maxwellii**
            6. Calyx 1.2-1.3 cm long; vexillum c. 2.5 cm long **6. A. candolleanus**
1. Leaf rachis terminating into aspine **7. A. grahamianus**

1. **A. concertus** Benth. in Royle, Ill. Bot. Himal. 199, 1835; Ali in Biologia 7 : 36, 1961; Stewart, l.c. 388. *A. vicioides* Grah. ex Baker in Hk.f., l.c. 131. **Fig. 56**

Erect, 30-60 cm tall, herbaceous undershrub; stems fistular, glabrous. Leaves 9-18 cm long, 15-23-foliolate; stipules membranous; leaflets oblong or ovate-oblong, 2-3.5 cm long, obtuse, glabrous above; beneath glaucous and thinly pubescent. Flowers yellow or with purplish tips, 1.5-1.7 cm long, in axillary, long-peduncled, dense racemes. Calyx membranous, 7-8 mm long, scattered with a few black hairs or glabrous; teeth hairy. Pod bilocular, turgid, 1.2-2 cm long excl. stipe and persistent style, curved.

Occ.: Restricted to wet gravelly islands in stream beds; not common.

Fls. July-Aug.; *Frts*. Aug.-Sept.

Gagarpathri near Liddarwat (3000 m) *B.M. Sharma* 14148 & 14697; near Liddarwat (3000 m) *P.S. Jamwal* 15683.

2. **A. chlorostachys** Lindl. in Trans. Hort. Soc. 2 : 249, 1830; Baker in Hk.f., l.c. 128.

Erect, up to 1.6 m tall, herbaceous shrub; stems pubescent or glabrous. Leaves 8-15 cm long, imparipinnately 13-17 foliolate; stipules lanceolate, spreading; leaflets oblong, 1.2-3×0.5-1.3 cm, obtuse, pubescent beneath, above glabrous or nearly so. Flowers yellow-green, frequently lilac-tinged, 1.2-1.5 cm long, in axillary, peduncled, 5-12 cm long racemes elongating to 20-25 cm in fruit; calyx *c.* 6 mm long, pubescent. Pod bilocular, oblong, 1-1.2 cm long, pointed. narrowed into a stipe longer than the calyx.

Occ. : Shady hillsides along Pehalgam-Aru road. also on wooded and open slopes in the areas around.

Fls. July-Aug.; *Frts*. Aug.-Sept.

Below Chumanaisar (2800 m) *B.M. Sharma* 14097; Lidder Vy. above Pehalgam (8-9000′) *Duthie* 13470; Lidder Vy. Chatponsal nallah (9-10,000′) *Duthie* 13172; roadside 3 km short of Aru (2300 m) *P.S. Jamwal* 15686.

3. **A. bakeri** Ali in Kew Bull. 1958: 307, 1958; Stewart, l.c. 386. *A. longicaulis* Baker in Hk.f., l.c. 129.

Reported by stewart (l.c.) from Pehalgam. Different from the preceding species in characters of the key and the calyx being covered by persistent, appressed, black hairs and longer, white, deciduous ones; leaflets narrower.

4. **A himalayanus** Kl. in Kl. & Garcke, Bot. Ergebn. Reise Wald. 160, t. 4, 1862; Baker in Hk.f., l.c. 126.

Diffuse or semierect perennial herb; stems slender, up to 45 cm long, adpressedly white-pubescent when young. Leaves 3-5 cm long, imparipinnately 19-25-foliolate; stipules small, lanceolate; leaflets oblong, 0.5-1.2 cm long, obtuse or subobtuse or retuse, adpressedly white-hairy. Flowers lilac, 1.2-1.5 cm long, in axillary heads; peduncles like calyx clothed with mixed black and white hairs; calyx 3-4 mm long. Pod oblong, 1-1.3 cm long, finely black-pubescent, narrowed into a stipe longer than calyx.

Occ. : Waysides about Liddarwat-Kolahoi route and open slopes of Pisu Ghati and Sant singh Pari.

Fls. June-Sept.; *Frts*. Aug.-Sept.

Satlanjan (3200 m) *B.M. Sharma* 14303; above Pisu Ghati (3450 m) *B.M. Sharma* 14641; Kolahoi Vy. (11-12000′) *Duthie* 13508; Dandabari-Seikwas route (3300 m) *P.S. Jamwal* 15687.

5. **A. maxwellii** Royle ex Benth. in Royle, Ill. 198, 1835; Stewart, l.c. 392; Ali in Nasir & Ali, Fl. W. Pak. 100 : 181, 1977. *A. ciliolatus* Benth. ex Bunge, Astrag. 1 : 27, 1868; Hk.f., l.c. 129.

Perennial herb with several, pubescent, up to 40 cm long, hollow, procumbent to semierect stems from woody base; hairs black and white. Leaves 5-11 cm long, imparipinnately 11-19-foliolate; stipules lanceolate or deltoid, free. Leaflets oblong-elliptic to oblong, 0.6-2.2×0.2-0.7 cm, obtuse or retuse, green and glabrous above, beneath glaucous and grey-pubescent. Flowers yellow or with the vexillum lilac tinged, 1.5-1.8 cm long, in axillary racemes, peduncle 5-10 cm long, clothed with black and white hairs, the former shorter and upwards predominant. Calyx tube black-pubescent, the teeth densely so and 1-3 mm long. Pod densely black or black-, and white-hairy, 2.5-2.6 cm long excl. stipe, the latter about equalling the calyx length; seeds 8-10.

Occ. : On banks and moist gravelly ground in an island in the torrent bed.

Fls. June-July.; *Frts.* July-Aug.

Bara Khalen (2900 m) *B.M. Sharma* 14152 & 14692; Lidder Vy. above Kainmul (12000′) *Duthie* 13121.

Treatment of above cited specimens under *A. maxwellii* is provisional since the former is stated (c.f. Hk.f., l.c. & Ali, l.c.) to have stems glabrous, leaves 15-23 cm long and calyx teeth 4.5-5 mm long. Although several characters are suggestive of closeness to *A. tibetanus* Benth., the latter (c.f. Hk.f., l.c. 124) is distinct from its zigzag slender stems, 21-25-foliolate leaves, capitate heads of purple or lilac flowers and the stipe shorter than calyx tube (leaflets 17-29, corolla azure violet; Ali, l.c. 180)

6. **A. candolleanus** Royle ex Benth. in Royle, Ill. Bot. Himal. 199, 1835; Hk. f., l.c. 132.

A shrub, likely to be met with in the area because of its having been cited by Ali (l.c.) from Zojipal, 11000′. In Dehradun Herbarium there is no specimen from the Lidder valleys.

7. **A. grahamianus** Royle ex Benth. in Royle, Ill. Bot. Himal. 199, t. 36, f. 2, 1835; Ali in Biologia 7 : 63, 1961; Stewart, l.c. 390. *A. cicerifolius* Royle ex Fisch. in Bull. Soc. Nat. Moscou 24(2) : 404, 1853; Parker, For. Fl. 141, 1918.

Spreading or semierect low shrub, with glabrous or pubescent young twigs. Leaves 2.5-7 cm long, pinnately 8-14-foliolate, with stramineous, pilose, stiff, rachises; stipules amplexicaul. scarious, with thickened bases, prominently nerved. Leaflets obovate, deciduous, 3-7.5 mm long, obtuse or retuse, mucronate, pubescent beneath. Flowers yellow, 2.5-3.2 cm long, usually 2 in an axil; calyx thinly silky, 1.2-1.5 cm long; standard panduriform. Pod turgid, 1.5-1.8 cm long, adpressedly silky, 15-18-seeded.

Occ. : Open dry slopes along Pehalgam Phaslan route; localised in small patches.

Fls. May-June; *Frts.* Sept.

Pehalgam-Phaslan route (2400 m) *B.M. Sharma* 14633 and *P. S. Jamwal* 15682.

8. OXYTROPIS DC.

O. cachemirica Camb. in Jacq. Voy. Bot. Inde. 38, t. 44, 1844; Baker in Hk.f., l.c. 139.

Perennial, 6-10 cm high hairy herb with ascending, short stems from a woody rootstock. Leaves odd-pinnate, 2.5-5 cm long; rachis densely pilose with white and a few black hairs; stipules connate below. Leaflets 11-21, opposite, elliptic, 0.5-1 × up to 0.4 cm, sessile, acute, densely white pilose on either side. Flowers purple. *c.* 1.5 cm long, in dense heads on blackish peduncles clothed with spreading black and white hairs; calyx campanulate, copiously pilose with largely white hairs; 0.8-1 cm long, with teeth *c.* half as much. Pod orbicular-oblong, membranous, inflated. 1.1-1.3 cm long, clothed with loose white and appressed black hairs.

Occ. : Sporadically distributed on usually dry, open alpine hillsides.

Fls. July-Sept.

Santsingh Pari (3900 m) *B.M. Sharma* 14630; Lidder Vy. near Sheshnag (12-13000′) *Duthie* 13337; Panchtarni (12000′) *K.M. Vaid*, DD no. 118084; above Pisu Ghati (11500′) *K.M. Vaid*, DD. no. 118083; Zojipal-Sheshnag route (3550 m) *P.S. Jamwal* 15678.

9. LESPEDEZA Michx.

L. juncea Pers., Syn. Pl. 2 : 318, 1807 var. **variegata** (Camb.) Ali in Biologia 12 : 41, 1966; Stewart, l.c. 411. *L. variegata* Camb. in Jacq. Voy. Bot. Inde. 42, t. 50, 1843; For. Fl. Sgr. 140.

Deciduous undershrub giving off 25-90 cm long, trailing or decumbent, striate, grey-pubescent shoots from a woody stock. Leaflets oblanceolate, mostly 0.7-1.5 × 0.25-0.5 cm, mucronate, glabrous above, densely grey-silky beneath, terminal shortly petioluled, lateral sessile or nearly so. Flowers pale purple, 6-7 mm long, in axillary, short (up to 5 mm)-peduncled, 2-4-flowered umbels. Calyx grey-silky, *c.* 4 mm long; teeth much longer than the tube. Pod silky, ovate; indehiscent, 1-seeded.

Occ. : Very common along roadsides and on dry scrubby places on the slopes between Mundlan and Aru.

Fls. July-Nov.

Beyond Mundlan (2300 m) *B.M. Sharma* 14632; Aru road (2350 m) *P.S. Jamwal* 19301.

10. HEDYSARUM Linn.

H. cachemirianum Benth. ex Baker in Hk.f., FBI 2 : 146, 1876.

Perennial, robust herb with 30-60 cm high, erect or ascending, grooved stems, upwards at least thinly patent-hairy. Leaves imparipinnate, 7-18 cm long, 19-27-foliolate; leaflets linear-oblong, 1.2-2.5 cm long, obtuse, ultimately glabrescent on either face. Flowers bright red, 2-2.5 cm long, in dense, 4-7.5 cm long racemes; bracts setaceous. Calyx 0.9-1.1 cm long, pubescent, partly coloured; standard longer than the keel and the wings. Pod oblong, segments with an entire wing on either suture.

Occ. : Sparingly distributed on alpine grassy slopes and in open places in stony nallahs.

Fls. July-Sept.

Saswarian (3400 m) *B.M. Sharma* 14178; slope below Tarsar (3700 m) *B.M.*

Sharma 14642; Lidder vy. above Kainmul (11-12000′) *Duthie* 13108; near Sheshnag (*c.* 3700 m) *P.S. Jamwal* 15690.

11. VICIA Linn.

Style conspicuously bearded on the lower side beneath the stigma; racemes nearly sessile to very shortly peduncled, 2-6-flowered **1. V. sepium**
Style finely and equally downy all round beneath the stigma; racemes long-peduncled, 6-20-flowered **2. V. rigidula**

1. **V. sepium** Linn., Sp. Pl. 737, 1753; Baker in Hk f., l.c. 179. **Fig. 57**

Usually simple, semierect or scandent, perennial herb; stem up to 80 cm long, upwards pubescent. Leaflets 4-7 pairs, ovate to ovate-oblong, 1.2-3.5×0.5-1.5 cm, thinly pubescent, rounded or truncate at apex; upper stipules entire. Flowers bluish purple, 1.4-1.7 cm long, in secund racemes; calyx pubescent, 8-9 mm long, unequally toothed. Pod glabrous, linear-oblong, rostrate, 4-7-seeded.

Occ. : Among Viburnum shrubs in Liddarwat; uncommon.

Fls. July-Aug.

Liddarwat (2950 m) *B.M. Sharma* 14058; Lidder Vy. Chatponsal nallah (10-11000′) *Duthie* 13175.

2. **V. rigidula** Royle, Ill. Bot. Himal. 200, 1835; Baker in Hk.f., l.c. 178.

Stewart (l.c. 425) has reported its occurrence in Pehalgam and Aru. It has creamy-white, orange or reddish flowers with glabrous, 4-6 mm long calyx.

12. LATHYRUS Linn.

1. Leaflets 2 **1. L. pratensis**
1. Leaflets 6-10:
   2. Rachis terminated by a short bristle **2. L. emodi**
   2. Rachis terminated by a long tendril **3. L. humilis**

1. **L. pratensis** Linn., Sp. Pl. 733, 1753; Baker in Hk.f., l.c. 180.

Perennial, branched, climbing or scrambling herb; stems angular, 30-75 cm long, hairy or glabrate. Leaflets lanceolate, 1.5-4.5×0.2-1 cm, acute, hairy beneath; stipules foliaceous, broadly lanceolate, 1-3.5×0.2-1 cm from unequally sagittate base. Flowers yellow, 1-1.5 cm long, usually 2-10 in axillary, leaf-exceeding, long-peduncled, hairy racemes. Calyx 0.6-1 cm long; teeth lanceolate, the lower longer than the tube. Pod glabrous, 3-4× 0.5-0.6 cm, 3-many-seeded,

Occ. : Usually in shaded, grassy places along hillsides.

Fls. June-Aug.; *Frts.* Aug.-Sept.

Aru (2375 m) *B.M. Sharma* 19241; Chandanwari (2950 m) *B.M. Sharma* 14667; Armiun glen (2760 m) *P.S. Jamwal* 15681; Aru (2350 m) *P.S. Jamwal* 15696.

2. **L. emodi** (Wall. ex Fritsch) Ali in Biologia 11(2) : 4, 1965 & in Nasir & Ali, l.c. 277; Stewart, l.c. 409. *Orobus emodi* Wall. ex Fritsch in Sitzung. Akad. Wissen. Wein. Math.-nat. Classe 104; 489, 1896. *O. luteus* Linn., Sp. Pl. 728, 1753. *L. luteus* (Linn.) Peterm, Deutsch. Fl. 155, 1849; Baker in Hk.f., l.c. 180. **Fig. 58**

Glabrous, almost erect, simple or branched, 0.5-*c.* 1 m tall perennial herb from a rather stout stock bearing thick

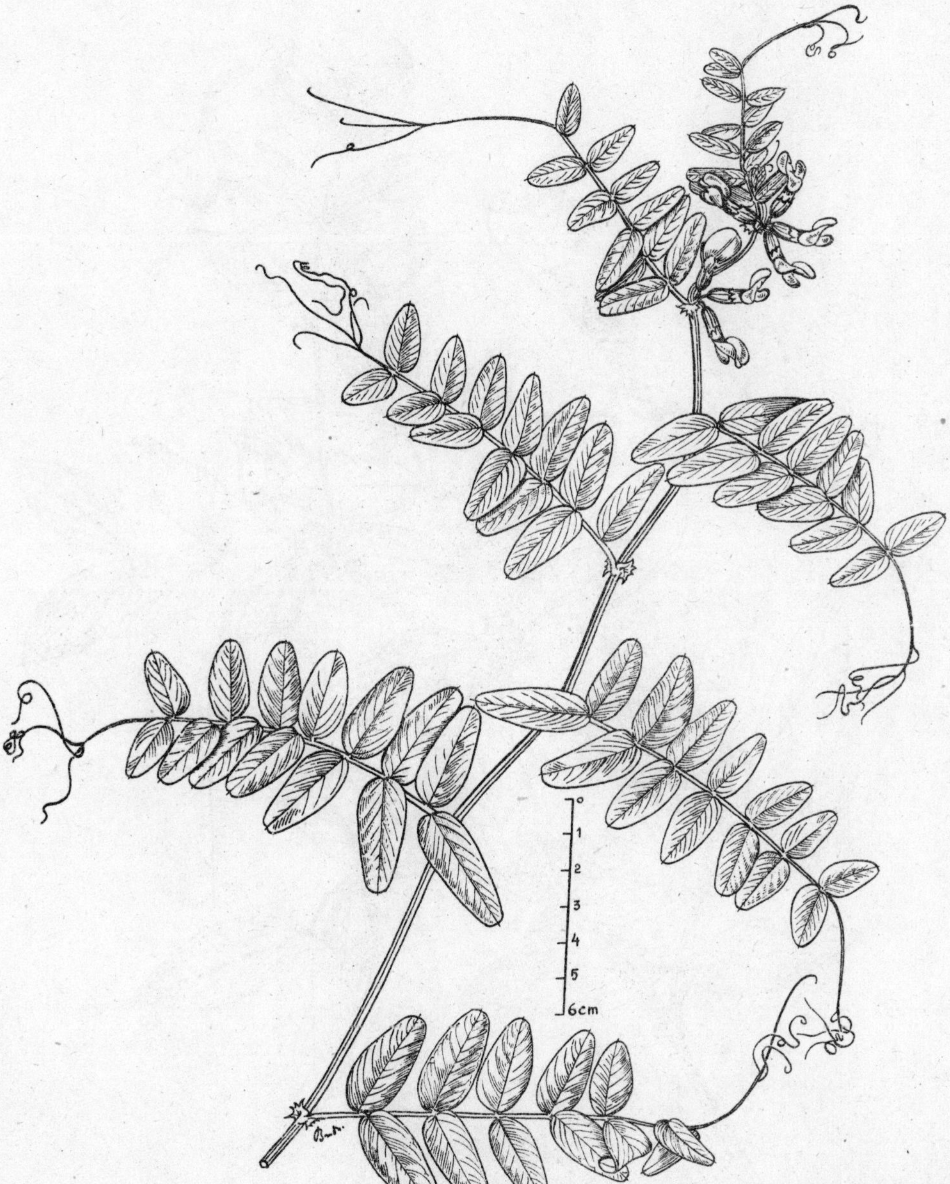

Fig. 57. *Vicia sepium* Linn.

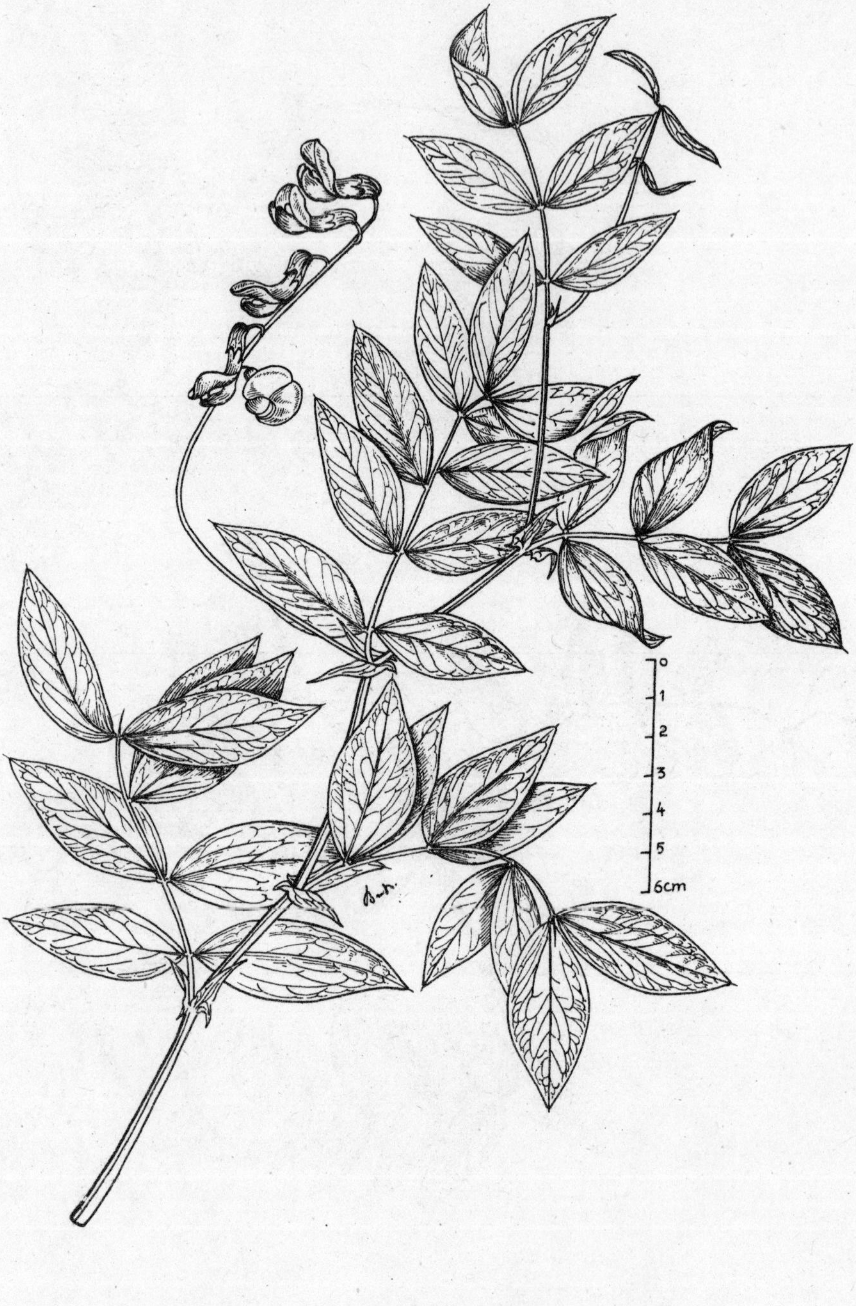

Fig. 58. *Lathyrus emodi* Ali

Fig. 59. *Prunus cornuta* Steud.

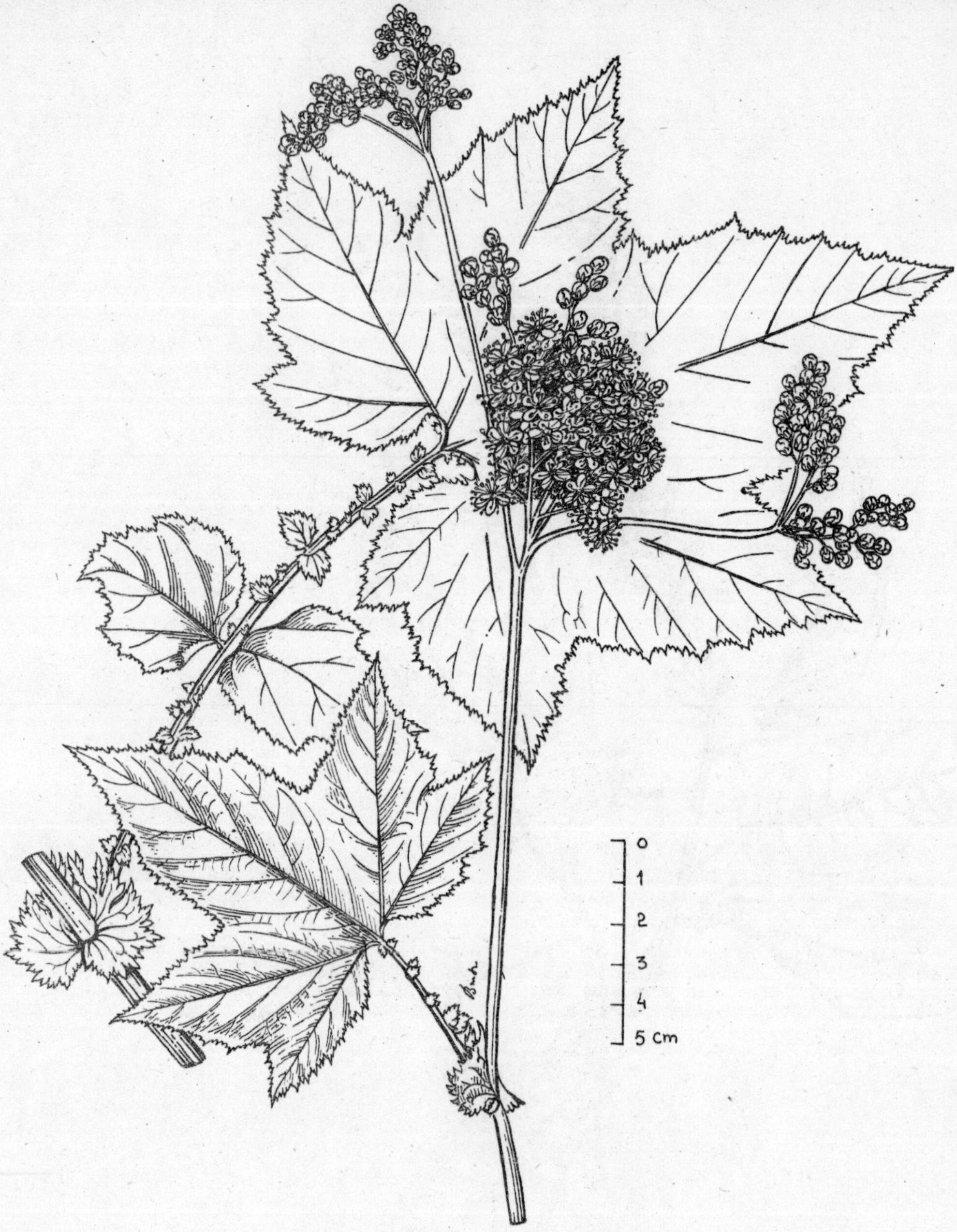

Fig. 60. *Filipendula vestita* Maxim.

root fibres. Leaves 5-14 cm to the bristle tip, 6-10-foliolate; leaflets ovate-lanceolate or elliptic, 3-7.5×1.2-3 cm, mucronate, puberulent beneath; stipules foliaceous, up to 3×1 cm. Flowers white, pinkish or yellow; 2-2.5 cm long, in axillary, 5-10-flowered racemes. Calyx 0.8-1.2 cm long; upper teeth deltoid, the lower longer and lanceolate. Pod beaked, 5-7.5×0.6-0.7 cm, glabrous.

Occ. : Sparingly distributed up to *c.* 3500 m, both in western and eastern Liddar valleys; usually among scrubs or in shade of rocks.

Fls. July.-Aug.; *Frts.* Aug.-Sept.

Satlanjan (3050 m) *B.M. Sharma* 14605; seen also at Aru (2400 m) and Zojipal (3450 m).

3. **L. humilis** (Ser.) Fisch. ex Spreng., Syst. Veg. 3 : 263, 1826; Ali in Biologia 11 (2) :. 5, 1965; Stewart, l.c. 410. *Orobus humilis* Ser. in DC., Prodr. 2 : 378, 1825. *L. altaicus* Ledeb. l.c. Pl. Fl. Ross. 1 : 13, t. 53, 1829; Baker in Hk.f., l.c. 180.

Glabrous erect 20-30 cm tall, perennial herb. Leaves paripinnate, 6-8-foliolate; leaflets elliptic or oblong-elliptic, 1.5-4×0.6-1.8 cm, glabrous, mucronate; stipules lanceolate, semisagittate, caudate, more or less toothed. Flowers red or purple, *c.* 1.8 cm long, in lax, 2-6-flowered, long-peduncled racemes; calyx *c.* 9 mm long, unequally toothed. Pod glabrous, 3-5 cm long.

Occ. : In shaded grassy places and woods throughout in Liddarwat-Pehalgam Zojipal tract; scarce.

Fls. June-Aug.

Aru (2375 m) *B.M. Sharma* 14650; Aru, Lidder Vy., *Inayat* 25561; Aru (2375 m) *P.S. Jamwal* 15685.

XXVI. Rosaceae

1. Petals absent — **11. Alchemilla**
1. Petals present :
    2. Carpels in fruit superior, i.e. neither enclosed within nor adnate to the calyx tube :
        3. Fruit dehiscent, follicular :
            4. Leaves simple; stipules absent — **4. Spiraea**
            4. Leaves imparipinnate; stipules persistent — **3. Sorbaria**
        3. Fruit indehiscent, composed of achenes or drupes :
            5. Trees; calyx deciduous; carpel solitary — **1. Prunus**
            5. Herbs or shrubs; calyx persistent; carpels more than 1 :
                6. Calyx ebracteolate :
                    7. Receptacle flat in fruit; fruit a head of achenes — **2. Filipendula**
                    7. Receptacle conical to globose in fruit; fruit a head of drupelets — **5. Rubus**
                6. Calyx with 5 bracteoles alternate with its lobes :
                    8. Achenes scattered on the surface of a brightly coloured, enlarged receptacle :
                        9. Bracteoles entire; petals white; receptacle pulpy — **8. Fragaria**

9. Bracteoles 3-toothed at the apex; petals yellow receptacle not pulpy **7. Duchesnea**

8. Achenes on a dry receptacle :

10. Stamens 5 **9. Sibbaldia**

10. Stamens 10 to many :

11. Styles elongated after anthesis **6. Geum**

11. Styles mostly very short, not elongating **10. Potentilla**

2. Carpels in fruit inferior, i.e., enclosed within the calyx tube :

12. Fruits small dry achenes, free from the calyx tube :

13. Unarmed herbs; achenes, 1 or 2; calyx hardened in fruit, crowned by a ring of hooked bristles outside the mouth **12. Agrimonia**

13. Armed shrubs; achenes numerous; calyx more or less fleshy in fruit, not as described above **13. Rosa**

12. Fruits fleshy, connate or adnate to the fleshy, enlarged calyx tube :

14. Fruit drupaceous, enclosing 1-5 bony stones :

15. Leaves entire; branches unarmed **17. Cotoneaster**

15. Leaves crenate, dentate, serrate or lobed; branches often armed with spines **16. Crataegus**

14. Fruit pomaceous (carpels with membraneous or leathery walls in fruit, the latter with fleshy endocarp :

16. Leaves (in our sp.) pinnate; flowers in compound corymbs **15. Sorbus**

16. Leaves simple; flowers in corymbs or umbelliform cymes **14. Malus**

1. PRUNUS Linn.

P. cornuta (Wall. ex Royle) Steud., Nom. Bot. ed. 2, 2 : 403, 1841; Rehder, Man. Cult. Tr. & Shrubs 479, 1940; Stewart, l.c. 362. *Cerasus cornuta* Wall. ex Royle, Ill. Bot. Himal. 207, t. 38, f. 2, 1835. *P. padus* Hk.f., FBI 2 : 315, 1878, non Linn. **Fig. 59.**

Deciduous, small or medium tree with rough, brown or blackish bark; young parts and inflorescence puberulent. Leaves elliptic to oblong-obovate or,-lanceolate, 10-15×3-7 cm from a mostly cordate base, acuminate, closely serrate, beneath glaucescent and bearded in nerve-axils; petioles red, usually 1-2-glandular near top. Flowers white, 6-8 mm across, in terminal and axillary, 8-16 cm long, pendulous racemes. Drupe 0.7-1 cm diam., red, on ripening black.

Occ. : Usually in association with *Sorbus lanata*, *Ulmus wallichiana*, *Acer caesium* and other deciduous trees both in the eastern and western Liddar valleys up to *c.* 3000 m.

Fls. May-June; *Frts.* July-Sept.

Aru ghasan (2500 m) *B.M. Sharma* 14321; Pehalgam (2250 m) *B.M. Sharma* 14677; Liddarwat (2900 m) *P.S. Jamwal* 15688.

The mawkish fruits are eaten by children. Not infrequently, the fruits develop a horn-like curved outgrowth-the 'kakarsingi'-as a result of the ovaries getting attacked by an insect. The leaves are dried and stored for use as cattle fodder. In fresh state they are believed to be poisonous.

2. FILIPENDULA Mill.

F. vestita. (Wall. ex G. Don) Maxim., Acta Horti Petrop. 6 : 248, 1879; Hand-Maz., Symb. Sin. Anth. 3 : 523, 1933; Stewart, l.c. 351; For. Fl. Sgr. 132. *Spiraea vestita* Wall. ex G. Don; Hk.f., l.c. 323. **Fig. 60**

Erect, pubescent or grey-tomentose, 20-90 cm tall herb from a perennial, rhizomatous base. Leaves 5-30 cm long, glabrous above, grey-felted or pubescent beneath, pinnatisect; lobes acutely toothed, the lateral small and very unequal, of one pair usually much larger than others (up to 5 cm long), terminal palmately 3-5 lobed, 5-17 cm across; stipules hemiorbicular. Flowers white, 5-8 mm diam., in terminal corymbose cymes; petals 4-5. Carpels villous.

Occ. : Common along drains and water channels in villages; also met with in damp meadows.

Fls. July-Sept.

Liddarwat (2900 m) *B.M. Sharma* 19382; Mundlan (2325 m) *B.M. Sharma* 14634; Aru (2400 m) *P.S. Jamwal* 19379.

3. SORBARIA A. Br.

S. tomentosa (Lindl.) Rehder, J. Arnold Arbor. 19 : 75, 1938; Stewart, l.c. 370; For. Fl. Sgr. 137. *Schizonotus tomentosus* Lindl., Bot. Reg. 26 : 71, 1841. *Spiraea lindleyana* Wall. ex Lindl. in Bot. Reg. 31, t. 33, 1841. *Spiraea sorbifolia* Hk.f., l.c. 324. **Fig. 61**

Deciduous, almost glabrous, 1.5-3 m tall shrub; older stems with rough, lenticellate, dark-brown bark. Leaves 15-30 cm long, pinnately 13-23-foliolate; leaflets lanceolate, 5-10×1-2 cm, acuminate, doubly serrate, glabrous or on the nerves beneath pubescent. Flowers white, 5-6 mm diam., in terminal 15-30 cm long panicles; stamens scarcely longer than petals. Follicles 5, glabrous,

Occ. : Common on open slopes and boulder ground about Liddar in Pehalgam, scarce along the roads to Aru and Chandanwari.

Fls. June-July; *Frts.* Aug.

Below Aru (2350 m) *B.M. Sharma* 14265; Aru (2400 m) *B.M. Sharma* 19305; Pehalgam (2160 m) *P.S. Jamwal* 15689.

4. SPIRAEA Linn.

Flowers bisexual; follicles villous, partially immersed in calyx tube **1. S. canescens**
Flowers unisexual; follicles glabrous but for ventral suture, not immersed **2. S. bella**

1. **S. canescens** D. Don, Prodr. Fl. Nep. 227, 1825; Hk.f., l.c. 325. **Fig. 62**

Deciduous, 2-3 m tall shrub; branches often arching; twigs leafy, reddish-, or greyish-brown, softly pubescent when young. Leaves pubescent, ovate, obovate or oblanceolate, 0.7-1.5×0.4-1 cm, from mostly cuneate base; apex toothed or entire. Flowers white, 5-7 mm diam., in dense, corymbose panicles terminal on the main shoot and short, leafy, lateral shoots, the latter usually all directed to one side. Calyx tomentose; tube glabrous within.

Occ. : On exposed scrubby slopes abutting on abandoned fields in Aru and Pehalgam.

Fls. May-June; *Frts.* Sept-Oct.

Aru (2400 m) *B.M. Sharma* 19380; Pehalgam (2160 m) *B.M. Sharma* 14646; Aru (2400 m) *P.S. Jamwal* 15691.

2. **S. bella** Sims, Bot. Mag. 50 : t. 2426, 1823; Hk. f., l.c. 324.

Deciduous shrub. Leaves membranous, ovate, elliptic or lanceolate 2-3.5 cm long, sharply and irregularly serrate in upper half at least, pale glaucous beneath. Flowers pink 5-7 mm diam., in terminal, pubescent, corymbose cymes Calyx-lobes glabrous; tube hairy within as well. Follicles brown, shining.

Fls. July.

Lidder vy. above Kainmul (10-11000′) *Duthie* 13087; Lidder vy. above Kainmul (11-12000′) *Duthie* DD, s.n.

5. RUBUS Linn.

1. Flowers white :
    2. Leaves white-tomentose beneath — **1. R. irritans**
    2. Leaves green beneath :
        3. Leaves pinnately 5-7-foliolate; calyx densely armed with prickles; carpels hairy — **2. R. pungens**
        3. Leaves pinnately 3-foliolate; calyx unarmed; carpels glabrous — **3. R. saxatilis**
1. Flowers pink :
    4. Leaves 5–7-foliolate; drupelets black — **4. R. niveus**
    4. Leaves 3-foliolate; drupelets red — **5. R. pedunculosus**

1. **R. irritans** Focke, Bibl. Bot. 72 : 192, 1910. Stewart, l.c. 368. *R. purpureus* Hk.f., l.c. 337 (non Bunge 1832).

Semierect, low shrub clothed with slender prickles and glandular bristles; young shoots brown or purplish, the older twigs with loose, brown or greyish papery bark. Leaves 6-16 cm long, mostly 3-, rarely 5-foliolate; petioles tomentose, armed and coloured like stem; leaflets doubly serrate or crenate, 3-10×1.5-8 cm, glabrous above, terminal ovate or near orbicular-cordate, the lateral sessile and obliquely ovate; stipules setaceous. Flowers drooping, white, *c.* 1.2 cm diam., in lower axils solitary, the upper in 2-4 flowered cymes. Calyx tomentose, armed with prickles and glandular bristles; lobes finely long-acuminate. Drupes many, red, tomentose.

Occ. : In open boulder strewn ground along nallah banks and as undergrowth in fir forests.

Fls. June-Aug.; *Frts.* Aug-Sept.

Bara Khalen (2950 m) *B.M. Sharma* 14272 & 19313; Chatponsal nallah, Lidder vy. (11-12000′) *Duthie* 13178; Aru-Chumanaisar route (2800 m) *P.S. Jamwal* 15692.

2. **R. pungens** Camb. in Jacq. Voy. Bot. Inde. 48, t. 59, 1844; Hk.f., l.c. 341.

Trailing shrub with long, weak, dull brown branches scattered with straight prickles and glabrous but for the hairy and sometimes glandular tips. Leaves 4-10 cm long including slender, prickly, pubescent petioles; leaflets ovate, 1-5× 0.7-3.5 cm; membranous, inciso-serrate, adpressedly pubescent or above nearly glabrous, terminal broadest, the lateral sessile; stipules lanceolate or setaceous. Flowers white, 2 cm diam., on slender 2.5-5 cm long pedicels, solitary or 2-3, axillary and terminal. Calyx tomentose and glandular; lobes caudate, patent in fruit. Drupelets pubescent.

Occ. : On shaded, humus laden rocks and forest floor about Liddarwat and Chandanwari.

Fls. May-June; *Frts*. July-Aug.

Chota Khalen (2900 m) *B.M. Sharma* 14273; Lidder vy. above Tanin (10-11000′) *Duthie* 14145; Lidderwat, Lidder vy. *Inayat* 25591; Nangneri nallah (2825 m) *P.S. Jamwal* 15693.

3. **R. saxatilis** Linn., Sp. Pl. 494, 1753; Hk.f., l.c. 233. **Fig. 63**

Minutely and sparingly armed, glabrous or hairy, herbaceous undershrub, throwing off slender, annual, up to 1 m long creeping vegetative shoots from a woody base, the latter sending forth ascending or semierect flowering stems next year. Leaflets rhombic-ovate or, -obovate from cuneate base, 3-8 × 2.5 cm, unequally toothed, glabrescent above, hairy beneath; stipules ovate, oblong or linear, Flowers white, 2-4 on short axillary peduncles; calyx lobes pubescent acuminate or mucronate, erect in fruit. Druplets few, 5-7 mm diam., red.

Occ. : Frequent in wet grassy places in alpine meadows.

Fls. May-June; *Frts.* July-Sept.

Yechi nar (3600 m) *B.M. Sharma* 16310; Lidder vy. below Tanin (8-9000′) *Duthie* 13443; above Pisu Ghati (3400 m) *P.S. Jamwal* 15695.

4. **R. niveus** Thunb., Dissert. de Rubo. 9, f. 3, 1813; Saxena, Indian For. 93 (12) : 798, 1967; Stewart, l.c. 368. *R. lasiocarpus* Sm. in Rees, Cycop. 30 : n. 6, 1819, Hk.f., l.c. 339.

Large shrub with long, glabrous, purple, arching, pendulous or spreading, pruinose branches often rooting at tips and scattered with stout, conical prickles. Leaves 8-28 cm long, sometimes 9-, or even 11-foliolate; leaflets ovate-lanceolate or ovate, 3-8 × 1.2-3.8 cm, serrate, glabrous above, white-, or grey-tomentose beneath, lateral sessile, terminal broadest and frequently lobed. Corymbs often panicled, terminal and axillary; calyx lobes lanceolate, tomentose on either face. Drupelets numerous, black, hoary.

Occ. : Frequent in forest clearings along the road from Pehalgam to Aru; less common in the forest.

Fls. May.-June

Aru (2350 m) *B.M. Sharma* 19297.

5. **R. pedunculosus** D. Don, Prodr. Fl. Nep. 234, 1825; Ohashi in Hara, Fl. East. Himal. 131, 1966; Stewart, l.c. 368. *R. gracilis* Roxb. ex DC., Prodr. 2 : 257, 1825. *R. niveus* Hk.f., l.c. 335.

Rambling shrub, much variable to the extent of indumentum, armature and colour of branches. Leaflets acuminate, simply or doubly serrate, 2.5-10 cm long, usually hairy at least along the nerves above, more commonly white-tomentose beneath, lateral mostly ovate or lanceolate, sessile or petiolulate, the terminal stalked, broadly ovate-cordate and frequently lobed. Flowers 0.8-2.5 cm diam., lower often solitary, the terminal in a few-flowered tomentose cluster; calyx lobes ovate lanceolate, tomentose on either face, acuminate. Drupelets numerous, hoary.

Frts. July-Aug.

Lidder vy. (6-7000') *Duthie* 13068; Lidder vy. above Pehalgam (8-9000') *Duthie* 14148.

6. **GEUM** Linn.

Calyx-lobes reflexed; style jointed in middle, in fruit hooked at the tip **1. G. urbanum**
Calyx-lobes spreading; style simple, not jointed, straight in fruit **2. G. elatum**

1. **G. urbanum** Linn., Sp. Pl. 501, 1753; Hk.f., l.c. 342. **Fig. 64**

Perennial, hairy herb with 20-75 cm tall, branched, leafy stems from a short rhizome. Leaves imparipinnate; leaflets coarsely toothed, of the upper cauline leaves mostly 3, at times united into a 3-lobed leaf, those of the radical and lower cauline in 2-5, alternately small and large, sessile pairs with odd one frequently lobed and largest, 2.5-7 cm across, suborbicular, rhombic or obovate-cuneate. Flowers pale yellow, 1.2-1.8 cm diam., in long-stalked terminal cymes. Achenes hairy, several in a globose head; style disarticulating at the joint about its middle to cast of the hairy, deciduous rostrum from the lower, persistent, glabrous part.

Occ. : Very common in hedges and along shady banks at altitudes mostly below 3300 m in both eastern and western valley.

Fls. May-July; *Frts.* July-Sept.

Liddarwat (2900 m) *B.M. Sharma* 14186; Pisu Ghati, on way to Amarnath (10,000') *K.M. Vaid*, DD no. 117559; Pehalgam (7200') *H.D. Thapliyal* 26513; Armiun glen (2750 m) *P.S. Jamwal* 16301.

2. **G. elatum** G. Don, Gen. Syst. 2 : 526, 1832; Stewart, l.c. 352. *G. elatum* Wall.; Hk.f., l.c. 343. *Acomastylis elata* F. Bolle in Fedde, Rep. Beih. 72 : 83, 1933; For. Fl. Sgr. 130. **Fig. 65**

Perennial herb, pubescent or clothed with patent, silky hairs. Stems sparingly leafy, 15-45 cm long, erect or arcuate from a stout, woody stock covered with fibrous remnants of old leaf-bases. Leaves chiefly radical, 10-30 cm long, pinnatisect into obovate or oblong, often lobed, crenate or serrate, sessile, up to 2.5 cm long, alternately large and small, hairy, broad-based segments progressively shorter downwards; terminal segment 3-lobed. Flowers bright yellow, 1.8-3 cm diam., solitary or a few together; style glabrous. Achenes ellipsoid, hairy.

Occ. : Common on alpine and subalpine pastures; mostly along streams.

Fls. June-Aug.

Bara Khalen (2920 m) *B.M. Sharma* 14145 & 14648; Rewil nallah (3200 m) *B. M. Sharma* 14651; Chandanwari (2700 m) *P.S. Jamwal* 16350.

7. **DUCHESNEA** Sm.

D. indica (Andr.) Focke in Engler & Prantl. Natural Pflanzenfam. 3(3) : 33, 1880; Juzz in Fl. URSS. 10 : 67,

t. 6-6, 1941; Stewart, l.c. 359. *Fragaria indica* Andr., Bot. Rep. t. 749, 1807; Hk.f., l.c. 343.

Low, perennial, silky herb, emitting long, creeping and rooting stolons. Leaves chiefly radical, rosulate, rather long-stalked, digitately 3-foliolate; leaflets obovate-cuneate, 1-2.5×0.7-2 cm, silky, crenate; stipules toothed. Flowers yellow, 1.2-1.5 cm diam., axillary, mostly solitary on nearly leaf-equalling stalks; bracteoles broad-obovate, longer than calyx-lobes. Achenes numerous, minute, on a bright red, globose 0.7-1 cm wide, insipid receptacle.

Occ. : Moister slopes and under shade of *Viburnum* and other bushes; common to *c.* 2400 m.

Fls. May-July; *Frts.* June-Aug.

Aru (2400 m) *B.M. Sharma* 14652; Gurkum (2300 m) *B.M. Sharma* 14653; Aru ghasan (2450 m) *P.S. Jamwal* 19377.

8. **FRAGARIA** Linn.

F. nubicola Lindl. ex Lacaita, J. Linn. Soc. 43 : 467, 1916; Stewart, l.c. 352. *F. vesca* Linn., Sp. Pl. 494, 1753 var. *nubicola* Lindl. ex Hk.f., l.c. 344; For. Fl. Sgr. 132. **Fig. 66**

Small, perennial, silky herb with erect, 6-18 cm long flowering scapes and filiform stolons from a stout rootstock. Leaves radical, rosulate, digitately 3-foliolate; petioles like scapes covered with appressed or spreading or both types of silky hairs. Leaflets ovate, obovate or rhombic, serrate, 1.5-6×1-4 cm, silky, usually silvery beneath. Flowers white, 1.5-2 cm diam., solitary or few in terminal cymes; bracteoles narrow. Achenes minute, scattered over and sunken in a red, globose, 0.7-1 cm wide, succulent receptacle.

Occ. : Abundant everywhere on the wooded slopes and humus rich soils.

Fls. May-Aug.; *Frts.* July-Aug.

Liddarwat (2750 m) *B.M Sharma* 14181; Pehalgam (2160 m) *B.M. Sharma* 14654; Chandanwari (2700 m) *P.S. Jamwal* 16351. Fruits are insipid or sweetish; eaten by children.

9. **SIBBALDIA** Linn.

S. cuneata Hornem., Ind. Sem. Hort. Hafn. 1842 et ex Kuntze in Linnaea 20 : 59, 1847; Wendelbo in Nytt. Mag. Bot. 1 : 40, 1952; Stewart, l.c. 369. *S. parviflora* Kitam., Faun. Fl. Nep. Himal. 160, 1955. *Potentilla sibbaldi* Hk.f., l.c. 345 (non Haller). **Fig. 67**

Perennial, tufted herb, with spreading or ascending, 5-25 cm long, silky hairy stems, from a branched woody rootstock. Leaves digitately 3-foliolate; leaflets obovate-cuneate, 0.5-2×0.3-1.5 cm, adpressedly siky, 3-5-toothed at the truncate apex. Flowers pale yellow, 5-6 mm diam., in dense, hairy cymes terminating leafy stems; bracteoles linear, narrow; petals, *c.* 2.5 mm long. Achenes ovoid, 1.5×1 mm, glabrous, obliquely striate.

Occ. : Abundant and often forming dense patches in wet places on sub-alpine and alpine meadows.

Fls. May-Aug.; *Frts.* June-Sept. Satlanjan (3100 m) *B.M. Sharma* 16378; Badzalkhod nallah, Lidder vy. (13000′) *Duthie* 14328; Pehalgam (7200′) *H.D. Thapliyal* 26517; Aru (2400 m) *P.S. Jamwal* 14397.

10. POTENTILLA Linn.

1- Leaves digitately compound :
  2. Leaves 3-foliolate :
    3. Undersurface of leaflets white, tomentose :
      4. Flowers yellow ... **1. P. argyrophylla**
      4. Flowers dark crimson ... **2. P. atrosanguinea**
    3. Undersurface of leaflets green, hairy or pubescent or glabrescent :
      5. Flowers 6 mm or less in diam.; receptacle glabrous ... **3. P. kleiniana**
      5. Flowers above 6 mm in diam.; receptacle villous or covered with long stiff hairs :
        6. Calyx glabrous; achenes completely concealed by receptacular hairs ... **4. P. eriocarpa**
        6. Calyx silky or pubescent; achenes not concealed :
          7. Leaflets truncated and 3-fid at apex; achenes 5 or 6, hairy ... **5. P. curviseta**
          7. Leaflets deeply or coarsely multicrenate; achenes many, glabrous :
            8. Leaflets deeply crenate; flowers 1.5-2.5 cm diam.; calyx lobes acute ... **6. P. gelida**
            8. Leaflets roundly and coarsely crenate at tip; flowers 0.8-1.2 cm diam.; calyx lobes obtuse ... **7. P. monanthes**
  2. Leaves all, or the lower at least, 5-foliolate :
    9. Flowers yellow :
      10. Flowers solitary, axillary ... **8. P. reptans**
      10. Flowers in terminal heads or cymes :
        11. Flowers in capitate, leafy heads ... **9. P. desertorum**
        11. Flowers in corymbose cymes :
          12. Flowers 6 mm or less in diam.; receptacle glabrous ... **3. P. kleiniana**
          12. Flowers *c.* 1.2 cm diam.; receptacle hairy ... **10. P. kashmirica**
    9. Flowers dark crimson or purple ... **11. P. nepalensis**
1. Leaves pinnately compound :
  13. Undersurface of leaflets white ... **12. P. multifida**
  13. Undersurface of leaflets green :
    14. Leaflets toothed; achenes glabrous ... **13. P. supina**
    14. Leaflets entire; achenes hairy ... **14. P. biflora**

1. **P. argyrophylla** Wall. ex Lehm., Pugill. 3 : 36, 1831 et Monogr. Pot. Suppl. 1 : t. 9, 1835; Hk.f., l.c. 356 p.p. **Fig. 68.**

Perennial herb from a short, rather thick rootstock; stems erect or ascending, 25-60 cm tall, like the undersurface of leaflets white-, or grey-tomentose. Leaflets ovate, obovate or suborbicular, 2-6 cm long, serrate, with green and hairy upper surface. Flowers 1.6-3.5 cm diam., on long pedicels, solitary or in panicled cymes; calyx lobes pointed, tomentose;

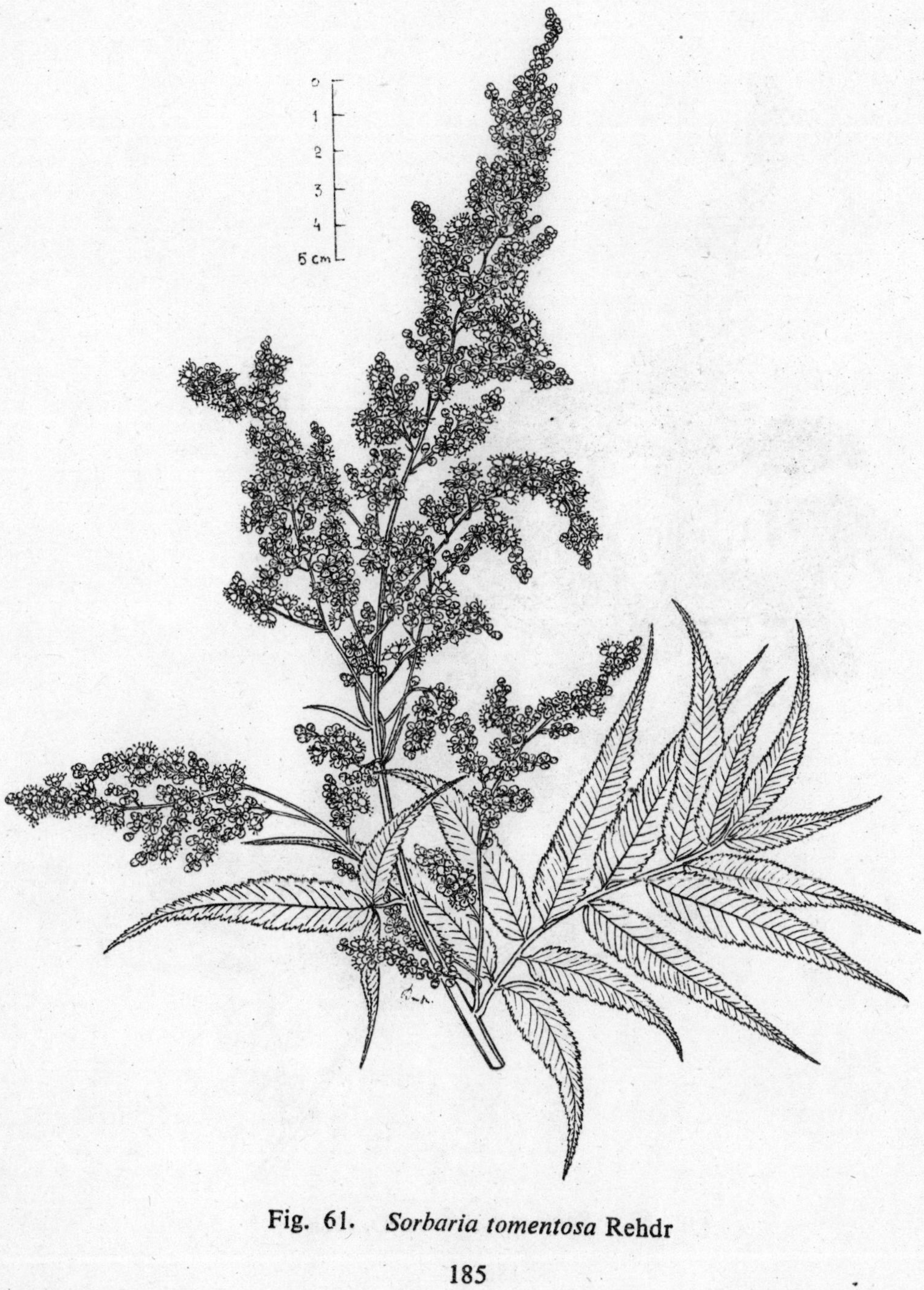

Fig. 61. *Sorbaria tomentosa* Rehdr

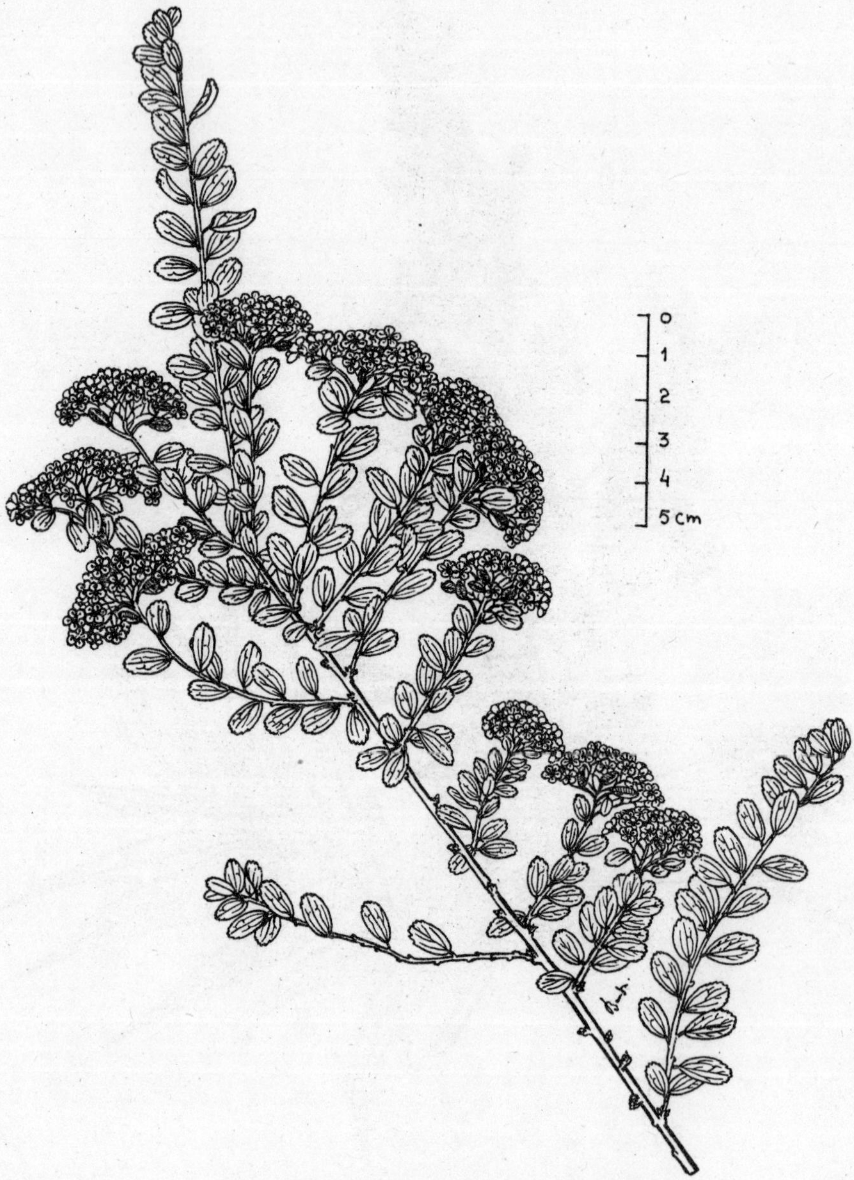

Fig. 62. *Spiraea canescens* D. Don

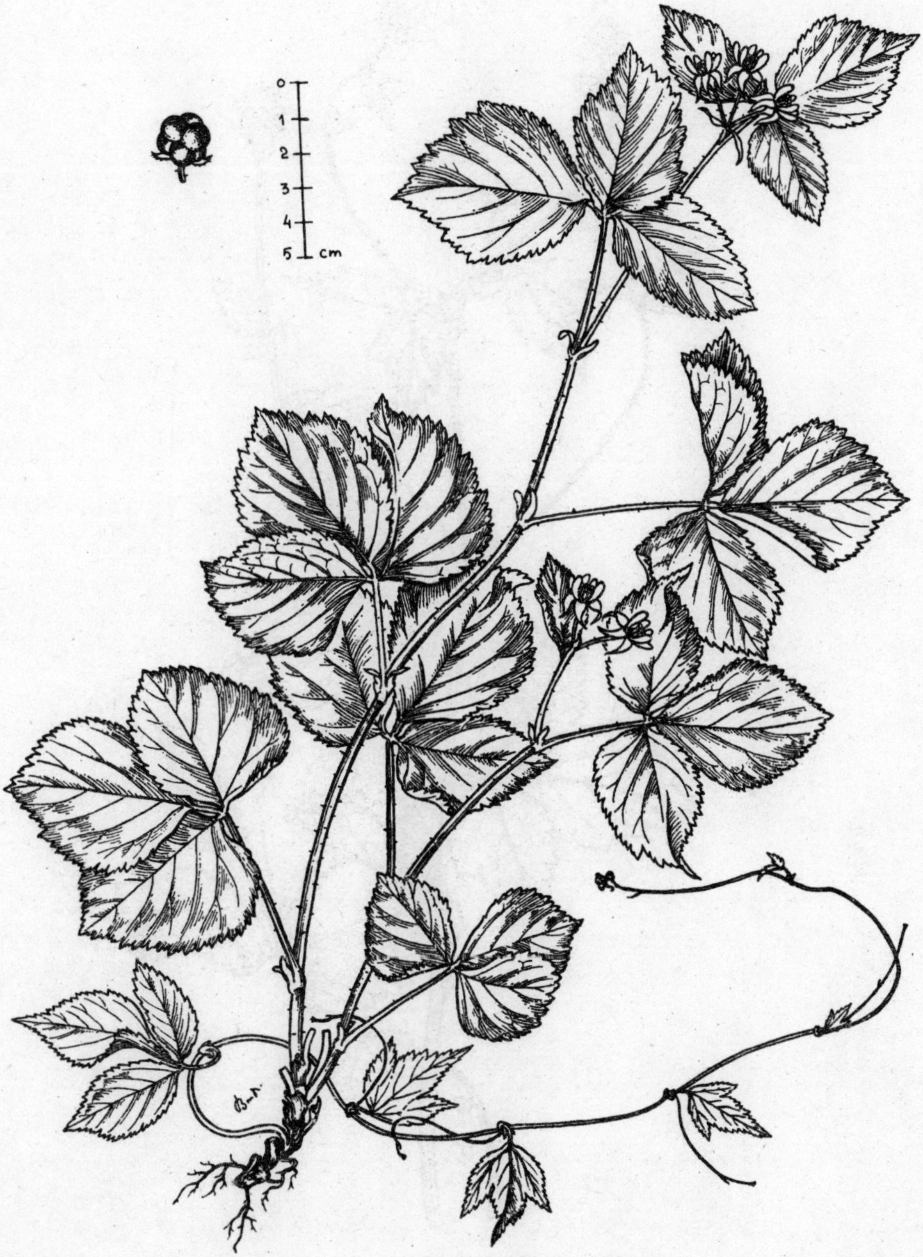

Fig. 63. *Rubus saxatilis* Linn.

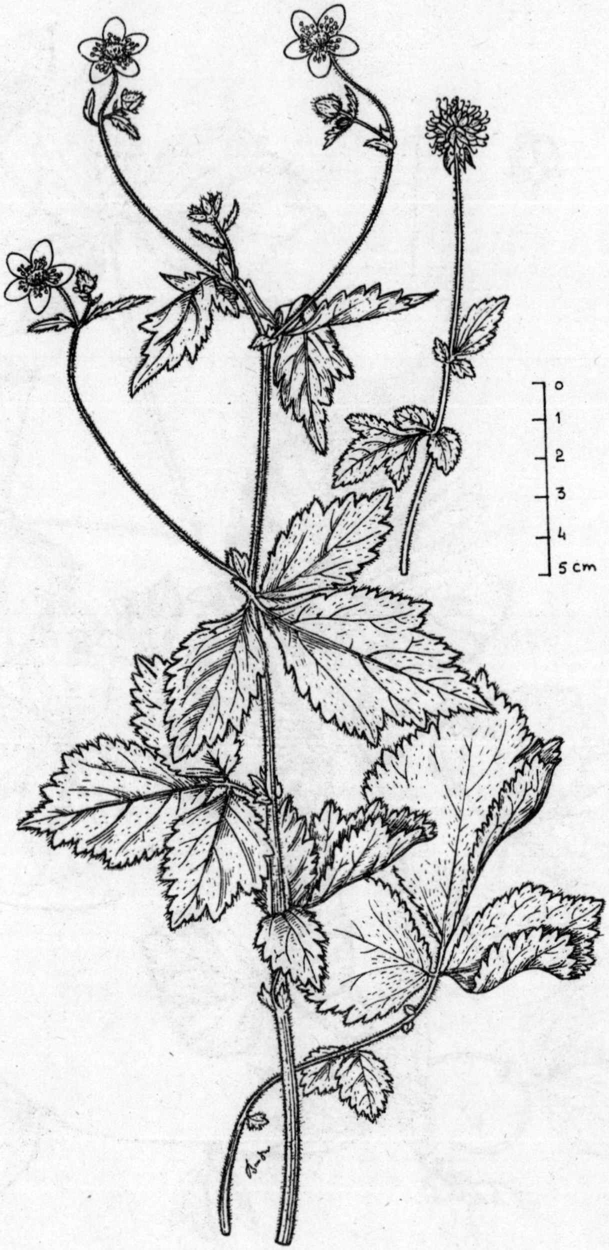

Fig. 64. *Geum urbanum* Linn.

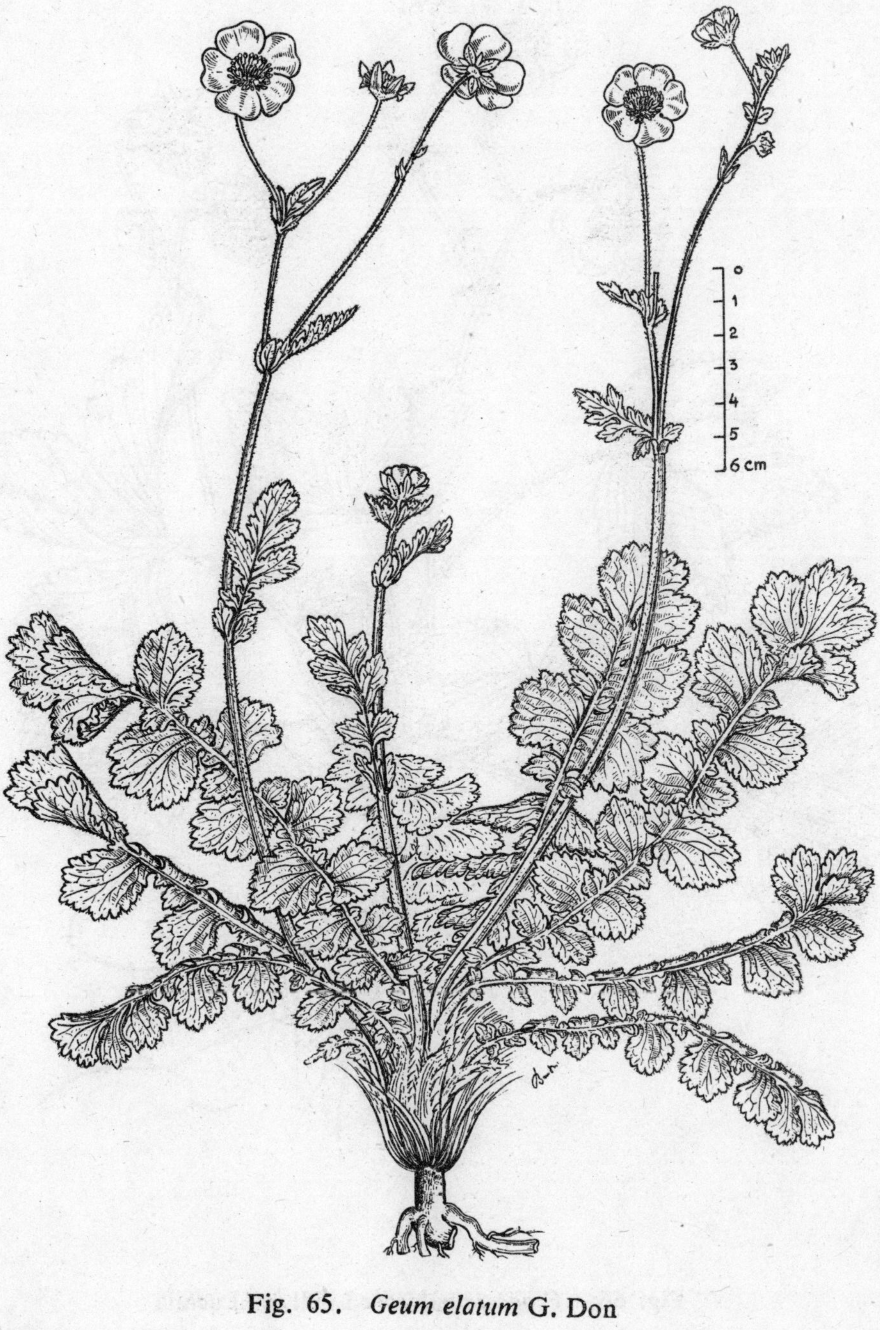

Fig. 65. *Geum elatum* G. Don

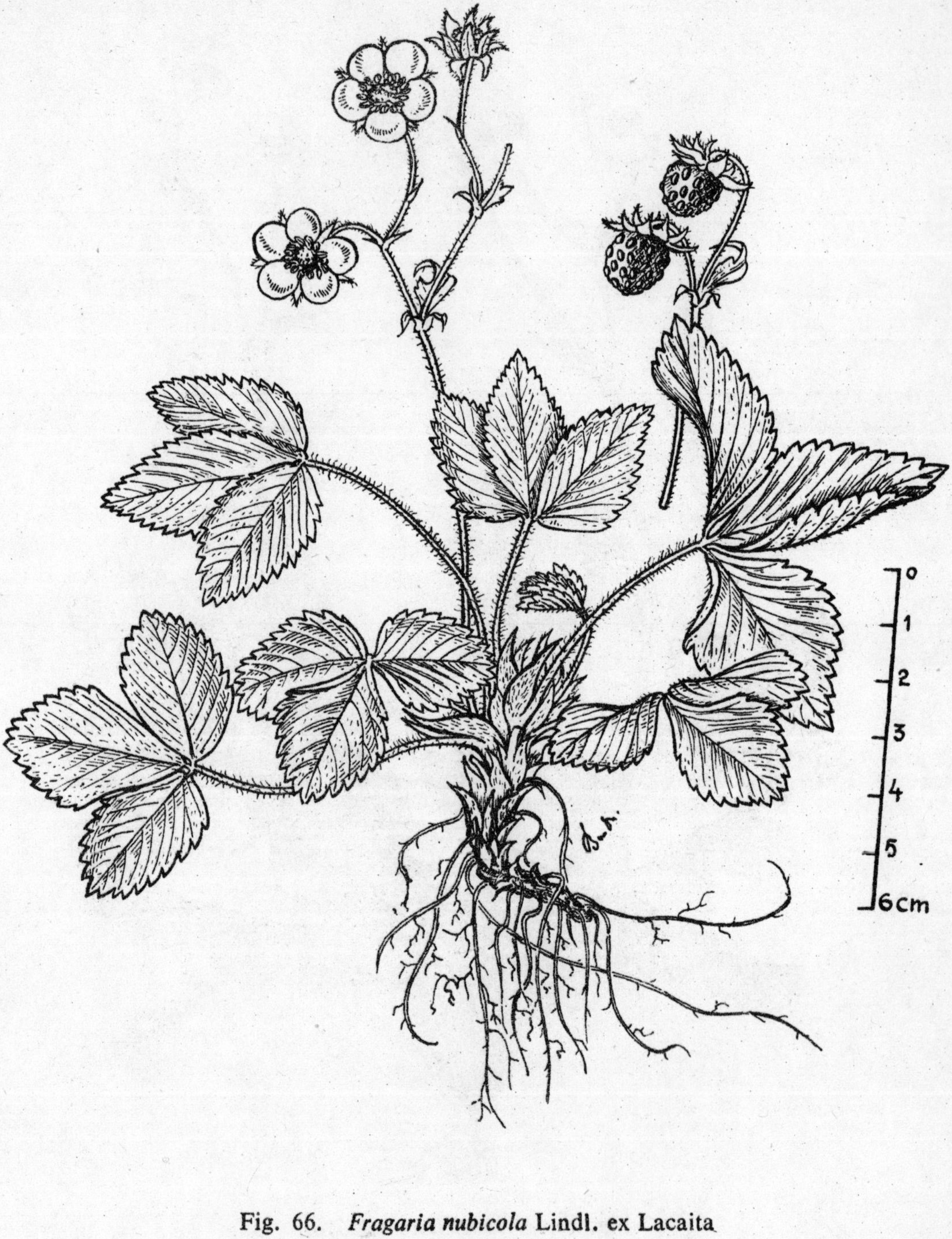

Fig. 66. *Fragaria nubicola* Lindl. ex Lacaita

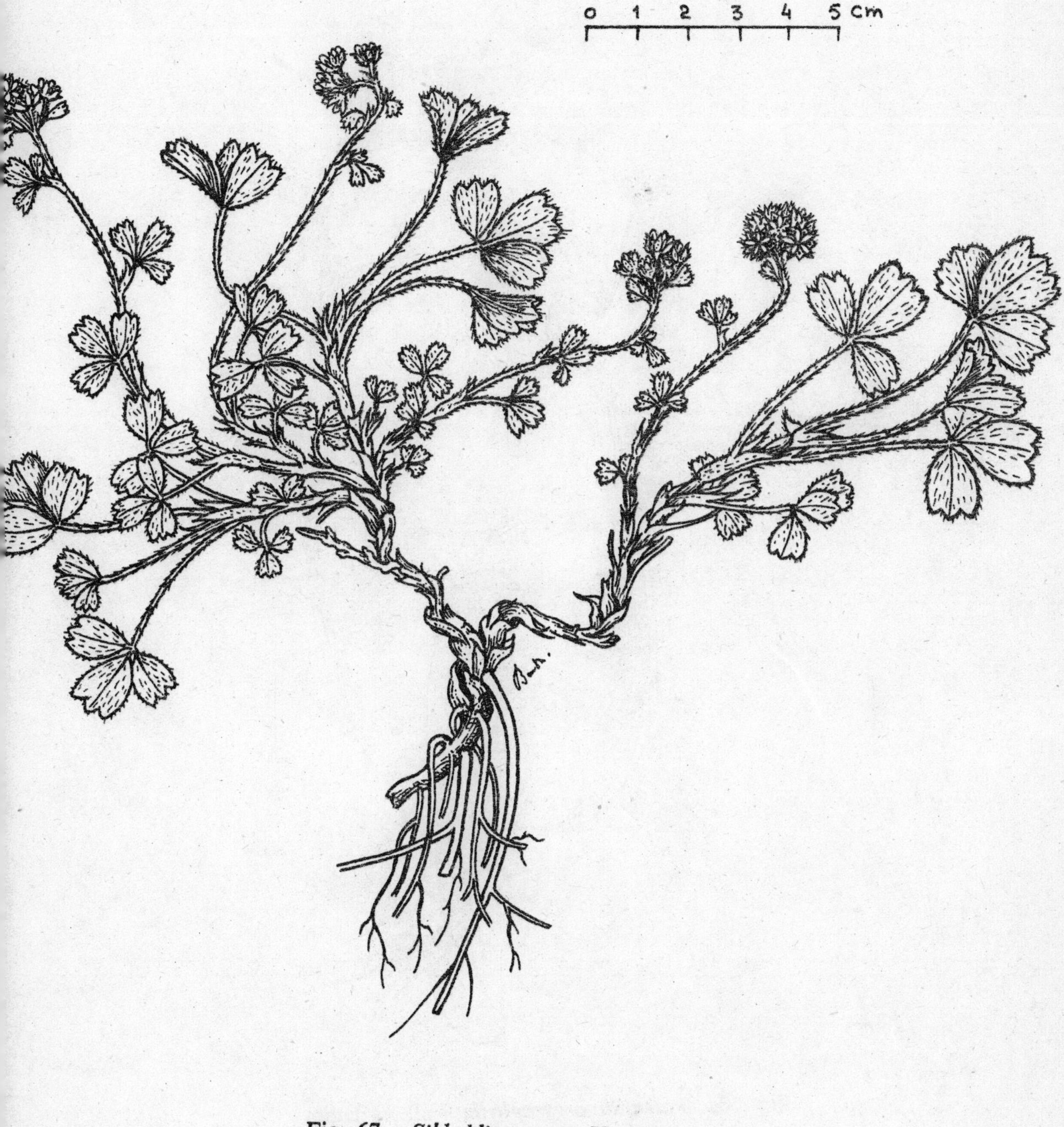

Fig. 67. *Sibbaldia cuneata* Hornem.

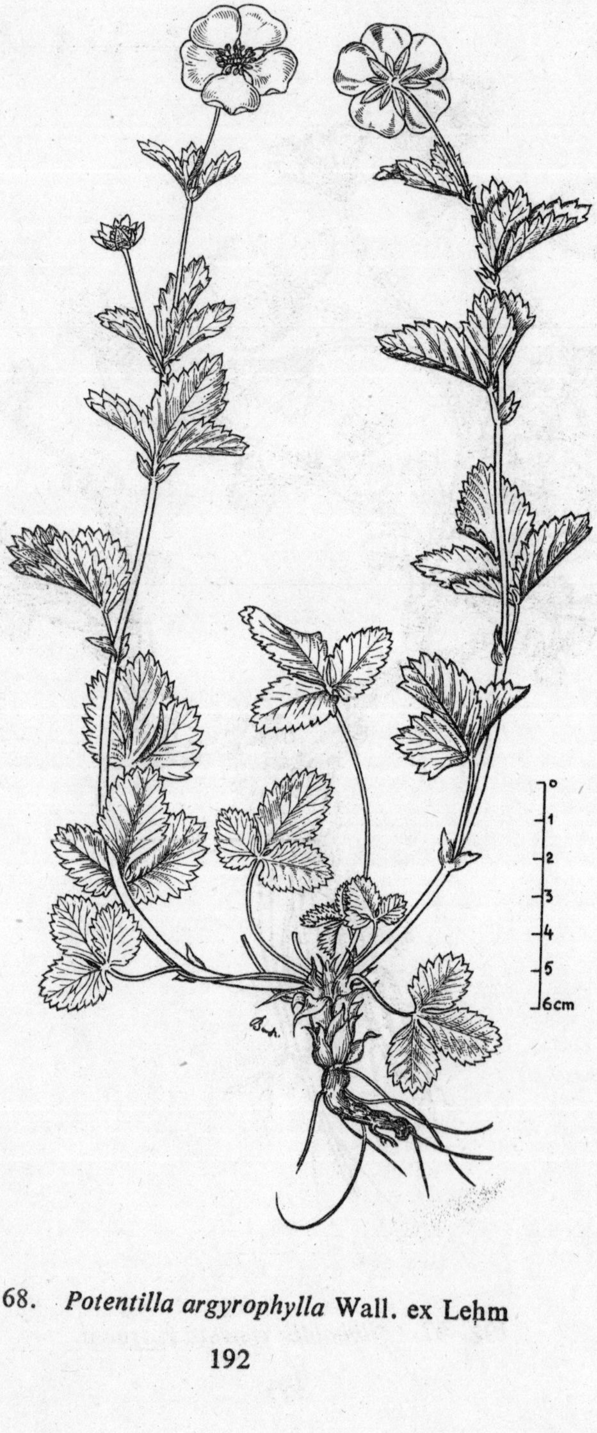

Fig. 68. *Potentilla argyrophylla* Wall. ex Lehm

petals obcordate. Achenes numerous, minute, glabrous.

Occ. : Abundant in the alpine region, especially in morainal ground throughout in Satlanjan-Kolahoi tract. Left alone by the grazing animals.

Fls. July-Aug.

Kolahoi (3700 m) *B.M. Sharma* 19376; Lidder vy. above Tanin (10-11000′) *Duthie* 13312; Sonsal nallah, Lidder vy. (13-14000′) *Duthie* 13372; Lidder vy. near Sheshnag (12-13000′) *Duthie* 14146/a; Mahagunus pass, on way to Amarnath (14000′) *K.M. Vaid* DD. no. 118078; Khalen, W. Lidder vy. (2900 m) *P.S. Jamwal* 14146.

2. **P. atrosanguinea** Lodd., Bot. Cab. 8 : t. 786, 1824; Stewart, l.c. 353; For. Fl. Sgr. 134. *P. argyrophylla* Wall. var. *atrosanguinea* Hk.f., l.c. 357.

Very like the preceding species in habit, habitat, altitudinal distribution and flowering period. Distinguishable alone from the crimson or purple colour of its petals Rather scarce.

Kolahoi (3750 m) *B.M. Sharma* 19375; Mount above Kolahoi (3900 m) *B.M. Sharma* 14696.

3. **P. kleiniana** Wt. & Arn., Prodr. 300, 1834; Hk.f., l.c. 359.

Annual, thinly silky herb with slender, 15-50 cm long, procumbent or diffusely spreading stems from the root. Leaflets unequal, 3-5, obovate, oblanceolate or obovate-oblong, mostly up to 4×2 cm, blunt, serrate except near the base, glabrous above, beneath silky especially on the nerves. Flowers 6 mm diam., in terminal corymbose cymes; calyx lobes and bracteoles thinly silky, the latter obtuse or acute, entire or 2-3 fid at the apex. Achenes minute, glabrous, deeply wrinkled.

Occ. : In depressed wet places; waysides and along paths.

Fls. & Frts. June-Sept.

Gha Shangan (2700 m) *B.M. Sharma* 14298; Pehalgam (2160 m) *B.M. Sharma* 14655; Mundlan-Aru road (2300 m) *P.S. Jamwal* 16315.

4. **P. eriocarpa** Wall. ex Lehm., Pugill. 3 : 35, 1831; Hk.f., l.c. 348.

Almost glabrous, perennial herb of alpine rocky cliffs; stems woody, rootstock-like, ascending, giving off semierect, herbaceous, 1-2-leaved to leafless flowering shoots from tips. Basal leaves few; leaflets cuneate, irregularly and sharply 3-many-toothed above. Flowers solitary, 2.5-3.8 cm diam., on naked, up to 7 cm long peduncles; petals orbicular-cordate. Achenes minute, long pilose.

Thuliyansar (12000′) *W.J. Lambert* DD. no. 44697, fls. 4.9. 1927; cited by Stewart (l.c.) also from Nafran (upper Lidder vy.) and Musjid valley.

5. **P. curviseta** Hk.f., FBI. 2 : 358, 1878.

Thinly silky-pubescent perennial herb of high level cliffs, distinguishable through characters of key and the characteristic sheath-like, narrow, brown petiolar wing formed by stipules adnate for half the length of the petiole. Leaves few; leaflets narrowly cuneate, 1.2-2.5×0.5-0.8 cm. Flowering stems 2-3-flowered.

Included after Stewart (l.c. 355) who states it to be common and cites specimens from Pehalgam and Mt. Kolahoi.

6. **P. gelida** C.A. Mey., Verz. Pfl. Cauc. 167, 1831; Hk.f., l.c. 357.

Softly hairy, sometimes glandular, tufted perennial herb with several procumbent or ascending, sparingly leafy flowering stems; rootstock mostly small. Basal leaves petiolate; leaflets obovate or cuneate, 0.7-1.8 cm long, entire for 1/2-1/3 of length from base; crenatures deep, irregular; cauline stipules ovate, 0.5-1 cm long, entire, rarely 1-2-toothed. Flowers 1-4 in a cluster; pedicels slender; bracteoles obtuse, shorter than the calyx lobes.

Occ. : Very common in open alpine pastures, preferably on morainal debris.

Fls. June-July.

Kolahoi (3700 m) *B.M. Sharma* 14312; Satlanjan (3300 m) *B.M. Sharma* 14695; Lidder vy. near Sheshnag (12-13000′) *Duthie* 14146; Lidder vy. above Kainmul 12000′) *Duthie* 13124; Kolahoi (3700 m) *P.S. Jamwal* 16337.

7. **P. monanthes** Lindl. ex Pugill. 3 : 33, 1831; Hk.f., l.c. 358.

Tufted, hairy or glandular perennial herb with ascending, 5-30 cm high, leafy stems. Leaves radical and cauline, petiolate; leaflets broad-obcuneate, 0.6-1.8 ×0.5-2 cm, glandular-pubescent; cauline stipules ovate, 0.5-1.2 cm long, rarely toothed. Flowers pedicellate and solitary in the axils, sessile or shortly stalked at the top of flowering stems; bracteoles orbicular-oblong, larger than the broadly obtuse calyx lobes. Achenes minute.

Occ. : In habitats similar for the preceding species but much less frequent.

Fls. June-July.

Kolahoi (3300 m) *B.M. Sharma* 19381; Tarsar (3750 m) *B.M. Sharma* 14694; Kolahoi (3650 m) *P.S. Jamwal* 16338.

8. **P. reptans** Linn., Sp. Pl. 499, 1753; Hk.f., l.c. 356.

Perennial herb from rather slender, woody rootstock; stems slender, creeping, glabrous to sparingly hairy, up to 60 cm long, rooting from leafy nodes. Leaflets obovate or oblanceolate, pubescent, in our specimens up to 3×1.3 cm, crenate-serrate, on up to 10 cm long common petiole; stipules oblong, adnate except for their lanceolate tips. Flowers 1.2-2 cm diam., on erect, slender, 2-8 cm long peduncles; bracteoles like the calyx lobes acute, silky and about their length. Achenes minute, ovoid, dark brown, glabrous.

Occ. : Common in wet depressed places, along paths and in shade of Viburnums in areas about Pehalgam and Aru.

Fls. June-Sept.

Aru (2450 m) *B. M. Sharma* 14297; Pehalgam (2150 m) *B.M. Sharma* 14656; Aru (2425 m) *P.S. Jamwal* 16302.

9. **P. desertorum** Bunge in Ledeb., Fl Alt. 2 : 257, 1830; Hk.f., l.c. 355.

Erect, perennial; stems tufted, leafy, 20-45 cm tall, clothed with soft, patent, simple and often glandular hairs. Radical leaves 5-foliolate, on 5-15 cm long petioles, the cauline 3-foliolate, short stalked, sessile upwards, with 1-2 cm long, ovate-lanceolate, entire or occasionally 1-2-toothed stipules. Leaflets obovate cuneate, adpressedly hairy, 2-5.5×1-2.5 cm, entire for 1/3-2/3 of length from base, coarsely crenate upwards. Flowers 1.2-2 cm diam.; bracteoles entire or 2-3-fid. Achenes minute, wrinkled, glabrous.

Occ. : Common in the alpine zone, on cliffs and in islands in mountain torrents.

Fls. June-July.

Hamwas (3200 m) *B.M. Sharma* 14274; Kolahoi (3650 m) *P.S.* Jamwal 16339.

10. **P. kashmirica** Hk.f., FBI. 2 : 355, 1878.

Erect, softly hairy, perennial herb. Leaves short-stalked; leaflets obovate oblong, hairy, cut 1/3 way down into lobes or serratures. Flowers in broad corymbose cymes; achenes deeply wrinkled.

Cited by Stewart (l.c. 356) from Aru, Lidder vy. However, in Dehradun Herbarium there is no specimen from our area, nor have we located any in the field.

11. **P. nepalensis** Hk., Exct. Fl. 2 : t. 88, 1824; Hk.f., l.c. 355.

Erect, 30-75 cm tall, hairy perennial herb; stem branched, leafy. Leaves long petioled, 5-foliolate, the upper often 3-foliolate; cauline stipules ovate or oblong, 1.8-2.5 cm long, upper often lobed. Leaflets unequal, oblong-ovate, obovate or oblanceolate, 2.5-7.5×0.6-3 cm, coarsely obtuse-, or acute-toothed, sparsely adpressed-hairy. Flowers 1.8-2.5 cm diam., in terminal dichotomous panicles. Achenes minute, glabrous wrinkled.

Occ. : Common on roadside slopes about Pehalgam town.

Fls. July-Aug.

Pehalgam (2160 m) *B.M. Sharma* 16364 & 16365.

12. P. **multifida** Linn., Sp. Pl. 496, 1753; Hk.f., l.c. 353.

Low perennial herb with ascending or prostrate, densely and softly white-hairy, rather slender flowering stems from a short, erect rootstock. Leaves chiefly radical, 2.5-5 cm long, pinnatisect into 5-9 linear-oblong, 0.5-2 cm long leaflets pinnatifid nearly to midrib into linear or lanceolate, ascending segments with green pubescent upper surface and recurved margins. Flowers 1-1.2 cm diam, on slender pedicels in open corymbose cymes with a solitary lower one; bracteoles and calyx-lobes densely white-silky; the former linear-oblong, obtuse, slightly shorter than or as long as the ovate, acute calyx lobes. Achenes minute, glabrous.

Fls. Sept.

Description based on a solitary specimen "Zojipal (3450 m) *B.M. Sharma* 14657" collected from a stony place along a nallah edge.

13. **P. supina** Linn., Sp. Pl. 497, 1753; Hk.f., l.c. 359.

Semiperennial, hairy herb, with several, branched, 15-40 cm long, prostrate stems radiating from the root. Leaves 1-7 cm long, 5-9-foliolate, the upper only 3-foliolate; leaflets oblong or obovate, up to 1.8×1 cm from a cuneate base, serrate, crenate or incised. Flowers 0.6-1 cm diam., on 0.4-1 cm long, solitary, axillary pedicels; calyx lobes and bracteoles densely silky, ovate, acute; petals oblong, shorter than the calyx. Achenes minute, glabrous, ridged.

Occ. : Common field weed in Aru and villages about Pehalgam; also in waste ground along roadsides.

Fls. & Frts. July-Aug.

Aru (2375 m) *B.M. Sharma* 14668; Mundlan roadside (2350 m) *B.M. Sharma* 14658; Aru (2400 m) *P.S. Jamwal* 14098.

14. **P. biflora** Willd. ex Schlecht., Gen. Naturf. Fr. Berl. Mag. 7 : 297, 1813; Stewart, l.c. 354. *P. inglisii* Royle, Ill. Bot. Himal. 207, t. 41, 1834; Blatter, Beaut. Fl. Kash. 108, pl. 21, f. 8. 1927. *P. fruticosa* Linn. var. *inglisii* (Royle) Hk.f., l.c. 348.

This dwarf alpine cushion-forming plant is included on the authority of Stewart (l.c.) who cites it from Pehalgam, Mt. Kolahoi, Zojipal and Harnag; leaflets 3-7, linear-lanceolate, *c.* 1 cm long: flowers usually solitary, terminal.

11. ALCHEMILLA Linn.

Stems densely clothed with patent hairs; calyx tube, lobes and bracteoles pilose **1. A. trollii**

Stems scattered with a few, patent or reflexed hairs; calyx tube, lobes and bracteoles glabrous **2. A. ypsilotoma**

1. **A. trollii** Rothm., Bull. bot. Gdns, Kew 1938: 272, 1938; Stewart, l.c. 346. **Fig. 69**

Perennial herb, flowering stems erect or ascending, 10-30 cm high from a creeping rootstock. Radical leaves orbiculate to subreniform, pilose, 2-4.5 cm along broadest width, shallowly 7-, rarely 9-lobed, with basal sinus open to nearly closed, on 2-12 cm long petioles clothed like the stem; teeth hairy-margined, tufted at tips; cauline leaves small, short petioled to nearly stalkless. Flowers greenish yellow, 3 mm diam., on very short, glabrous pedicels in dense, contracted or sparingly branched clusters; sepals ovate; bracteoles ovate-lanceolate. Achenes 1-2, glabrous.

Occ. : Common among grasses in Fodder Research Station in Aru.

Fls. June-Aug.

Aru (2450 m) *B.M. Sharma* 14669; Aru (2425 m) *P.S. Jamwal* 16303.

2. **A. ypsilotoma** Rothm., l.c. 273; Stewart, l.c. 346.

Much like the preceding but stems arcuate-ascending; leaves thinly pilose to near glabrous except the nerves beneath, reniform with more open basal sinus and divaricate, more deeply toothed lobes; teeth mucronulate. Petioles flexuous, hairy to almost glabrous; corymbs laxer.

Occ. : Abundant and often gregarious in wet places on alpine and sub-alpine pasture lands.

Fls. June-Aug.

Between Satlanjan and Kolahoi (3300 m) *B.M. Sharma* 14137; Pehalgam (7200′) *H.D. Thapliyal* DD. no. 26515/127881 (under *A. vulgaris* Linn.); Liddarwat (2800 m) *P.S. Jamwal* 16308.

12. AGRIMONIA Linn.

A. eupatoria Linn., Sp. Pl. 448, 1753; Hk. f., l.c. 361; Stewart, l.c. 346. *A. pilosa* Ledeb. var. *nepalensis* (D. Don) Nakai, Tokyo Bot. Mag. 47 : 247, 1933; For. Fl. Sgr. 130 **Fig. 70**

Stout, glandular, 40-90 cm tall, perennial herb from creeping, woody rootstock. Leaves pinnately divided into 5-9 major leaflets separated by much smaller, very unequal, paired or scattered intervening leaflets; main leaflets elliptic ovate to, -obovate or lanceolate, 1.5-5 × 1-2.2 cm, coarsely serrate, hairy on either face, glandular beneath; stipules foliaceous, toothed. Flowers yellow, 6-8 mm diam., in axillary and terminal spike-like racemes; calyx tube turbinate, grooved, villous and glandular, limb 5-lobed.

Occ. : Common in shade of bushes among forest clearings and on humus rich slopes along waysides.

Fls. June-Aug.; *Frts*. Aug.- Sept.

Liddarwat (2750 m) *B.M. Sharma* 14375; Aru (2400 m) *B. M. Sharma* 14325; Pehalgam (7200′) *K.M. Vaid* DD. no. 117511; Liddarwat (2750 m) *P.S. Jamwal* 16316.

13. ROSA Linn.

1. Habit upright or spreading, not climbing; styles free :
   2. Prickles straight, sometimes only slightly curved :
      3. Leaves 5-20 cm long; leaflets 7-11 **1. R. macrophylla**
      3. Leaves 2-4.5 cm long; leaflets (in our specimen) 5-7 **2. R. nanothamnus**
   2. Prickles stout and strongly hooked **3. R. corymbifera**
1. Habit climbing; styles united in a column much exserted beyond the mouth of the calyx tube **4. R. brunonii**

1. **R. macrophylla** Lindl., Monogr, Ros. 35, t. 6, 1820; Hk.f., l.c. 366.

Deciduous, 1.5-2.5 m high shrub with purplish brown, erect or arching, nearly smooth stems or armd with pale-brown, broad-based prickles. Leaves odd-pinnate; petiole tomentose, often prickly; stipules broad, with glandular margins. Leaflets elliptic-ovate, serrate, 1.2-3.5 cm long, glabrous above, usually pubescent beneath. Flowers bright pinkish red, 3.5-6 cm diam., solitary or in few-flowered subterminal corymbs. Calyx lobes long-caudate, often dilated at tips. Fruit oblong-ovoid, up to 5 cm long, red.

Occ. : Common as forest undergrowth and on open slopes.

Fls. May-July; *Frts*. Aug.-Oct.

Aru (2400 m) *B.M. Sharma* 19335; Pehalgam, Lidder vy., *Coventry* DD. nos 27449 & 27451; Pehalgam (7200′) *K.M. Vaid* DD. no 117528; Chandanwari (2950 m) *P.S. Jamwal* 16396.

Var. **minor** Lindl., l.c. 35; Hk.f., l c. 366.

This has smaller, generally obtuse leaflets, more or less pubescent petioles and white or pinkish white, 3-4 cm wide flowers.

Common on rocky hillsides and on humus covered precipitous slopes.

Fls. June-July.

Sawarian nalla base (2900 m) *B.M. Sharma* 14164; Nangneri (2850 m) *P.S. Jamwal* 16310.

2. **R. nanothamnus** Boulenger, Bull. Jard. bot. Brux. 13, fasc. 3 : 206, 1935; Stewart, l.c. 366. *R. webbiana* Wall. var. *microphylla* Crep., J. Linn. Soc., 19 : 161.

Habit of preceding, species; prickles generally more numerous, yellow. Leaves. odd-pinnate; petioles prickly, glandular bristly, glabrous or more or less pubescent; stipules small, with glandular margins. Leaflets obovate or suborbicular, 0.5-1 cm long, coarsely sharp-toothed, glabrous except the petiolules. Flowers pink or pinkish-white, 3.5-5 cm diam., solitary; calyx-lobes caudate-acuminate, dilated at tips, like the tube and peducle bristly-glandular. Fruit red, globoid, 1.2 cm long.

Occ. : In open stony ground and steep rocky slopes.

Fls. June

Pehalgam-Chandanwari road (2250 m) *B.M. Sharma* 14673; Pehalgam (7500′) *H.D. Thapliyal* 26500.

3. **R. corymbifera** Borkh., Vers. Forst. bot. Beschr. Holzart. 319, 1790; Klastersky. in Fl. Eur. 2 : 30, 1968; Stewart, l.c. 365. *R. dumetorum* Thuill. Fl. Par. (ed. 2) 250, 1799. *R. jacquemontii* Crep. mss.

Spreading shrub with scattered prickles; leaflets 5-7, serrate, broadly elliptic to oval-orbicular, hairy or at least such on veins beneath. Flowers white or pale pink, corymbose; sepals deciduous. Fruit ovoid, 1.25-1.8 cm long, orange red. Lidder valley below Tanin (8-9000′) *Duthie* 13437; fruiting 2-8-1893. Cited from Pehalgam also by Stewart, l.c.

4. **R. brunonii** Lindl. Ros. Monogr. Pl. 14, 1820; Rehder, Man. Cult. Tr. & Shrubs, 446, 1940; Stewart, l.c. 365. *R. moschata* auct., non J. Herrm., Diss. de Rosa, 15, 1762; Hk.f., l.c. 367 pro parte.

This tall, white-flowered, temperate zone shrub is conspicuous from its climbing habit and is easily hold apart owing to the styles being connate into a column protruding much beyond the mouth of the calyx tube.

14. MALUS Mill.

M. pumila Mill., Gard. Dict. ed. 8, no. 3, 1768; Stewart, l.c. 352. *Pyrus malus* Linn., Sp. Pl. 479, 1753; Hk.f., l.c. 373.

Deciduous, 5-7 m tall tree with blackish bark and tomentose young twigs, petioles and inflorescences. Leaves ovate, elliptic or obovate, 3-8.5 × 1.5-3.5 cm, acuminate, obtusely serrate or crenate, above glabrous but for midrib, beneath more or less permanently tomentose. Flowers pinkish white, on short, lateral, leafy shoots; pedicels 1.5-3 cm long. Fruits densely fascicled, pale green, globose,

3.5-4 cm diam., tipped with persistent, tomentose calyx.

Occ.: A few trees on foot of slope abutting on abandoned fields about Aru.

Fls. April ; *Frts.* July.-Sept.

Aru (2375 m) *B.M. Sharma* 14670; Aru 2375 m) *P.S. Jamwal* 16304.

Wild apple has an insipid sourish taste and is occasionally eaten by children.

15. **SORBUS** Linn.

Leaves simple	**1. S. lanata**
Leaves pinnate	**2. S. cashmiriana**

1. **S. lanata** (D. Don) S. Sch., Uebers. Arb. Schles. Ges. Vaterl Kult. 1847; 292, 1848; Stewart, l.c. 371. *Pyrus lanata* D. Don, Prodr. Fl. Nep. 237. 1825; Hk.f., l.c. 375. **Fig. 71**

Moderate, deciduous tree with rough, dark-brown bark and white-floccose young twigs and inflorescence. Leaves ovate or elliptic, lobulate, 7-18×4-10 cm, prominently nerved, above ultimately glabrous but for midrib, beneath white-floccose; lobules serrate. Flowers white, scented, 1.2-1.5 cm diam., in woolly corymbs; calyx lobes woolly within. Fruit pyriform, 1.2-2.5 cm long, yellowish red, tipped with calyx lobes.

Occ. : Locally common in Aru and Phraslan.

Fls. April-May; *Frts.* June-Aug.

Aru (2400 m) *P.S. Jamwal* 14344; Aru (2400 m) *B.M. Sharma* 14678 & 14679.

The fruits are stated to be sweet and edible. But they become nearly half rotten by the time they are ripe.

2. **S. cashmiriana** Hedl., Monog. d. Gatt. Sorbus 35, 1901; Stewart, l.c. 370. *Pyrus foliolosa* Wall., Pl. As. Rar. 2; 81, t. 189, 1831 var. *cashmiriana* (Hedl.) Parker, For. Fl. 229, 1918. **Fig. 72**

Deciduous shrub or treelet with smooth, white-lenticellate twigs. Leaves 10-16 cm long, imparipinnately 13-19-foliolate; petiole tomentose beneath with rusty and grey hairs; leaflets oblong-lanceolate, 1-4×0.5-1.5 cm, serrate, glabrous and green above, beneath pale and pubescent or tomentose on the midribs. Flowers white or pink, 1.2-1.5 cm diam., in dense, compound corymbs; inflorescence branches scattered with loose white hairs to nearly glabrous. Fruit globose, red or covered with blue bloom, 0.8-1 cm diam.

Occ. : Rare Associated with *Rhodendron campanulatum* shrubs on Dudhsar slope; also a solitary specimen among junipers on the mount opposite.

Fls. May-June; *Frts.* Aug.-Sept.

Dudhsar (3900 m) *B.M. Sharma* 19280; Mt. opposite Dudhsar (4200 m) *P.S. Jamwal* 16349.

16. **CRATAEGUS** Linn.

C. **songarica** C. Koch in Verh. Verz. Beford Gartenb. Preuss. n.s. 1 : 287, 1853; Stewart, l.c. 351. *C. oxyacantha* Hk.f., l.c. 383 (non Linn.). *C. monogyna* auct.; For. Fl. Sgr. 131. **Fig. 73**

Deciduous shrub or small tree up to *c.* 3.5 m high, with greyish, often armed twigs; young shoots and petioles with scattered, long hairs. Leaves 2.5-7×3-8 cm from a cuneate base, irregularly pinnatifid into 3-7 lobes sharply incise-dentate at tips, beneath long-haired on nerves especially towards base, above thinly so or glabrescent; stipules foliaceous, the upper entire and falcate or semicircular. Flowers white, 1.2-1.3 cm diam., in terminal corymbose cymes; styles mostly 2-3. Fruit red, ovoid, *c.* 1 cm long, glabrous, crowned by reflexed, obtuse, subacute or obtusely acuminate calyx lobes.

Occ. : Common in and around Pehalgam, in both E. & W. Liddar valleys; also a few specimens on scrubby slope near Kolahoi-n-kot.

Fls. April-May; *Frts.* June-Aug.

Hoksar (3450 m) *B.M. Sharma* 14385; Pehalgam (7200′) *H.D. Thapliyal* 26509; Pehalgam (2200 m *P.S. Jamwal* 16305.

Liquid extract of fruits is reported to be a useful remedy for organic and functional heart diseases.

17. COTONEASTER Medic.

1. Habit erect; leaves deciduous, 1.2 cm and more in length; flowers in 3-many-flowered cymes:
   2. Petals pink, erect; ripe fruits red :
      3. Leaves glabrous or hairy beneath, not tomentose; calyx-tube and,-lobes glabrous without **1. C. rosea**
      3, Leaves tomentose beneath; calyx-tube and,-lobes tomentose without **2. C. obovata**
   2. Petals white, patent; ripe fruits black **3. C. affinis**
1. Habit procumbent; leaves evergreen, up to 8 mm in length; flowers mostly solitary **4. C, microphylla**

1. C. **rosea** Edgew., Trans. Linn. Soc. 20 : 46, 1851; Rehder, Man. Cult. Tr. & Shrubs, 354; Stewart, l.c. 350. *C. multiflora* Hk.f., l.c. 386 pro parte (non Bunge). **Fig. 74**

Slender shrub, 2.5-3 m high; twigs brown, arcuate, with thin papery bark, scantily hairy only when young. Leaves elliptic, ovate or obovate, above dark green and glabrous with slightly hairy midrib, pale beneath, obtuse, acute or rounded at tip, mucronate. Flowers 6-7 mm diam., in 3-10 flowered, glabrous cymes; calyx woolly within. Fruit ovoid, 7-8 mm long.

Occ. : Open, scrubby, humus rich slopes about Aru and Pehalgam.

Fls. May.-June; *Frts.* Aug.

Aru (2400 m) *B.M. Sharma* 1443; Pehalgam (2160 m) *B.M. Sharma* 14671; Aru (2400 m) *P. S. Jamwal* 16336.

2. C. **obovata** Wall. ex Dunn in Kew Bull. 1921: 119, 1921; Parker, For. Fl. 232, 1918; Stewart, l.c. 350.

Shrub, with densely hairy young twigs; leaves usually rounded at both ends, up to 11×7.5 cm, snow-white on the undersurface. Cited by Stewart (l.c.) from Pehalgam. To distinguish *C. obovata* from other related species Parker (l.c.) has recorded the following interesting field observation.

"Usually found as a small shrub 2-3′ high with a single vigorous shoot and one or more dead or dying shoots

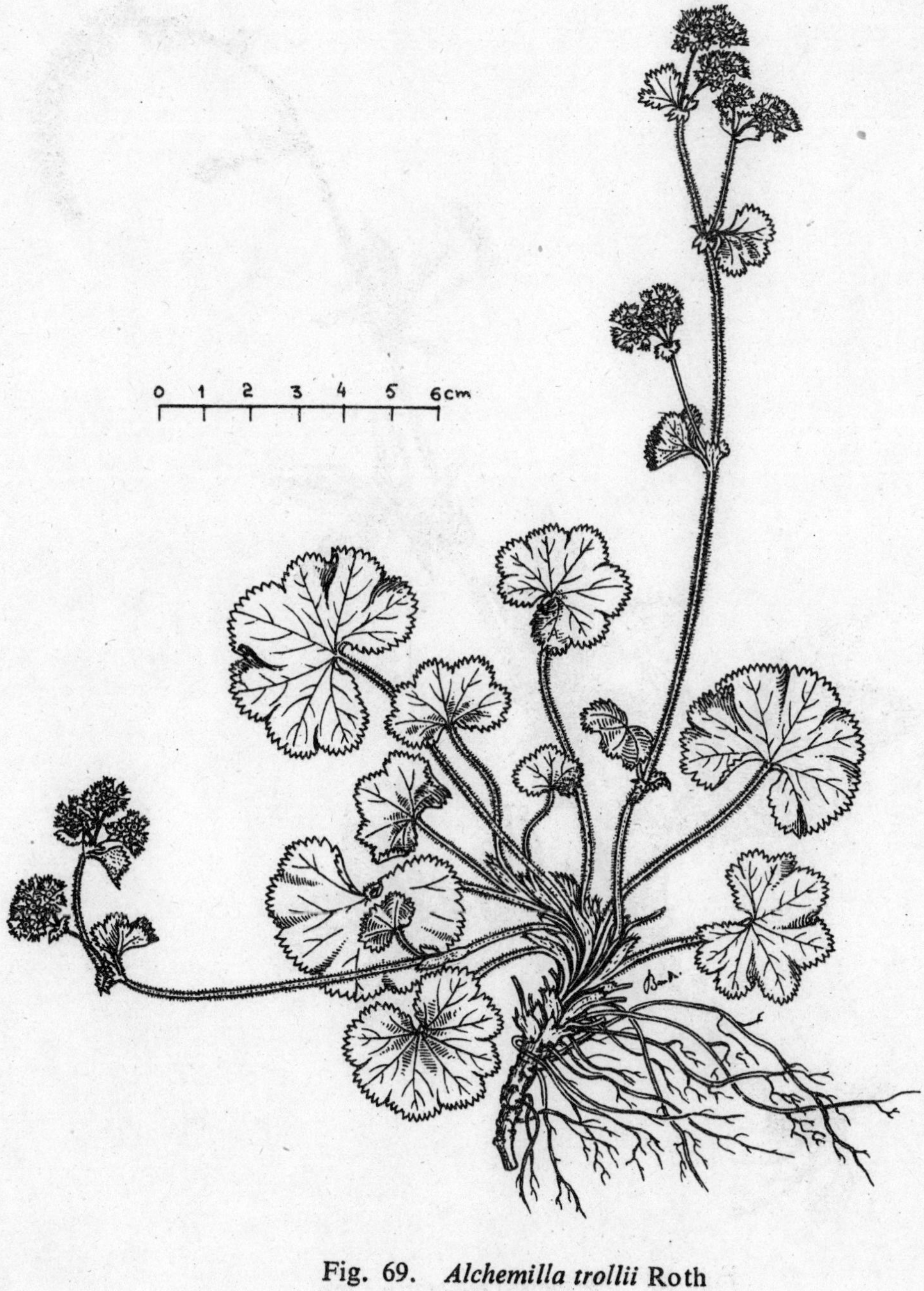

Fig. 69. *Alchemilla trollii* Roth

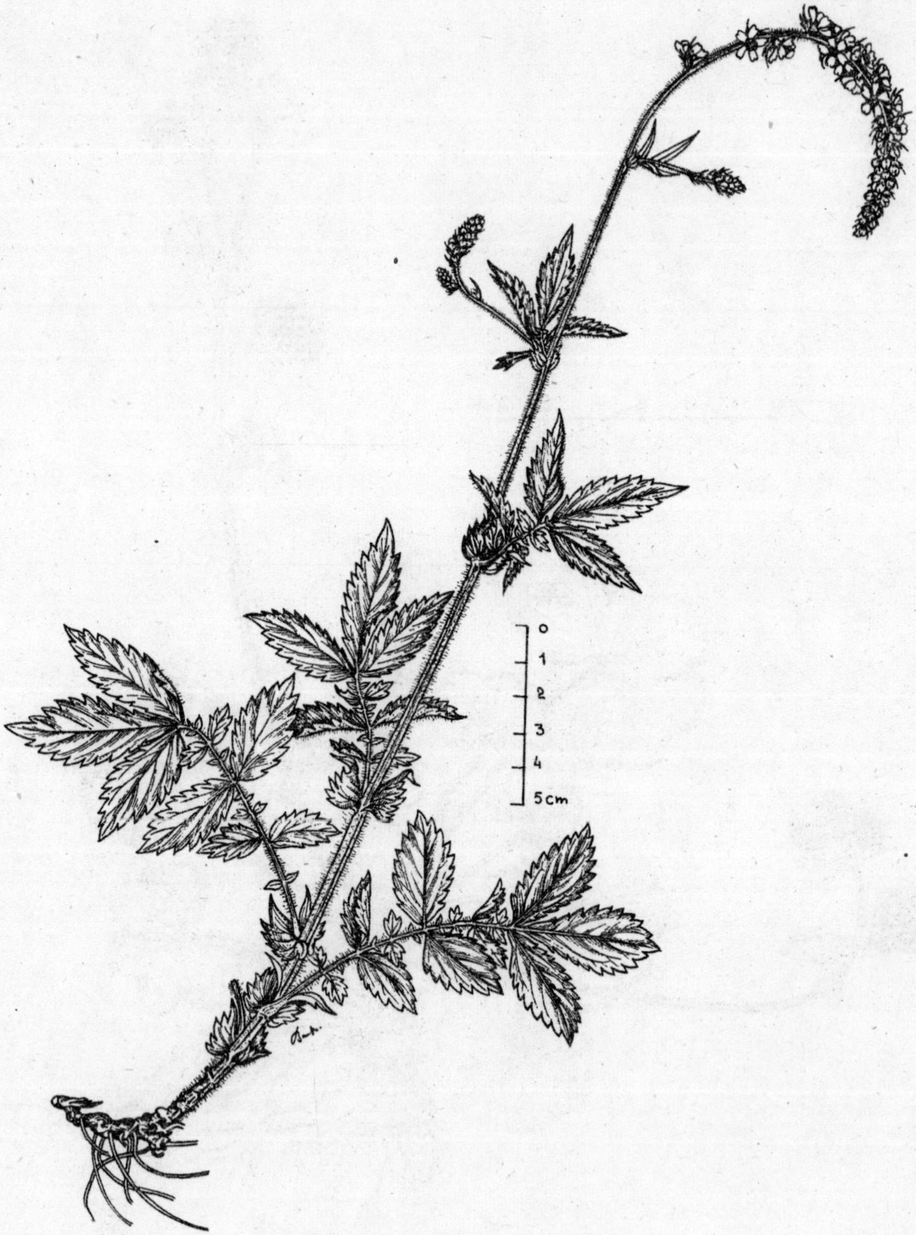

Fig. 70. *Agrimonia eupatoria* Linn.

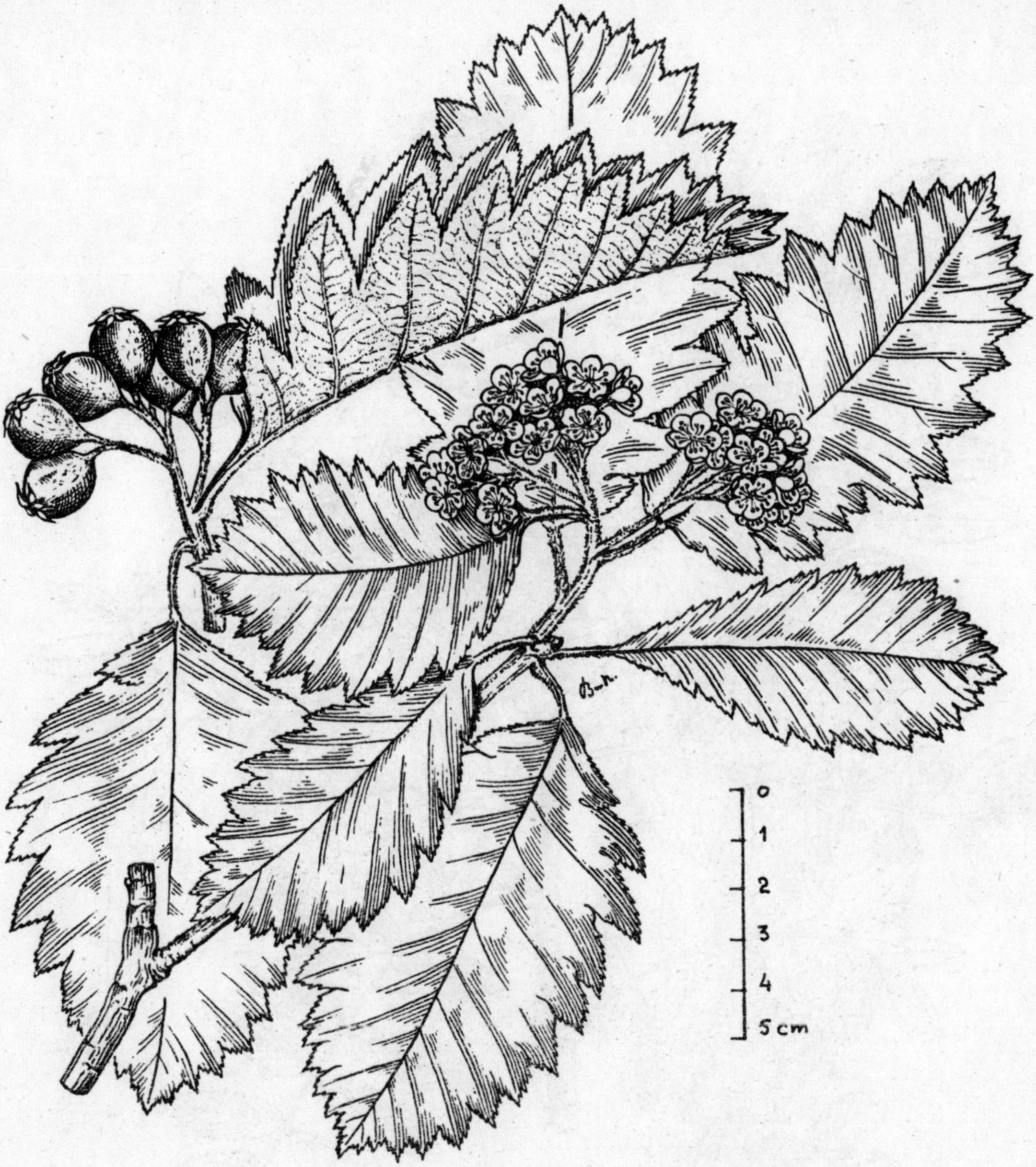

Fig. 71. *Sorbus lanata* S. Sch.

Fig. 72. *Sorbus cashmiriana* Hedl.

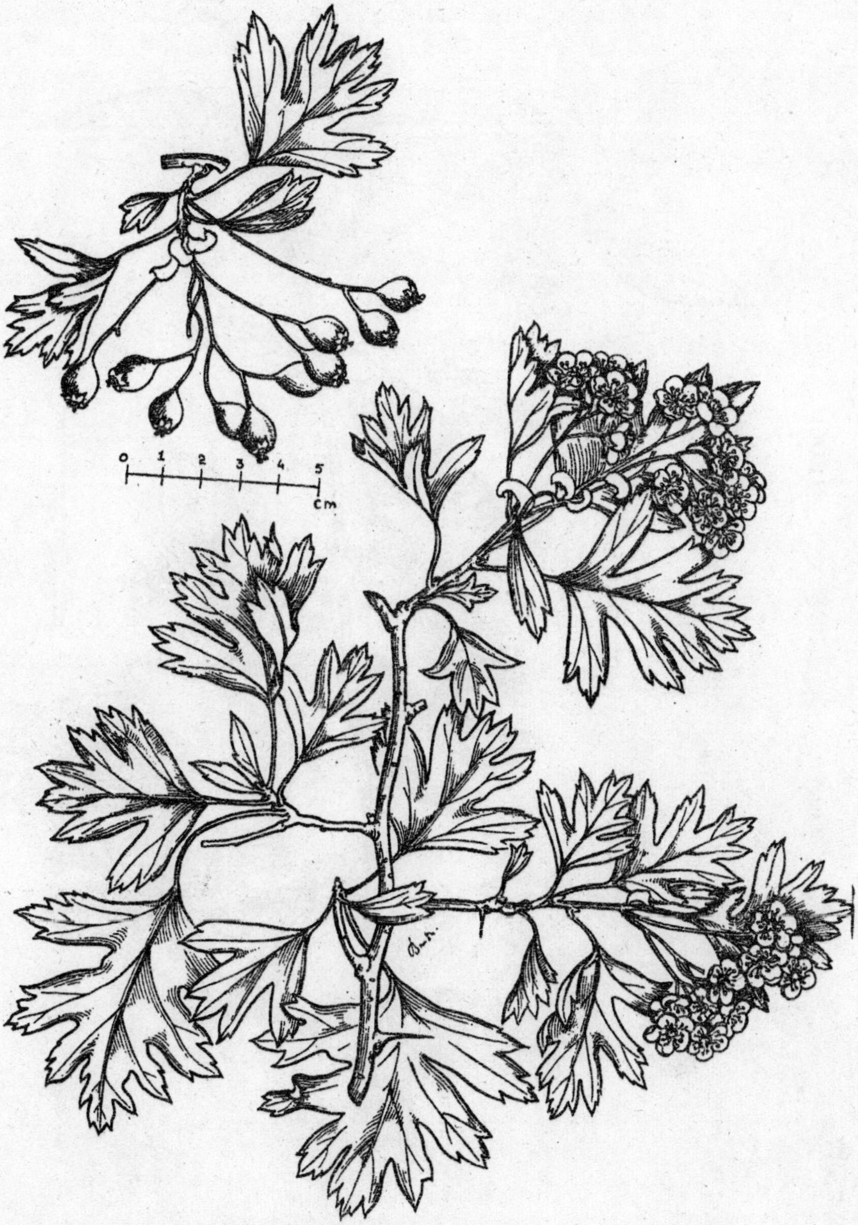

Fig. 73. *Crataegus songarica* C. Koch

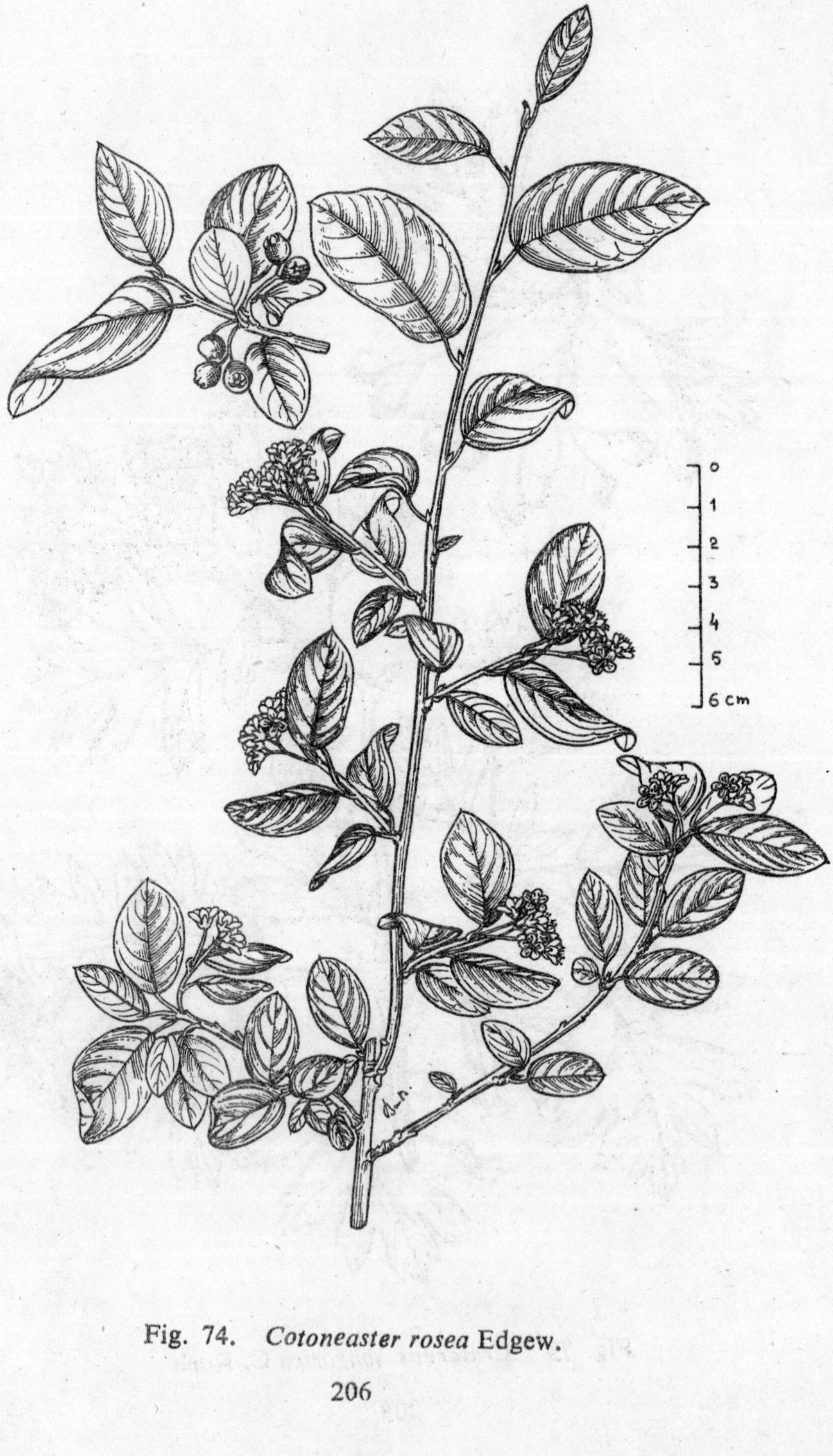

Fig. 74. *Cotoneaster rosea* Edgew.

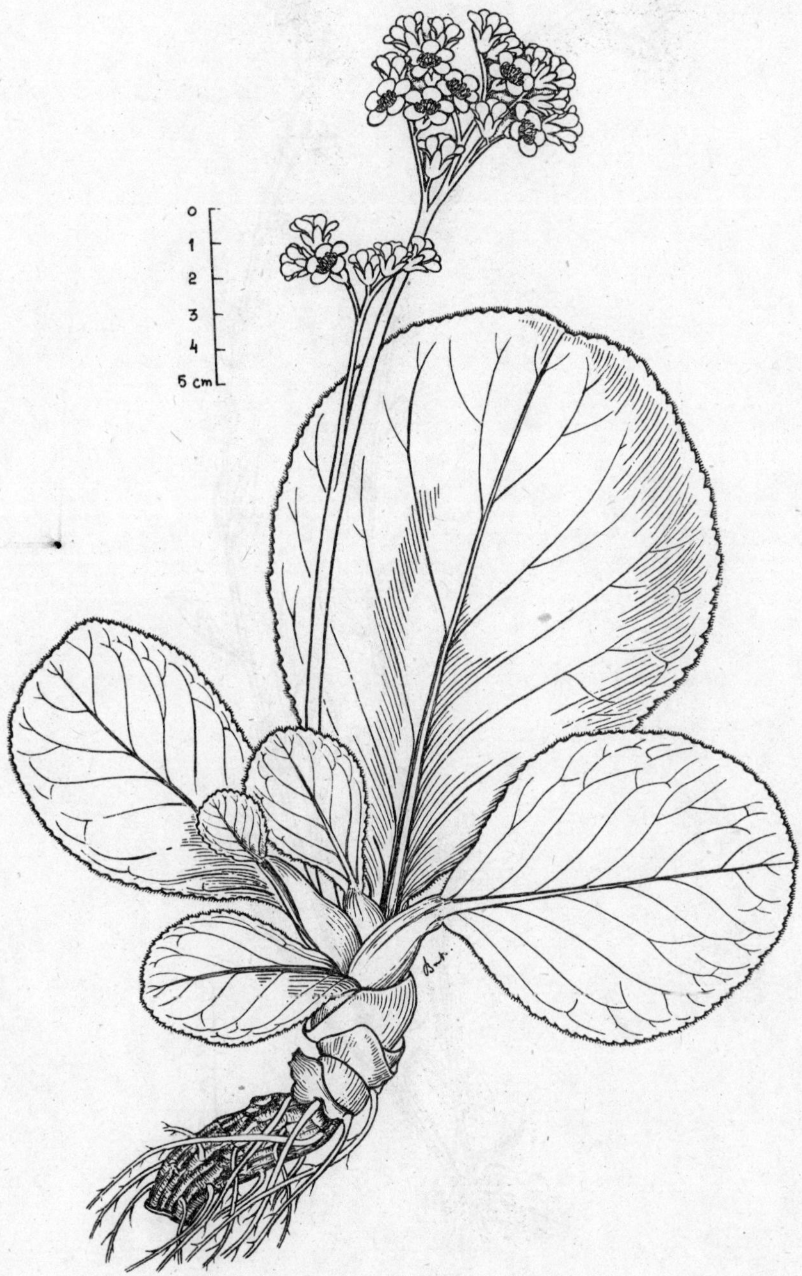

Fig. 75. *Bergenia stracheyi* Engl.

Fig. 76. *Saxifraga moorcroftiana* Wall. ex Sternb.

separate from the ground level. The first shoot gets pressed down by snow and rapidly loses vigor, being replaced by a succession of others which share the same fate, the plant thus appears to be single-stemmed and does not form a clump like *C. bacillaris*."

3. C. **affinis** Lindl., Trans. Linn. Soc. 13 : 101, 1821 var. **bacillaris** (Lindl.) Schneid., Ill. Handb. Laubh. 1 : 757, 1906; Fl. Nainital. 124; Stewart, l.c. 347. *C. bacillaris* Wall. ex Lindl., Bot. Reg. Pl. 15, 1229, 1829, var. *affinis* Hk.f., l.c. 385.

Shrub or small tree, with smooth, reddish brown branches; petioles, young shoots and inflorescence grey-hairy or, -tomentose. ovate, obovate or ovate-lanceolate, 2.5-7×1.8-3.5 cm, glabrous above, beneath pale green and generally pubescent; apex rounded or acute, often mucronate. Flowers 7-8 mm diam., in few-, or many flowered, compound cymes terminal on lateral branchlets; calyx grey pubescent or, -woolly. Fruit 6-8 mm diam., globose.

Occ. : In scrub forests near villages.

Fls. May-June; *Frts.* Sept.-Oct.

Aru (2400 m) *B.M. Sharma* 14672; Aru (2400 m) *P.S. Jamwal*, 14350; Pehalgam (7000') *H.D. Thapliyal* 26508.

The wood is prized for making walking sticks.

4. C. **microphylla** Wall. ex Lindl., Bot. Reg. 13 : t. 1114, 1827 var. **thymifolia** (Baker) Koehne; Stewart. l.c. 349.

Low, intricately branched, rigid shrub; young branches purplish brown, ultimately glabrous. Leaves leathery, obovate-, or oblong-cuneate, 5-8×3-5 mm, with rotund, retuse or mucronate apex, above glabrous or on the depressed midrib ciliate, beneath pale and thinly hairy to almost tomentose; margins recurved. Flowers white, *c.* 8 mm diam., Fruit globoid, 5-6 mm diam., scarlet.

Occ. : Forming dense, stiff, interlaced patches on rocks in forest blanks or trailing on open grassy slopes.

Fls. May-June; *Frts.* June-Aug.

Aru Ghasan (2550 m) *B.M. Sharma* 14372; above Liddarwat (2980 m), *B.M. Sharma* 14693; Liddarwat-Tarsar route (2980 m) *P.S. Jamwal* 16309.

XXVII. SAXIFRAGACEAE

Plants rhizomatous; rhizomes stout, fleshy; leaves punctulate with immersed glands **1. Bergenia**
Plants non rhizomatous or with thin or very short rhizomes; leaves glandular, if glandular then the glands superficial or on hairs **2. Saxifraga**

1. BERGENIA Moench

Leaf base cuneate; petiole sheathing nearly throughout its length; calyx and pedicels pubescent **1. B. stracheyi**
Leaf base cordate or rounded; petiole sheathing at base alone; calyx and pedicels glabrous **2. B. ciliata**

1. B. **stracheyi** (Hk.f. & Th.) Eng. in Bot. Zeitg. 26 : 842, 1868; Blatter, l.c. 126, pl. 23, f. 8; Yeo, Kew Bull. 20 (1) : 128, 1966; Stewart, l.c. 341. *Saxifraga stracheyi* Hk.f. & Th., J. Linn. Soc. Bot. 2 : 61, 1858; Cl. in Hk.f., FBI 2 : 398, 1878. **Fig. 75**

Perennial herb; rhizome covered with dead, blackish remains of leaf bases. Leaves rosulate, fleshy, erect, obovate or oblong-obovate, 6-17×4-12.5 cm, glabrous, crenate or crenate-dentate, ciliate, with rounded or, in young leaves truncate apex. Flowers white or pink-tinged, campanulate, 1.5-2 cm diam., in a rather compact, spreading panicle on a leafless, usually 12-30 cm long flowering stem; inflorescence branches glandular-hairy. Calyx and hypanthium together 0.8-1.2 cm long; teeth oblong, with rounded apex. Petals obovate-spathulate, *c.* 1× 0.5 cm. Capsule (incl. styles) 1-1.3×0.5-0.6 cm.

Occ. : Gregarious and forming exclusive patches on steep rocky banks and cliffs at alpine altitudes. The leaves turn beautiful red during autumn.

Fls. June-Aug.

Rewil (3200 m) *B.M. Sharma* 14157; Chandanwari (10,000′) *H. D. Thapliyal* 26539; Kolahoi (3450 m) *P.S. Jamwal* 16343.

2. **B. ciliata** (Haw.) Sternb., Revis. Saxifrag Suppl. 2 : 2, 1831 forma *ligulata* Yeo, l.c. 134. *Saxifraga ligulata* Wall. in Asiat. Res. 13 : 398, 1820; Cl. in Hk.f., l.c. 398; Coventry, Wild Fl. Kash. 1 : 39. t. 40, 1923. *B. ligulata* sensu Blatter, l.c. 125. *B. himalaica* Boriss., Acta Inst. bot. Acad, Sci. URSS. ser. 5, 4 : 322, 1956.

Habit of *B. stracheyi* Leaves spreading suborbicular to orbicular or broad obovate or, -ovate, usually 3-10×2.5-8.5 cm at flowering, enlarging subsequently in autumn, glabrous, with entire or shallowly sinuate-crenate or, -denticulate, ciliate margins, rounded at apex; petioles naked above, their sheath-apices free and ciliate. Flowers rose-pink, 1.5-2 cm diam., corymbose or in a 1-sided raceme subtended by a reddish or greenish scape often bearing an ovate, leafy bract; inflorescence branches glabrous. Calyx and hypanthium together 0.6-1.2 cm long; teeth ovate, with rounded apex. Petals orbicular, clawed, 1.1-1.4×0.7-0.9 cm.

Occ. : Common on rocky slopes about nallahs; often gregarious in patches.

Fls. April-May.

Aru (2400 m) *B.M. Sharma* 16691; Mundlan (2300 m) *B.M. Sharma* 16693; Mundlan-Aru route (2350 m) *P.S. Jamwal* 16331 & 16332.

2. SAXIFRAGA Linn.

1. **Plants emitting from base leafless, filiform stolons terminating in a leafy rosette** — **1. S. flagellaris**
1. **Plants non stoloniferous :**
    2. **Upper leaves tipped with a solitary, white, lime-encrusted pit** — **2. S. pulvinaria**

2. Leaves not pitted :
    3. Bulbils present in the axils of basal or cauline leaves :
        4. Bulbils confined to the axils of basal leaves — **3. S. sibirica**
        4. Bulbils distributed in the axils of cauline leaves — **4. S. cernua**
    3. Bulbils wanting :
        5. Habit caespitose by virtue of dense crowding of leafy shoots :
            6. Petals equalling the length of sepals or wanting — **5. S. microphylla**
            6. Petals at least one and a half times as long as the sepals :
                7. Flowers yellow; leaves entire — **6. S. jacquemontiana**
                7. Flowers white; basal leaves often apically 3-lobed — **7. S. androsacea**
        5. Habit not caespitose :
            8. Sepals free; petals 0.6-1.1 cm long, yellow — **8. S. moorcroftiana**
            8. Sepals basally united; petals up to 5 mm long, not yellow :
                9. Flowers 1 (-2-3), reddish — **9. S. pseudo-pallida**
                9. Flowers many, white — **10. S. pallida**

1. **S. flagellaris** Willd. ex Sternb. Revis. Saxifrag. 25, 1810 ssp. *mucronulata* (Royle) Engl. & Irm. in Pflanzenr. 4 : 117. 1, 163, 1916; Hulten, Svensk Bot. Tidskr. 58 (1) : 98, 1964; Stewart, l.c. 342. *S. mucronulata* Royle, Ill. Bot. Himal. 227, t. 50, f. 2, 1835. *S. flagellaris* var. *mucronulata* Cl. in Hk.f., l.c. 397.

Erect, 5-12 cm tall, simple, perennial herb; stem leafy, glandular-pilose; stolons red, glandular all along. Leaves sessile, stiffly ciliate-margined, spinulose-mucronulate, the basal ones rosulate, ovate-lanceolate to obovate, 0.5-1 × 0.15-0.4 cm, the cauline alternate, lanceolate or oblong lanceolate; cilia of basal leaves long and pointed, those of the cauline mostly gland-tipped. Flowers yellow, 0.8-1 cm diam., solitary or few in a corymb; sepals oblong, acute, basally adnate, like the pedicels glandular pilose; petals oblong obovate, 5 × *c.* 2.75 mm, like the sepals 5-nerved.

Occ. : In damp, stony localities at alpine altitudes.

Fls. July-Aug.

Tarsar (4050 m) *B.M. Sharma* 19383; Mt. Kolahoi (4500 m) *B.M. Sharma* 16616; Lidder vy. above Kainmul (12000′) *Duthie* 13119; Sangam vy. above Lidderwat (13000′) *Duthie* 14151; Mt. opposite Dudhsar (4200 m) *P.S. Jamwal* 16346.

ssp. **stenophylla** (Royle) Hulten, l.c. 194. *S. stenophylla* Royle, l.c., f. 1.

Cited by Stewart, l.c., from Kolahoi. Stolons glandular only near the base; leaf cilia neither long nor pointed; sepals free (Hulten, l.c.)

2. **S. pulvinaria** H. Smith, Bull. Br. Mus. nat. Hist. 2(4) : 105, 1958; Stewart, l.c. 344; S. Ghaz. in Nasir et Ali, l.c. 108 : 25, 1977. *S. imbricata* Royle, l.c., t. 49, f. 1; Cl. in Hk.f., l.c. 394.

Perennial herb forming dense tufts through overcrowding of 1.5-2.5 cm long, erect or ascending, closely leafy shoots of much branched, 3-6 cm long, decumbent stems. Leaves densely imbricated, oblong, 2.5-3.5 × 1-1.25 mm, obtuse, glabrous or on lower margins sparingly ciliate, keeled beneath, the upper perforated at the tip with a white pit. Flowers small, white, solitary, terminal, sessile; sepals ovate, obtuse; petals suborbicular from an unguiculate base, about twice as long as the sepals.

Occ. : Common on alpine slopes about Amarnath cave.

Fls. July.

Near Amarnath (3900 m) *B.M. Sharma* 16677; Sangam vy. (14000′) *Duthie* 13300; near Amarnath (3900 m) *P.S. Jamwal* 16392.

3. S. **sibirica** Linn., Sp. Pl. ed. 2, 577. 1762; Cl. in Hk.f., l.c. 390.

Glandular-villous herb, perennating through ovoid bulbils in basal leaf-axils; stems weak, leafy, erect or ascending, 7-20 cm high, often flexuous or bent. Leaves sparingly pubescent to almost glabrous, the basal ones reniform, on 1.5-4 cm long, glandular-villous petioles, with margin cut into 5-10, broadly triangular-ovate, acute or obtuse lobes; cauline leaves short-stalked, with usually acute, oblong or ovate lobes, the uppermost nearly sessile, 3-lobed or oblong and entire. Flowers white, solitary or 2-6 to each stem; pedicels glandular-villous. Sepals oblong, obtuse or acute, 3-4 mm long, glandular-pubescent; petals obovate-cuneate, up to 1.3×0.5 cm, 3-7-nerved, slightly notched at apex.

Occ. : Common alpine of shady nooks and open, moist places along streams.

Fls. July-Aug.

Seikwas (3400 m) *B.M. Sharma* 14015; near Sheshnag (*c.* 3600 m) *P.S. Jamwal* 16320.

4. S. **cernua** Linn., Sp. Pl. 403, 1753; Cl. in Hk.f., l.c.

Stem simple, bearing a solitary, terminal, white flower, or the latter as well rudimentary, rarely upper part of stem with a few branches each terminating in a normal or rudimentary flower. Leaves all with axillary reddish bulbils; basal leaves palmately lobed, long stalked, the upper ones short stalked to almost sessile and few-lobed, the uppermost entire.

This sp. and the next included after Stewart (l.c. 342 & 343) who has cited these from above Pehalgam and from Mt. Kolahoi respectively.

5. S. **microphylla** Royle ex Hk.f. & Th., J. Linn. Soc. bot. 2 : 72, 1858; Cl. in Hk.f., l.c. 395.

Caespitose herb. Flowers small, solitary on puberulous pedicels; sepals adnate below; petals 0 or as long as the sepals.

6. S. **jacquemontiana** Decne. in Jacq. Voy. Bot. Inde. 68, t. 78, f. 2, 1844; Cl. in Hk.f., l.c. 395.

Low, caespitose, perennial herb; branches densely clothed with close-set, sessile leaves. Leaves orbicular or broad-elliptic, 3-6 mm long, the uppermost obovate-oblong and forming rosettes at branch tips. Flowers yellow, solitary, terminal, sepals glandular-pubescent, about half the length of 5-6 mm long, broadly elliptic petals; stamens equalling the length of petals.

Fls. Aug.

Sangam vy. above Lidderwat (13-14000′) *Duthie* 13561.

7. S. **androsacea** Linn., Sp. Pl. 339, 1753 var. **tridentata** Gaud. in Allg. Schweiz. Ges. 4 : 67, 1818; Stewart, l.c. 341.

Caespitose. Scapes aphyllous or 1-2-leaved, sparingly glandular, 1-3-flowered, arising from an apparently basal rosette of spathulate to oblanceolate, apically 3-toothed leaves. Flowers white, campanulate; sepals adnate below; petals *c.* 2½ times as long as the sepals.

Cited by Stewart, l.c., from Mt. Kolahoi.

8. S. **moorcroftiana** Wall. ex Sternb., Revis. Saxifrag. Suppl. 2 : 28, 1831; Stewart, l.c. 344. *S. diversifolia* Cl. in Hk.f., l.c. 393 pro parte (non Wall. ex Sternb.) **Fig. 76**

Perennial, 20-45 cm tall herb; stem leafy, glabrous in lower part, glandular-hairy upwards. Basal and lower cauline leaves long-stalked, elliptic to lanceolate, 3-10 (incl. petiole)×0.5-1.5 cm, acute, attenuate, with ciliate petiolar-bases; upper cauline leaves progressively smaller upwards, ovate or oblong, 1-4×0.5-1.8 cm, sessile, at least the few uppermost glandular on margins. Flowers yellow, 1.2-1.8 cm diam., pedicels glandular. Sepals 4-7 mm long, glandular especially on margins, with nerves confluent at the tip; petals *c.* 1½ times the length of the sepals, oblong-obovate or, -ovate. Capsule ovoid, *c.* 7 mm long, tipped with 2, short, divergent styles.

Occ. : In wet open spots at alpine altitudes; often several plants at the same spot.

Fls. July-Sept.

Yechi Nar, Kolahoi (3600 m) *B.M. Sharma* 19323; Yamher Pass (13000′) *Duthie* (13579; Seikwas, Kashmir (12000′) *M.V. Laurie* 16781; Seikwas, (3600 m) *P.S. Jamwal* 16321 & 16322.

9. S. **pseudopallida** Engl. & Irm. var. **bellidifolia** Engl. & Irm. in Engl., Pflanzenr. Saxi. iv. 117 : 53, 1916; S. Ghaz., l.c. 14.

Erect herb with ovate, entire to crenate, petiolate basal leaves and 0-2, sessile, lanceolate cauline leaves; petals as long as the sepals.

Reported from Nafran, Upper Lidder vy. by Stewart, l.c.

10. S. **Pallida** Wall. ex Ser. in DC. Prodr. 4 : 38, 1830; Stewart, l.c. 344. *S. micrantha* Edgew. in Trans. Linn. Soc. 20 : 50, 1846; Cl. in Hk.f., l.c. 394.

Perennial herb, with erect or ascending, 8-20 cm tall, white-pubescent flowering stems from a rather lax, basal rosette of up to 3 cm long, spathulate, ovate, oblong-, or rhomboid-ovate, dentate-crenate, glabrous or hairy leaves from a cuneate, truncate or cordate base; petiole flattened, ciliate towards the base at least. Cauline leaves few, sessile. Flowers many, small, white with purplish calyx, in lax cymes; petals broad-ovate to almost orbicular, 2.5-4×2-3 mm, hardly exceeding the triangular to triangular-ovate, basally united sepals.

Occ. : On wet rocks and among morainal debris at high altitudes.

Fls. June-July; *Frts.* Aug.

Kolahoi (3450 m) *B.M. Sharma* 14202; Lidder vy. above Kainmul (10-11000′) *Duthie* 13081; Badzulkhod nallah Lidder vy. (13000′) *Duthie* 13418; Yamher Pass overlooking Sind vy. (13-14000′) *Duthie* 13577 & 14152; Kolahoi (3450 m) *P.S. Jamwal* 16344.

XXVIII. Parnassiaceae

PARNASSIA Linn.

Staminodes deeply lobed; lobes cylindrish; divaricately disposed **1. P. laxmanii**
Staminodes shallowly or obscurely lobed; lobes neither cylindrish nor divaricately disposed **2. P. nubicola**

P. laxmanii Pallas ex Schultes, Syst. Veg. 6 : 696, 1820; Kitamura, Pl. W. Pak. & Afgh. 3 : 79, 1964. *P. ovata* auct. non Ledeb., Clarke in Hk.f., l.c. 403.

Perennial, 7-15 cm tall, glabrous herb supporting a solitary, sessile leaf on the lower half of slender, angled, 1-flowered scape. Radical leaves long-stalked, ovate or ovate-oblong from cordate base, 0.8-2 cm long; apex rotundate. Flowers white, *c.* 1.5 cm diam., petals spathulate-obovate, 1-1.2×0.4-0.5 cm, obtuse or subacute; staminodes upwards dilated and deeply 3-lobed. Capsule obconic below, subhemispheric upwards.

Occ. : Frequent in wet open places at alpine and subalpine altitudes.

Fls. Aug.-Sept.

Kolahoi (3450 m) *B.M. Sharma* 16614 & 16615; Dudhsar (3750 m) *P.S. Jamwal* 19276.

P. nubicola Wall. ex. Royle ssp. **occidentalis** Schonbeck-Temesy in Rech. f., Fl. Iran. Pt. 20 : 3, 1966; Stewart, l.c. 340.

Similar to the preceding but more robust and larger in all its parts; scapes 15-30 cm tall; radical leaves 2-6 (excl. petiole)×1.2-3 cm, with acute, obtuse or rotundate apex, cauline leaf about or below the middle of the scape. Flowers 1.5-2.5 cm diam., staminodes shallowly 3-lobed.

Occ. : In habitats similar to those of *P. laxmunii.*

Fls. Aug.-Sept.

Near Amarnath (3650 m) *B.M. Sharma* 16678; Amarnath (12,500′) *K.M. Vaid* DD. no. 117991; Sangam vy. (3600 m) *P.S. Jamwal* 16393.

XXIX. Grossulariaceae

RIBES Linn.

Plants prickly. Flowers solitary **1. R. alpestre**
Plants unarmed. Flowers racemed **2. R. himalayense**

1. **R. alpestre** Decne. in Jacq. Voy. Bot. Inde. 64, t. 75, 1844; Stewart, l.c. 337. *R. grossularia* auct. non Linn.; Cl. in Hk.f., l.c. 410, *R. himalensis* Royle, Ill. Bot. Himal. 225, 1835 nom. nud.

Deciduous, erect shrub, armed at each node with usually 3, straight, up to 1.8 cm long prickles; twigs scattered with prickly bristles or inconspicuous gland-tipped hairs. Leaves slightly hairy, orbicular, obtusely 3-5-lobed, 3-7-nerved at base; lobes inciso-crenate; petioles pubescent, sometimes glandular. Flowers 2-sexual, 5-7 mm diam., usually solitary on short, pubescent, bracteate peduncles; calyx bristly -glandular, adnate to ovary. Berry ovoid, 1.2 cm long, glandular-hairy, tipped with withered calyx and corolla.

Fls. June

Chandanwari (9500'. Rare) *H.D. Thapliyal* 26531.

2. **R. himalayense** Decne. in Jacq. Voy.Bot. Inde. 66, t. 77, 1844. *R. rubrum* auct. non Linn.; Cl. in Hk.f., l.c. 411. *R. emodense* Rehder in J. Arnold Arbor. 5 : 161, 1924, nom. illegit. **Fig. 77**

Deciduous shrub, 2-3 m high; twigs glabrous or nearly so, blackish or dark brown. Leaves cordate, 2.5-6.5 cm long and about as wide, 3-5-lobed, above glabrous or scattered with pale hairs, beneath hairy in the nerve axils; lobes inciso-serrate; petiole 1.5-5 cm long, pubescent and scattered with gland-tipped hairs, broadened and fimbriate at base. Racemes many-flowered, pubescent, eglandular; bracts shorter than pedicels. Berries black, 5-7 mm diam.

The calyx tube produced above the fruit and the black colour of the berries distinguish this from *R. glaciale Wall.* which possesses red fruits with calyx tube scarcely produced above. In the specimens cited below, the petioles scattered with gland-tipped hairs are reminiscent of *R. glaciale* but their broadened fimbriate bases is a point in departure.

Occ. : Among the scrubs on rocky slopes about stream banks in Liddarwat and Aru.

Frts. Aug.

Liddarwat (2900 m) *B.M. Sharma* 19327; Aru (2400 m) *P.S. Jamwal* 16345.

XXX. CRASSULACEAE

Rhizome stout; radical leaves persistent, scale like; flowers unisexual **1. Rhodiola**

Rhizome wanting or abortive; radical leaves absent; flowers bisexual **2. Sedum**

1. RHODIOLA Linn.

1. Flowers yellow :
   2. Leaves ovate, obovate or broad oblong; styles outcurved in ripe fruit **1. R. heterodonta**
   2. Leaves linear; styles not outcurved in fruit **2. R. crassipes**
1. Flowers red or pink-red **3. R. tibetica**

1. **R. heterodonta** (Hk.f. & Th.) A. Boriss. in Kom., l.c. 9 : 476, 1939 (in obs.). *Sedum heterodontum* Hk.f. & Th. in J. Linn. Soc. Bot. 2 : 95, 1858; Cl. in Hk.f., l.c. 417. *S. roseum* Scop. var. *heterodontum* Fedtsch. in Frod., The Genus Sedum, pt. 1 : 40, 1930, Stewart, l.c. 335.

Glabrous, glaucous, perennial herb from a thick, fleshy, almost erect rootstock; stems simple, annual, erect, 30-70 cm tall, leafy. Leaves alternate, ovate, obovate or broad oblong, 1.2-3.5 × 0.7-2.3 cm, rather loosely imbricate, dentate, white-margined especially towards the base. Flowers congested in terminal cymes; sepals narrow, oblong. Follicles 4-5, erect, red, 6-7 mm long, with shortly outcurved styles. Seeds oblong, reddish brown.

Occ. : On rocks and slopes in the juniper zone; common.

Fls. June; *Frts*. July.

Saswarian (3300 m) *B.M. Sharma* 14171; Kolahoi (3600 m) *B. M. Sharma* 14278; Hamwas (3250 m) *P.S. Jamwal* 16361.

2. **R. crassipes** (Wall. ex Hk.f. & Th.) A. Boriss. in Kom., l.c. 41. *Sedum crassipes* Wall. ex Hk.f. & Th. in J. Linn. Soc. Bot. 2 : 99, 1858; Stewart, l.c. 333. *S. asiaticum* Cl. in Hk.f., l.c. 419.

Glabrous, erect, perennial herb; stems leafy, unbranched, 12-30 cm tall, several from a thick, woody rootstock. Leaves alternate, linear, 1.2-2.5×0.1-0.2 cm, coarsely and remotely toothed. Flowers yellowish white, 0.8-1 cm diam., crowded in compact, terminal corymbose cymes; petals lanceolate-spathulate, about twice the length of the sepals. Follicles oblong-lanceolate; styles nearly straight.

Occ. : On open slopes, humus laden moist rocks and in wet gravelly ground by the torrents.

Fls June-July..

Kolahoi (4100 m) *B.M. Sharma* 16609; Gagar Pathri, Khalen (3000 m) *B.M. Sharma* 14141; Hamwas (3220 m) *P.S. Jamwal* 16360.

3. **R. tibetica** (Hk.f. & Th.) Fu in Acta Phytotax. 1 : 121, 1965; Ohba in J. Fac. Sci. Univ. Tokyo, 3. 12 : 190, 1978. *Sedum tibeticum* Hk.f. & Th., J. Linn. Soc. Bot. 2 : 96, 1858; Cl. in Hk.f., l.c. 418. Stewart, l.c. 335. *S. tibeticum* Hk.f., & Th. var. *stracheyi* Cl. in Hk.f., l.c. 418.

Glabrous, perennial herb from stout, erect, usually branched, up to 3 cm thick rootstock; stems erect, simple, 7-25×*c.* 0.2 cm, leafy, fastigiate. Leaves alternate, oblong or lanceolate-oblong, 0.4-1.2×0.15-0.2 cm, sessile, imbricated acute or obtuse, entire or upwards denticulate. Flowers often 5-merous, red, 4-5 mm diam., in terminal, bracteate, 2-10-flowered cymes; sepals oblong to triangular, acutish. Follicles 5-7 mm long, erect; seeds oblong-ellipsoid.

Fls. July.-Aug..; *Frts*. Aug.-Sept.

Amarnath (—) *S.L. Nayar*, 3 sheets under col. no. 244. RL. Herb.

***R. quadrifida** Fisch. & C.A. Mey. Syn. *Sedum quadrifidum* Pall. has also been reported from the area. According to *Flora Europaea*, 1 : 363, 1968, however the plants from the Himalaya often cited under this name appear to be very different. The 5-8 cm tall, few-, to many- stemmed, commonly occurring meadow herb with subterete, 5-6 mm long, close set leaves and deep red cymes of small flowers may belong here.

Fls. July.

Tarsar (3900 m) *B.M. Sharma* 14022.

2. SEDUM Linn.

Leaves opposite or uppermost few alternate; petals pink or rose-purple, free. **1. S. ewersii**
Leaves all alternate; petals yellow or yellowish white, connate at base **2. S. oreades**

1. **S. ewersii** Ledeb., Ic. Pl. Fl. Ross. 1 : 14, 1829 et Fl. Alt. 2 : 191, 1830; Cl. in Hk.f., l.c. 421. **Fig. 78**

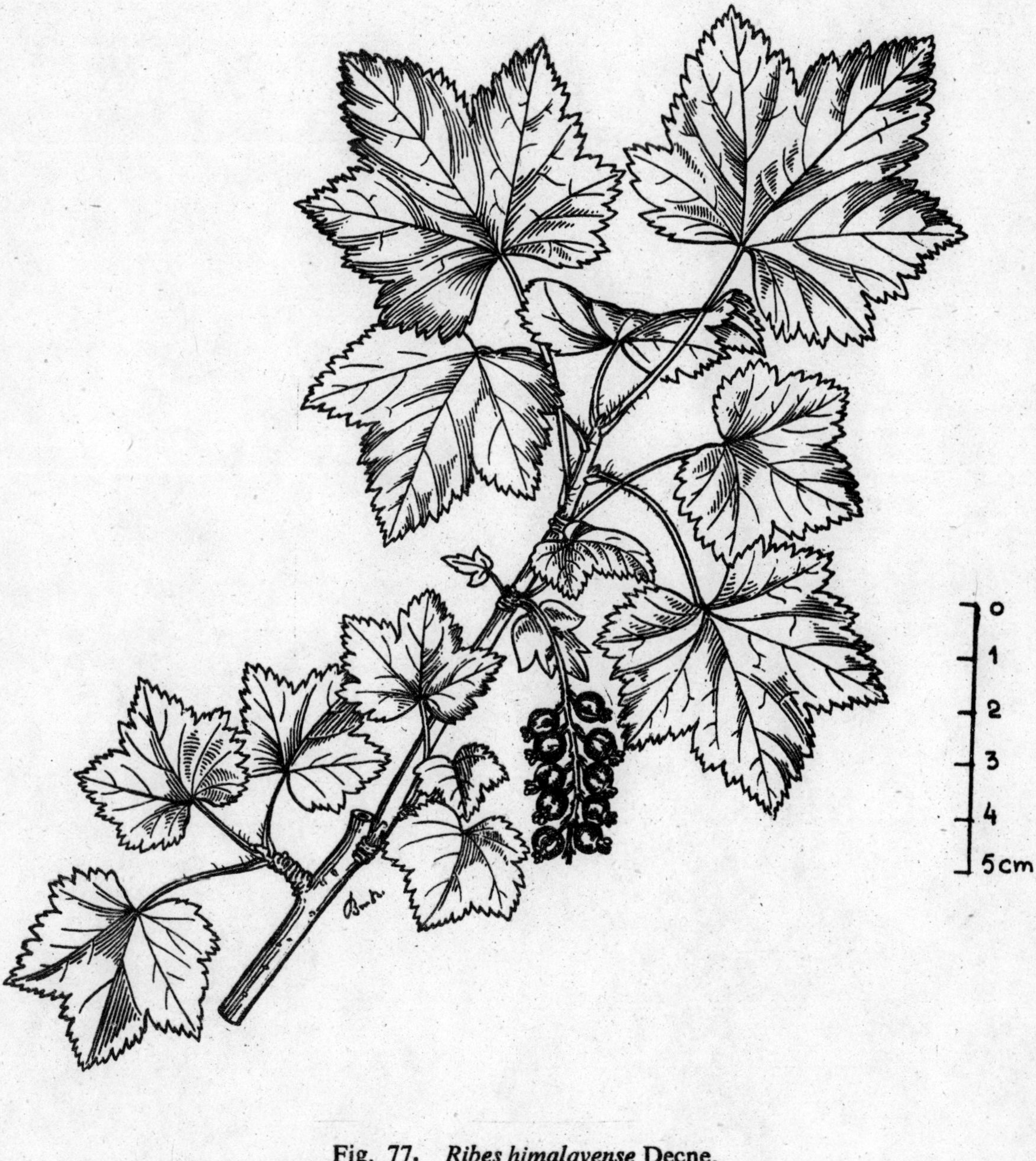

Fig. 77. *Ribes himalayense* Decne.

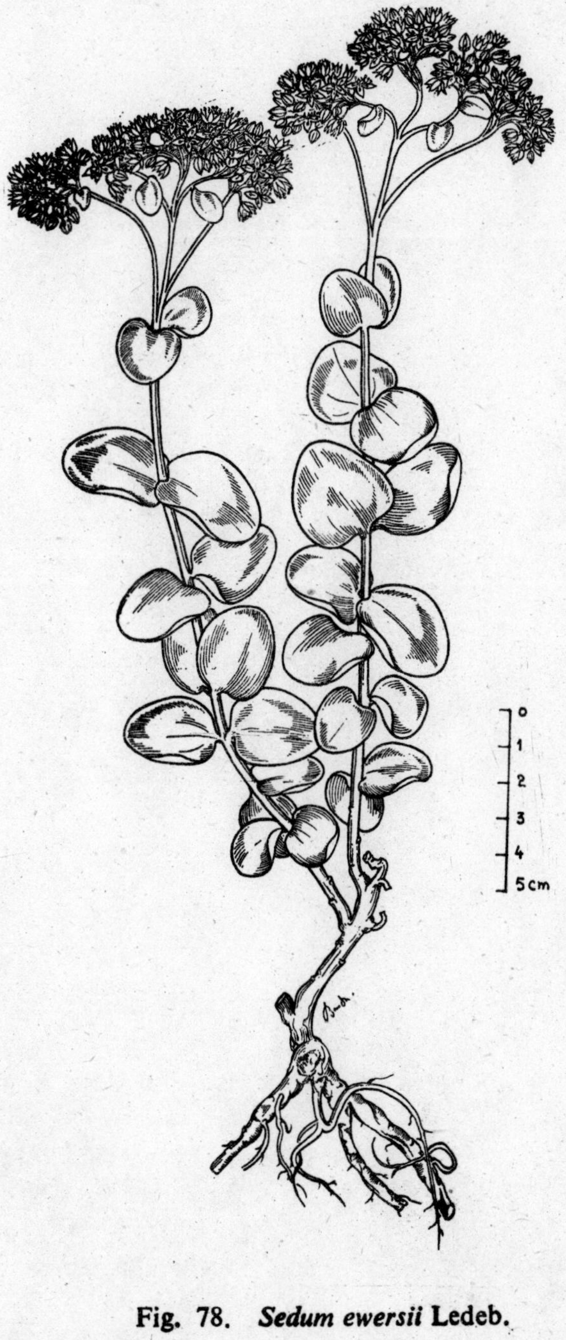

Fig. 78. ***Sedum ewersii*** **Ledeb.**

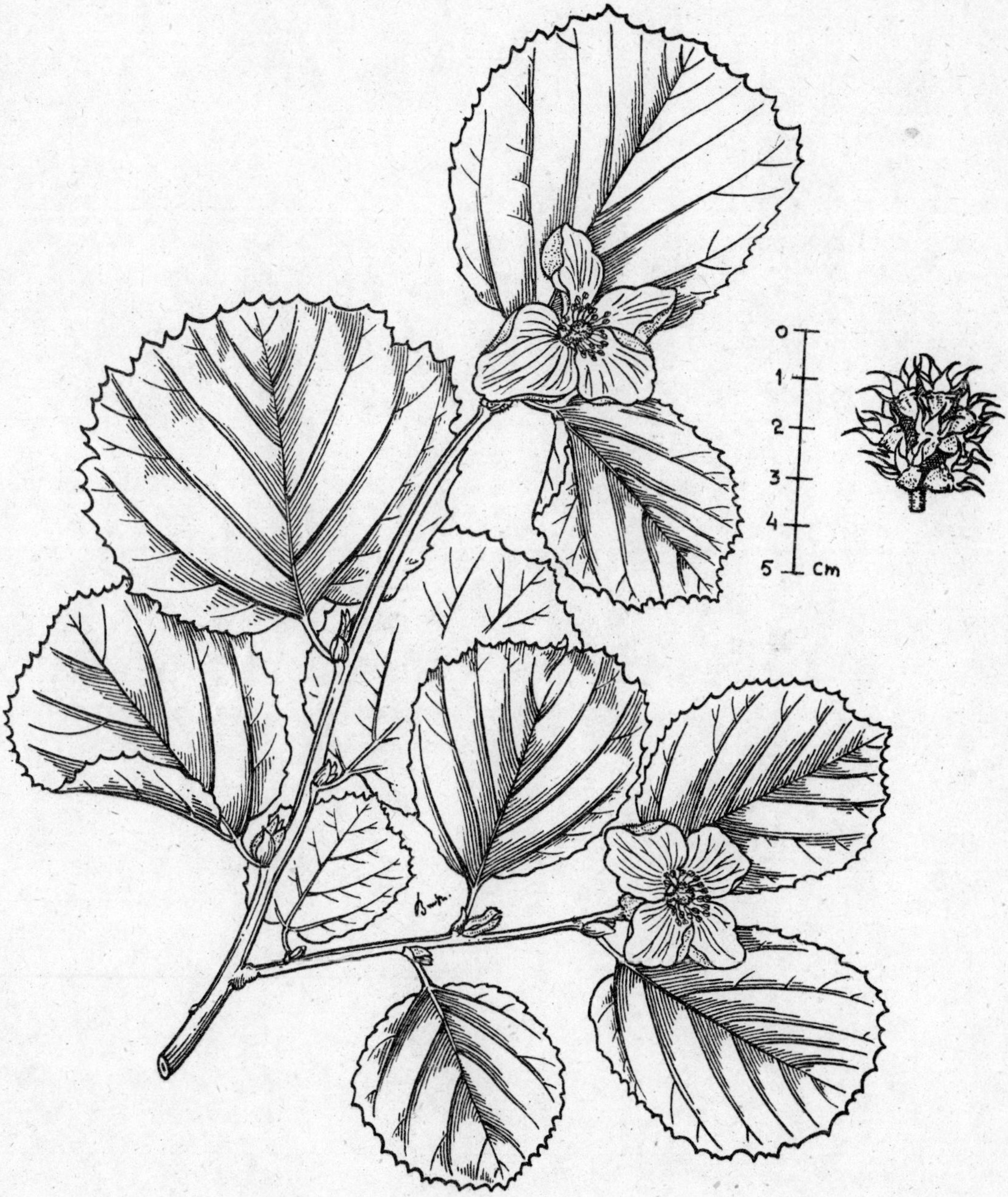

Fig. 79. *Parrotiopsis jacquemontiana* Rehder

Fig. 80. *Epilobium latifolium* Linn.

Glabrous, perennial herb; stems often reddish, 10-30 cm tall from a procumbent, woody basal part. Leaves glaucous, fleshy, orbicular, obovate or broad-ovate, 1-3 cm across, opposite, or the upper few alternate. Flowers pink or rose-purple, 6-8 mm diam., in dense carymbose cymes; petals 4-5; stamens 8-10. Follicles 4-5 erect.

Occ. : Common; forming patches mostly from crevices in moister rocks.

Fls. July-Sept.

Near Satlanjan (3100 m) *B.M. Sharma* 16620; above Pisu Ghati (3300 m) *B M. Sharma* 16674; Pehalgam (7200′) and Pisu Ghati (11500′) *K.M. Vaid*, both under DD. Herb. no. 117557; Panchtarni (—) *K.S. Ahluwalia*, RL. Herb. nos. 3852-3854,-56 &-57; Saswarian nallah (2950 m) *P.S. Jamwal* 19232.

2. S. **oreades** (Decne.) Hamet, Bull, bot. Soc. Fl. 56 : 571, 1909; Stewart, l.c. 334. *Umbilicus oreades* Decne., Jacq. Voy. Bot. Inde. 62, 1844. *Cotyledon oreades* Cl. in Hk.f., l.c. 416.

Glabrous, fleshy annual; stems several, 5-10 cm long, ascending, leafy. Leaves narrow, lanceolate, acute, 4-7 mm long, alternate. Flowers yellow, 7-8 mm long, solitary or up to 3 at the tips of the branches. Sepals oblong, acute, *c.* 6 mm long; petals united at the base; stamens 10-8 (-5). Follicles 5; seeds ellipsoid, microscopically tuberculate.

Occ. : In shaded, moist places at alts. 3300-4200 m.

Fls. July-Aug.

Liddarwat-Tarsar route (3300 m) *B.M. Sharma* 14006; Sangam vy, (13-14000′) *Duthie* 14154; Sangam vy. above Lidderwat (11-12000′) *Duthie* 14155; Kolahoi, Lidder vy., *Inayat* 25602: Mt. opposite Dudhsar, Kolahoi (4000 m) *P.S. Jamwal* 16359.

XXXI. HAMAMELIDACEAE

PARROTIOPSIS Schn.

P. **jacquemontiana** (Decne.) Rehder, J. Arnold Arbor. 1 : 256, 1920; Hutchinson, Gen. Fl. Pl. 2 : 101, 1967; Stewart, l.c. 345. *Parrotia jacquemontiana* Decne., Jacq. Voy. Bot. Inde. 4 : 73, t. 82, 1844; Cl. in Hk.f., l.c. 426. **Fig. 79.**

Deciduous, 2.5-5 m high, shrub or tree with grey-silvery bark; young shoots, petioles and stipules stellately hairy. Leaves crenate-serrate, suborbicular, obovate or broadly so, 3-8 cm long and nearly as broad. Flowers appearing prior to young leaves, minute, greenish yellow, apetalous, forming 0.5-1 cm wide heads supported by broad, 4, white 2-2.5 cm long bracts; calyx densely woolly, truncate, adnate to the ovary base; stamens *c.* 15. Capsules clustered, 2-celled, 1.2-1.3 cm long, stellately tomentose, in the lower half adherent to the calyx-tube; seeds ellipsoid, brown, *c.* 6 mm long, polished.

Occ. : Common and often gregarious in patches along roadside slopes between Mundlan and Aru.

Fls. May-June; *Frts.* Aug.

Aru (2375 m) *B.M. Sharma* 14318; Pehalgam (7200′) *H.D. Thapliyal* 26511; Pehalgam (7200′) *K.M. Vaid*, DD. no. 132211; Mundlan (2300 m) *P.S. Jamwal* 15669.

XXXII. LYTHRACEAE

LYTHRUM Linn.

L. salicaria Linn., Sp. Pl. 447, 1753; Stewart, l.c. 500; For. Fl. Sgr. 177.

Perennial, 0.6-1.2 m tall, sparingly branched, minutely grey-pubescent herb; stem with 4-6 raised lines running node to node. Leaves sessile, lanceolate or lanceolate-oblong, 2-5×0.7-1.2 cm, acute, lower in whorls of 3 or simply opposite, the upper subopposite or alternate sometimes. Flowers heteromorphous, bluish-purple, fascicled in axils of leafy bracts and forming long, terminal, interrupted spikes. Calyx tubular, 4.5-5×1.5-2 mm, 12-ribbed; teeth 6, deltoid, alternate with as many narrow, lanceolate, accessory teeth. Stamens 12, all or at least some exserted.

Occ. : Riverside and wet borders of fields; not common, although gregarious in a few localities only.

Fls. July.-Aug.

Pehalgam (2100 m) *B.M. Sharma* 14665; Ganeshbal, Lidder vy. (2150 m) *P.S. Jamwal* 15699; Bawan, Lidder vy. *Inayat* 25615.

XXXIII. ONAGRACEAE

Sepals and petals 4 each; stamens 8; fruit a many-seeded loculicidal capsule; seeds comose **1. Epilobium**

Sepals and petals 2 each; stamens 2; fruit indehiscent; 1-2-seeded; seeds not comose **2. Circaea**

1. EPILOBIUM Linn.

1\. Flowers zygomorphic; stamens deflexed on one side; style bent on one side at least before the anthers dehisce; stigmas distinctly 4-lobed, spreading :
- 2\. Leaves prominently reticulately nerved; flowers subtended by narrow, caducous bracts **1. E. angustifolium**
- 2\. Leaves obscurely nerved; flowers subtended by ordinary foliage leaves **2. E. latifolium**

1\. Flowers actinomorphic; stamens and style erect; stigmas combined or capitate :
- 3\. Middle cauline leaves sessile; amplexicaul **3. E. laxum**
- 3\. Middle cauline leaves evidently petiolate :
  - 4\. Middle and lower internodes glabrescent, with distinct, glabrous or puberulous, elevated ribs running along their length from petioles; leaves subcoriaceous, narrow lanceolate, up to 1.2 cm broad **4. E. cylindricum**
  - 4\. Middle and lower internodes pubescent all round or glabrescent, with hairy lines unelevated when present; leaves flaccid, broad-lanceolate or elliptic, 1.2-2.5 cm broad **5. E. royleanum**

1\. **E. angustifolium** Linn., Sp. Pl. 347, 1753; Cl. in Hk.f., l.c. 582, 1879. *Chamaenerion angustifolium* (Linn.) Scop., Fl. Carn. ed. 2, 1 : 271, 1772.

Erect, almost glabrous, 0.6-1.2 m tall, perennial herb; stems stout, angular, usually simple. Leaves pale beneath, alternate, opposite or subopposite, lanceolate, 8-15×1-1.8 cm, almost sessile, acute, entire or minutely denticulate, narrowed at the base. Flowers rose-purple, 2.2-2.5 cm diam., spreading, in long, terminal spike-like raceme. Calyx tube closely white-tomentose; teeth lanceolate, purplish, acuminate. Petals clawed, rounded at the tip. Capsule linear, 6-8.5 cm long, pubescent.

Occ. : Common on rocky hillsides above 2400 m in W. Liddar vy.

Fls. Mid July-Mid Aug.

Nangneri (2800 m) *B.M. Sharma* 14119; Satlanjan (3000 m) *P.S. Jamwal* 16372.

2. **E. latifolium** Linn., Sp. Pl. 347. 1753; Cl. in Hk.f., l.c. 583. **Fig. 80**

Erect, 20-40 cm tall, branched, perennial herb from a woody base; stems puberulous to almost glabrous. Leaves scattered and opposite, elliptic, 1.5-5 × 0.5-1.5 cm (in our specimens), obtuse to ± acute, glabrous or obscurely pubescent. Flowers showy, rose-purple, 3-4 cm diam., solitary, axillary on distant stalks. Calyx lobes purplish, broadly lanceolate, acuminate; petals obovate. Capsule linear, *c.* 7.5 cm long, closely pubescent; seeds narrowly obovoid, smooth.

Occ. : Common and gregarious in gravelly and stony islands in alpine stream beds, particularly near Kolahoi glacier.

Fls. July-Aug.; *Frts.* Aug.

Bara Khalen (3100 m) *B.M. Sharma* 14144; Kolahoi, near glacier (3725 m) *B.M. Sharma* 16617; Kolahoi, above Lidderwat (11-12000′) *Duthie* 13496; Kolahoi (3700 m) *P.S. Jamwal* 16373.

3. **E. laxum** Royle, Ill. Bot. Himal. 211, t. 43, f. 2 1835; Raven in Bull. Br. Mus. nat. Hist. (Bot.) 2 : 10, 369, 1962. *E. amplectans* (Cl.) Benth. ex Hausskn. Monogr. Epilob. 208, 1884.

Erect, generally 12-40 cm tall herb, sometimes attaining up to 60 cm; stems 2-angled, with a strongly marked line of crisped hairs running along each angle. Leaves ovate or oblong, 1.5-4 × 0.8-2 cm, largely opposite, acuminate, closely coarse-serrulate, glabrous except on the distinctly raised nerves beneath. Flowers pinkish-purple, 0.5-1.5 cm long, solitary, axillary and in racemes approximated towards the ends of the branches. Calyx lobes lanceolate, hairy; petals obovate, notched at tips. Capsule linear, 4-5 cm long; seeds narrow elliptic, minutely papillose.

Occ. : Along the edges of drains and on open, moist slopes; common.

Fls. July-Sept.

Between Kolahoi and Dudhsar (3750 m) *B.M. Sharma* 19228; Aru-Chummanaisar route (2600 m) *B.M. Sharma* 16636.

4. **E. cylindricum** D. Don, Prodr. Fl. Nep. 222, 1825; Raven, l.c. 355. *E. roseum* Schreb. var. *cylindricum* (D. Don) Cl. in Hk.f., l.c. 585.

Perennial, 30-90 cm tall herb, with stems and upper surface of leaves often purple; upper part of stem pubescent all round, the lower almost glabrous with puberulent or glabrous, raised lines decurrent from petioles. Leaves pale beneath, narrow-lanceolate, on the margins and mid vein puberulent. Flowers pale pink, 5-6 mm long, in several pubescent racemes. Capsules linear 3.5-6.5 cm long, on 1-2 cm long peduncles; seeds obovoid, minutely papillose.

Occ. : Usually along streams up to *c.* 3000 m.

Fls. July.-Sept.; *Frts.* Aug.-Sept.

Gha Shangan (2900 m) *B.M. Sharma* 19334 & 19229; Armiun nalla (2750 m) *B.M. Sharma* 16633.

5. **E. royleanum** Hausskn. in Oesterr. Bot. Zeitschr. 29 : 55, 1879; Raven, l.c. 358. *E. roseum* Schreb. var. *indicum* Cl. in Hk.f., l.c. 584. *E. roseum* var. *dalhousieanum* Cl. in Hk.f., l.c.

Perennial, up to 60 cm tall herbs; stems round, pubescent all round or with well marked hairy lines decurrent from the leaf bases. Leaves pale beneath, at least the middle cauline opposite, lanceolate, elliptic or elliptic-lanceolate, 3-6.5 × 1.2-2.5 cm, denticulate, acute, on margins and nerves pubescent, narrowed at either end; petiole sometimes up to 6 mm long. Flowers pink, 6-8 mm long, in pubescent racemes. Capsules linear, 4-6 cm long, on peduncles 0.4-1 cm long; seeds obovoid, minutely papillose.

Occ. : Habitats of preceding species; common.

Fls. : July.-Sept.; *Frts.* Aug.-Sept.

Pehalgam (2200 m) *B.M. Sharma* 16627; Armiun nallah (2750 m) *B.M. Sharma* 16631 & 16632.

Excluded Species

E. amurense Hausskn. ssp. **laetum** Raven and **E. rhynchospermum** Hausskn., reported from area of present work (Stewart, l.c. 505 & 507) could not be studied because of inavailability of their materials.

2. **CIRCAEA** Linn.

Ovary unilocular; fruit 1-seeded	**1.** C. **pricei**
Ovary bilocular; fruit 2-seeded	2. C. **cordata**

1. C. **pricei** Hayata, Ic. Pl. Formosa 5 : 72, 1915; Stewart, l.c. 505. *C. alpina* Linn. ssp. *imaicola* (Asch. & Magn.) Kitam., Fl. Afgh. 279, 1960, For. Fl. Sgr. 177. *C. alpina* var. *himalaica* Cl. in Hk.f., l.c. 589. **Fig. 81**

Erect, 15-28 cm tall, pubescent, usually unbranched herb. Leaves opposite ovate 1.5-4.5 × 1.2-2.5 cm, sinuate dentate, obtuse, acute or acuminate, from rounded, acute or truncate base; petioles up to 2.5 cm long. Flowers very small, white or pinkish white, in terminal and lateral, glandular-hairy racemes. Calyx lobes glabrous. petals notched. Fruit clavate, *c.* 2 × 1 mm, glabrous or clothed with hooked hairs.

Occ. : Common and often gregarious along water drains and in shady places under wet rocks.

Fls. & Frts July-Sept.

Liddarwat (3050 m) *B.M. Sharma* 14192; Lidder vy. (6-7000′) *Duthie* 13064; Pisu Ghati, on way to Amarnath (10,500′) *K.M. Vaid* DD. Herb. no 117553.

2. C. **cordata** Royle, Ill. Bot. Himal. 211, t. 43, f. 1, 1834; Cl. in Hk.f., l.c.

Erect. 30-60 cm tall, stout, hairy herb. Leaves ovate-lanceolate, acuminate,

7-10×4-6 cm, more or less pubescent, sinuate-denate to subentire, from mostly cordate base; petioles up to 6 cm long. Flowers small, white, in terminal glandular-hairy racemes. Calyx lobes hairy, reflexed; petals obcordate, *c.* 2 mm long, deeply notched. Fruit ovoid from an unequal base, 3 mm long, deflexed, clothed with hooked hairs.

Occ. : Under forest shade in Pehalgam and places around.

Fls. & Frts. July-Aug.

Pehalgam (2200 m) *B.M. Sharma* 16626; Lidder vy. near Pehalgam (7-8000′) *Duthie* 13168; Lidder vy. above Pehalgam (8-9000′) *Duthie* s.n. DD.

XXXIV. Apiaceae

1. Leaves simple, undivided, linear to lanceolate or ovate, with entire margins — **3. Bupleurum**
1. Leaves compound, variously divided :
  - 2. Leaves palmately 3-5 partite — **1. Sanicula**
  - 2. Leaves pinnately divided :
    - 3. Fruit with epicarp transversely plicate-rugose, the transverse folds occupying the whole back except the ridges (stylosum) **15. Pleurospermum**
    - 3. Fruit not as described above :
      - 4. Dorsal, intermediate and lateral ridges on the fruit wall prominent or not prominent, never winged :
        - 5. Plants partially submerged in water or denizens of marshy habitats; fruit ridges prominent, all corky — **6. Sium**
        - 5. Plant neither submerged nor inhabiting marshy places; fruit ridges prominent or not, not corky :
          - 6. Fruit 6-12 mm long :
            - 7. Fruit glabrous, narrowed at apex: ridges prominent, broad, obtuse — **9. Chaerophyllum**
            - 7. Fruit tuberculate to hispid or minutely bristly, rarely at length glabrate; ridges not prominent — **10. Anthriscus**
          - 6. Fruit at the most up to 5 mm long ;
            - 8. Fruit prickly with long, dense, curved bristles — **17. Torilis**
            - 8. Fruit glabrous or pubescent, never clothed with prickles :
              - 9. Root swollen into a globoid tuber — **5. Bunium**
              - 9. Root not tuberous or globoid :
                - 10. Fruit glabrous :
                  - 11. Vittae in fruit furrows obscure — **8. Aegopodium**
                  - 11. Vittae in fruit furrows evident;
                    - 12. Fruit furrows 1-vittate; inner face of seed plane — **4. Carum**
                    - 12. Fruit furrows 2 to multi-vittate; inner face of seed grooved — **2. Vicatia**
                - 10. Fruit hispid, pubescent or pilose :
                  - 13. Calyx-teeth linear, deciduous. Fruit subterete to dorsally compressed; fruit furrows 1-vittate — **11. Seseli**

13. Calyx-teeth obsolete. Fruit laterally compressed; fruit furrows 2-3-vittate **7. Pimpinella**

4. Dorsal and intermediate ridges prominent or not prominent, winged or not winged, the lateral ones invariably winged:

14. Bracteoles 10-15, broadly white-margined, oblong to obovate or elliptic, obtuse; inner face of the seed deeply grooved **15. Pleurospermum** (candolii)

14. Bracteoles not white-margined; if white-margined then only 5-10 and linear to filiform with sometimes lobed top; inner face of seed plane or concave :

15. Vittae evident, extending only to two-third of the length of the fruit **16. Heracleum**

15. Vittae obsolete, or evident and then extending to the base of the fruit :

16. Fruit 6-1.5 mm long; wings of the lateral ridges of the two mericarps separate at margins **13. Angelica**

16. Fruit 3-5 mm long; wings of the lateral ridges of the two mericarps closely appressed at the margins ;

17. Calyx-teeth linear; dorsal fruit furrows 1–vittate or vittae obsolete **14. Selinum**

17. Calyx-teeth obsolete; dorsal fruit furrows 2-3-vittate **12. Ligusticum**

1. SANICULA Linn.

S. **elata** Buch.-Ham. ex D. Don, Prodr. Fl. Nep. 183, 1825; Shen et Constance, Gen. Sanicula in Univ. Calif. Pub. Bot. 25 : 16, 1951; Backer et Bakh. f., Fl. Jav. 2 : 173, 1965. *S. europaea* Cl. in Hk.f., l.c. 670 (non Linn.)

Glabrous, 20-75 cm tall perennial, dichotomously branched upwards. Radical and lower cauline leaves palmate, long petioled; segments ovate to rhomboid, lobed or incised, 2.5-5 × 1.5-3 cm, serrate; teeth mucronate. Upper leaves sessile. Umbels falsely compound; rays simple or forked. Bracts leafy, few, lobed, toothed or entire; bracteoles minute. Flowers minute, white; calyx teeth acute. Fruit ovoid, 2.5-3 mm long, clothed with hooked bristles.

Occ. : In gravelly, shady and moist places along banks; infrequent.

Fls. & Frts. June-July.

Hamwas (3220 m) *B.M. Sharma* 14039; Lidder Valley above Pehalgam (8000′) *Duthie* 13459.

2. VICATIA DC.

V. **coniifolia** DC., Prodr. 4 : 243, 1830; Cl. in Hk.f., l.c. 671.

Glabrous, 15-50 cm high perennial, branching from near the base. Leaves 2-3-pinnate, finely dissected into lanceolate to linear, acute ultimate segments. Umbels unequally 6-12-rayed; bracts 1-2, or 0; bracteoles linear, longer than the umbellules. Flowers minute, dark red; calyx teeth 0. Fruit oblong to ovoid, 3.5-4 mm long, narrowed towards the apex, bluntly ridged; ridges slender but distinct.

Occ. : Common in forest clearings and in open stony and grassy places at 3000-4000 m.

Fls. June-July; *Frts.* July-Aug.

Below Dudhsar (3600 m) *B.M. Sharma* 19386.

3. BUPLEURUM Linn.

1. Middle and upper cauline leaves with 13 to 25 or even more veins :
   2. Flower dark purple to almost black: bracteoles longer than the flowering umbellule, 5-7 veined **1. B. longicaule**
   2. Flowers yellow; bracteoles wanting, or as long or shorter than the flowering umbellule and then 3–5-veined **2. B. jucundum**
1. Middle and upper cauline leaves with 11 or less veins :
   3. Plants decumbent; stems many from the base; flowers dark purple to almost black **3. B. atroviolaceum**
   3. Plants erect; stems solitary or few from the base; flowers yellow :
      4. Leaves gradually long–tapering at the apex; leaf-bases completely engirdling the stem **4. B. marginatum**
      4. Leaves suddenly acute; leaf-bases only semi-amplexicaul, not completely engirdling the stem **5. B. hoffmeisteri**

1. **B. longicaule** Wall. ex DC., Prodr. 4 : 131, 1830. var. **himalayense** (Kl.) Cl. in Hk.f., l.c. 677. *B. himalayense* Kl. in Kl. et Garcke, Bot. Ergeb. Reise, Prinz. Waldem. 146, t. 51, 1862. **Fig. 82**

Erect, 15-60 cm tall perennial; stem generally solitary, sometimes 2 or 3 from the base. Lower leaves linear-lanceolate, upper lanceolate, amplexicaul, acuminate. Umbels 2-6 rayed; rays up to 6 cm long. Bracts 1-3 (-4), unequal; bracteoles foliaceous, 6-10, ovate-lanceolate, prominently nerved. Flower ± 2 mm across; carpel ridges narrowly alate.

Occ. : Common in alpine pastures about Kolahoi, Rewil and Seikwas.

Fls. July-Aug.

Dudhsar (4000 m) *B. M. Sharma* 14033, 15616 & 15617; Sonsal Nallah, Lidder Valley (13-14000′) *Duthie* 13346; Amarnath Cave (12500′) *K. M. Vaid* DD no. 117984 & 131904.

'Above Seikwas, Kashmir (12000′) *M.V. Laurie 18779*' is a mixture with var. **ramosum** Nasir. The latter, reported also from near Mt. Kolahoi (13000′) in Stewart, l.c., differs from var. **himalayense** in decumbent habit, many stems rising from base and fewer, linear to oval bracteoles slightly united at base.

2. **B. jucundum** Kurz in Seem. Journ. Bot. 5 : 240, 1867. var. **cachemirica** Cl. in Hk.f., l.c. 675; Stewart, l.c. 513. **B. swatianum** Nasir in Univ. Calif. Pub. Bot. 27 : 425, 1954 et in Nasir & Ali, Fl. W. Pak. 20 : 50, 1972.

Up to 60 cm tall perennial with fistular; 1-few stems from a decumbent base. Leaves mucronate, white margined, lower linear to lanceolate, amplexicaul, the upper ovate to ovate-oblong or nearly orbicular from a deeply cordate base; transverse nerves reticulate. Umbels 6-8-rayed; rays angled. Bracts solitary, foliaceous or like the bracteoles lacking; flowers very small. Fruit ridges subalate; furrows 3-vittate.

Occ. : In shady localities; not common.

Fls. June

Aru (2400 m) *B.M. Sharma* 14341 & 14601; water mill near Aru (± 2400 m) *P.S. Jamwal* 15675.

3. **B. atroviolaceum** (O.E. Schulz) Nasir in Nasir et Ali, l.c. 54. *B. aitchisonii* (Boiss.) Wolff, f. *atroviolacea* O.E. Schulz, Notizbl. bot. Gart. Mus. Berl. 11 : 876, 1933; Nasir in Univ. Calif. Pub. Bot., l.c. 441; Stewart, l.c. 512.

Included on authority of R.R. Stewart, l.c. who has cited it from Kolahoi and Yamher. Stated to be rare.

4. **B. marginatum** Wall. ex DC., Prodr. 4 : 132, 1830; Stewart, l.c. 514; Nasir in Nasir et Ali, l.c. 60. *B. falcatum* Linn. var. *marginata* Wall. ex DC.; Hk.f., l.c. 676; For. Fl. Sgr. 183.

Fig. 83

Perennial, up to 1 m tall, with generally 1 or 2 stems from the base. Leaves linear to lanceolate, 4-15×0.4-0.8 cm, from rather broad, amplexicaul base, acuminate, mucronate; margins thickened, white. Umbel rays 3-10, slender, 0.6-2 cm long. Bracts generally 5, unequal; bracteoles about as long as the flowering umbellule. Fruit 3-4 mm long, distinctly ridged; ridges white.

Occ.: Generally on open rocky slopes and among stony places along the banks, less commonly among the scrubs.

Fls. July-Aug. ; *Frts.* Aug.

Satlanjan (3080 m) *B.M. Sharma* 19333 & 14647; Lidder Valley (5-6000′) *Duthie* 13061; Kolahoi, Lidder vy., *Inayat* 25617.

5. **B. hoffmeisteri** Kl. in Kl. et Garcke, Bot. Ergeb. Reise, Prinz. Waldem. 146, t. 52, 1862; Nasir in Nasir et Ali, l.c. 59. *B. falcatum* Linn. var. *hoffmeisteri* (Kl.) Cl. in Hk.f., l.c. 676.

More branched than the preceding species, which it closely resembles except for the characters of the key. Leaves up to 1.3 cm broad, the lower with mostly attenuate bases; involucre wanting or of 1-2 bracts.

Occ.: Common among scrubs along Mundlan-Aru road; also in rocky nooks between Liddarwat and Hamwas.

Fls. : June-Aug.; *Frts*: July-Sept.

Aru (2300 m) *B.M. Sharma* 19298; Above Liddarwat (2990 m) *B.M. Sharma* 14020; Gurkum (2260 m) *B.M. Sharma* 14602; between Gurkum and Aru (2310 m) *P.S. Jamwal* 15618.

4. CARUM Linn.

C. **carvi** Linn., Sp. Pl. 263, 1753; Cl. in Hk.f., l.c. 680. *C. gracile* Lindl. in Royle, Ill. Bot. Himal. 232, 1835.

Glabrous, erect or diffuse aromatic perennial, 30-60 cm high from an elongate fusiform root. Leaves finely 2-pinnatisect; lower with 3-4 mm long, lanceolate segments, those of the upper linear to filiform, umbel rays 4-9; bracts wanting or up to 3, linear, sometimes dissected; bracteoles few or 0. Flowers 3 mm across, white or pinkish; calyx-teeth 0. Fruits ovoid to oblong, 3-4 mm long, very unequally pedicelled, prominently ridged.

Occ.: Common along the drains in villages; sometimes found also in meadows and open damp places along the nallahs.

Fig. 81. *Circaea pricei* Hayata

Fig. 82. *Bupleurum longicaule* Wall. ex DC. var. *himalayense* Cl.

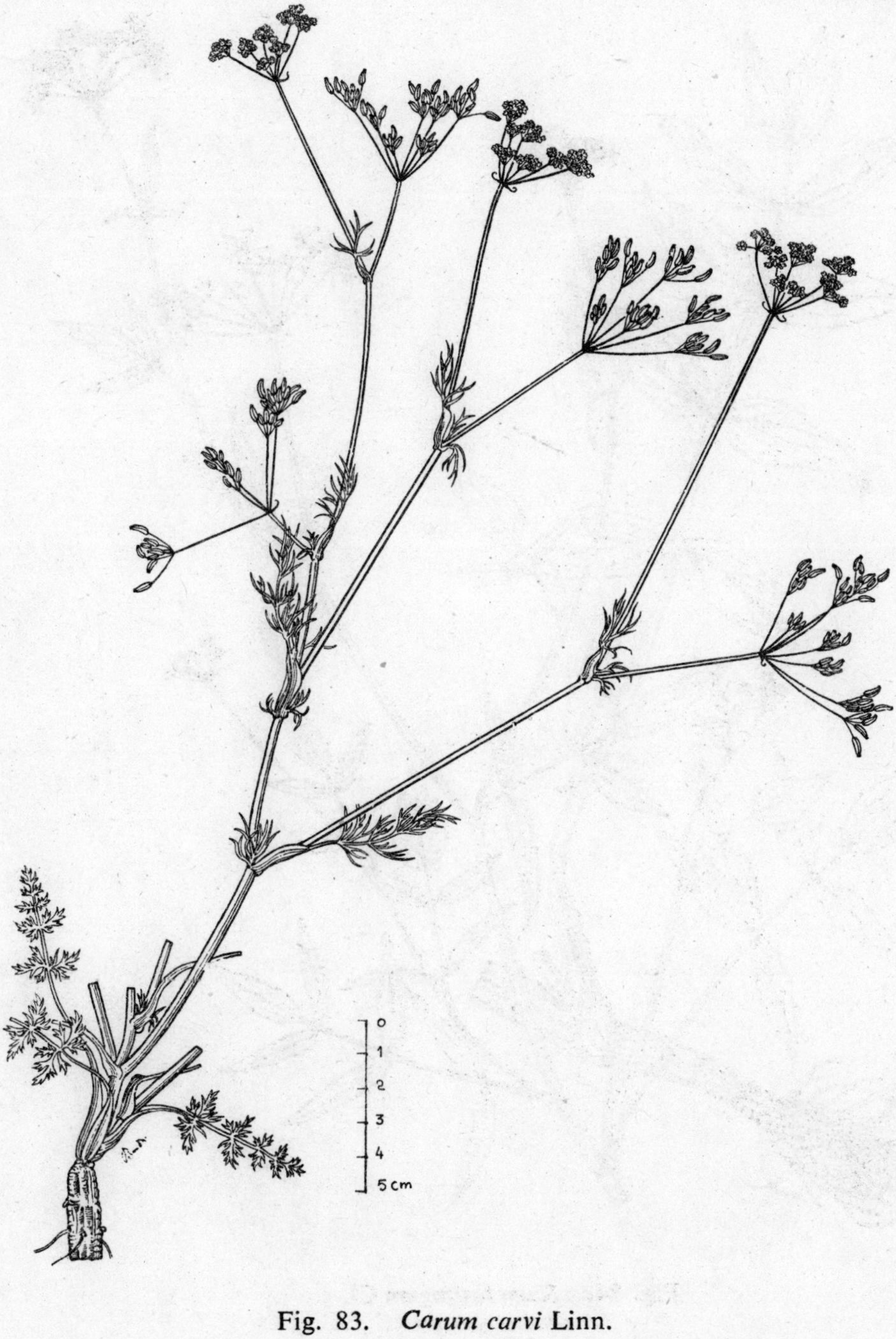

Fig. 83. *Carum carvi* Linn.

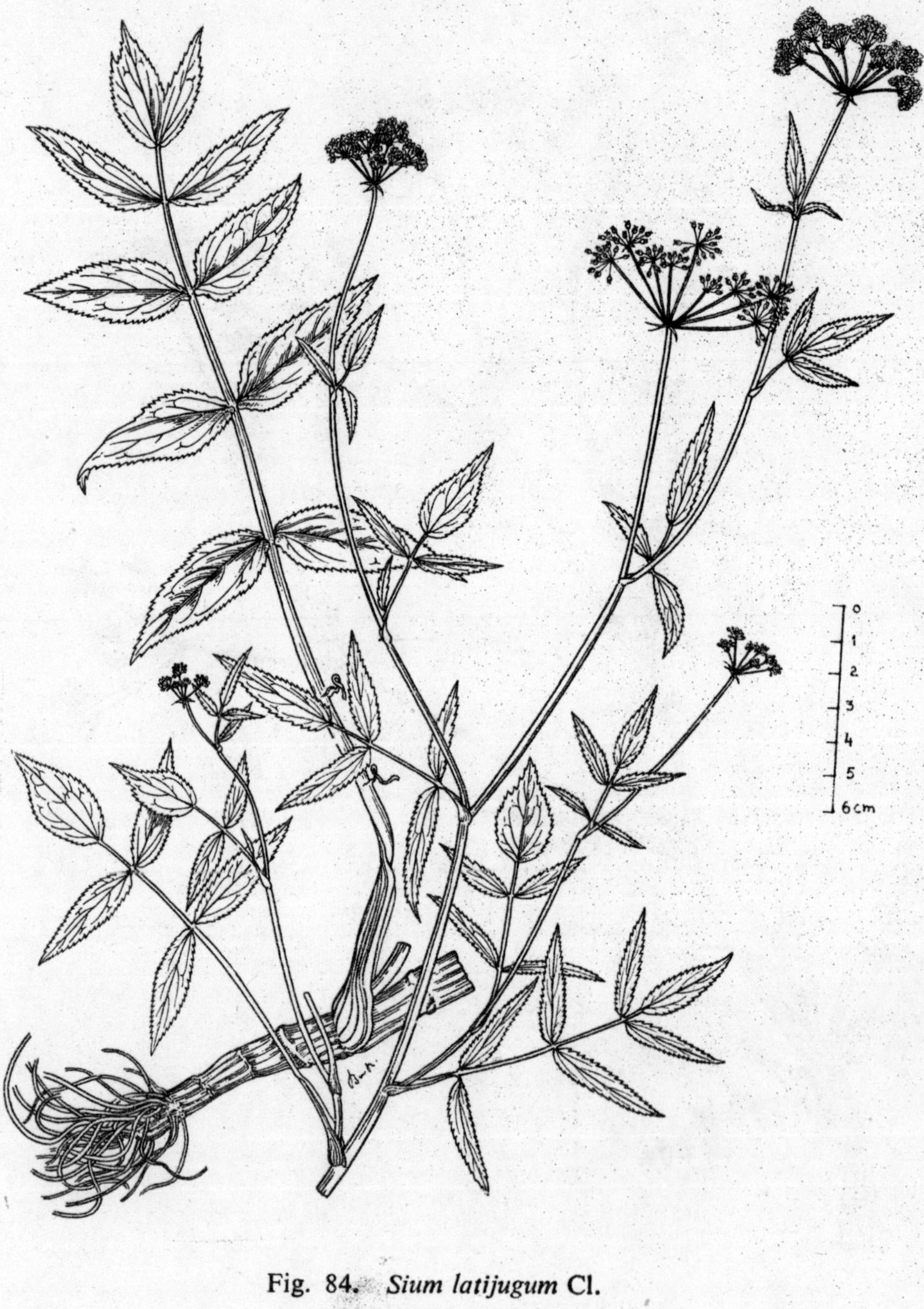

Fig. 84. *Sium latijugum* Cl.

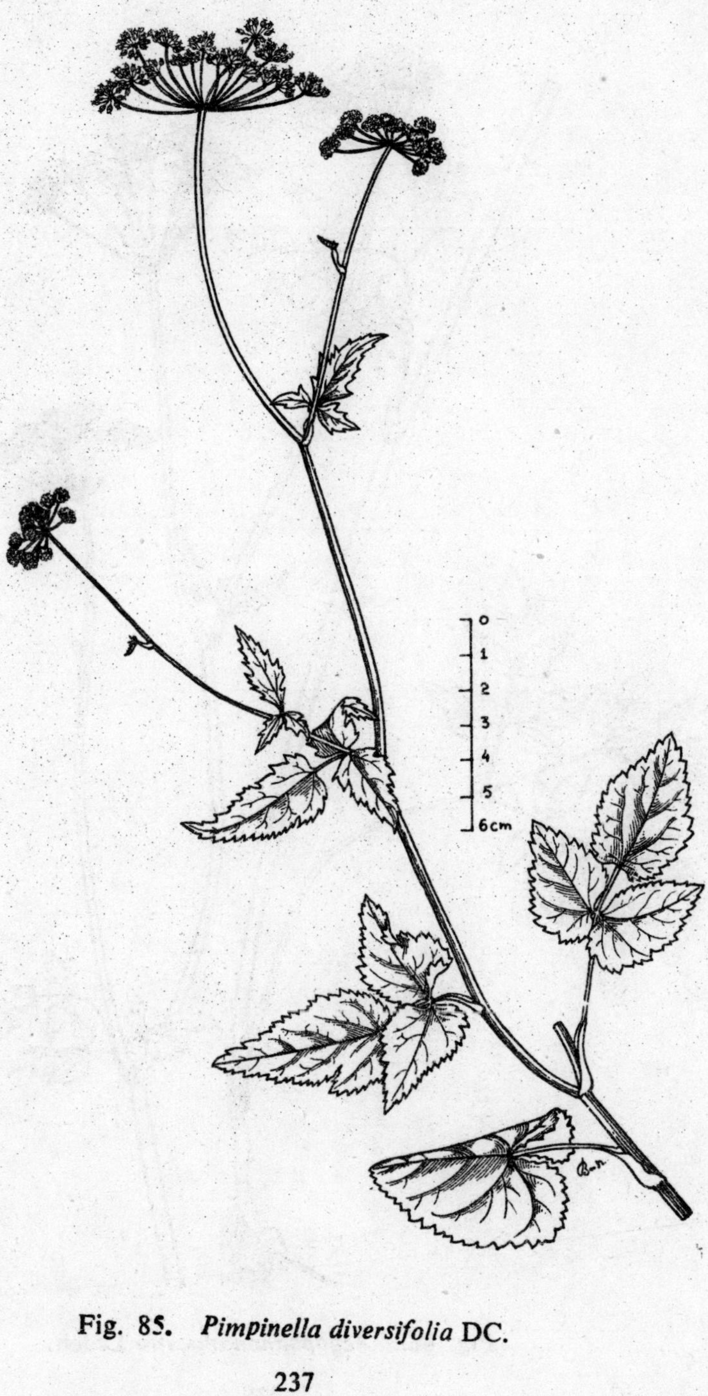

Fig. 85. *Pimpinella diversifolia* DC.

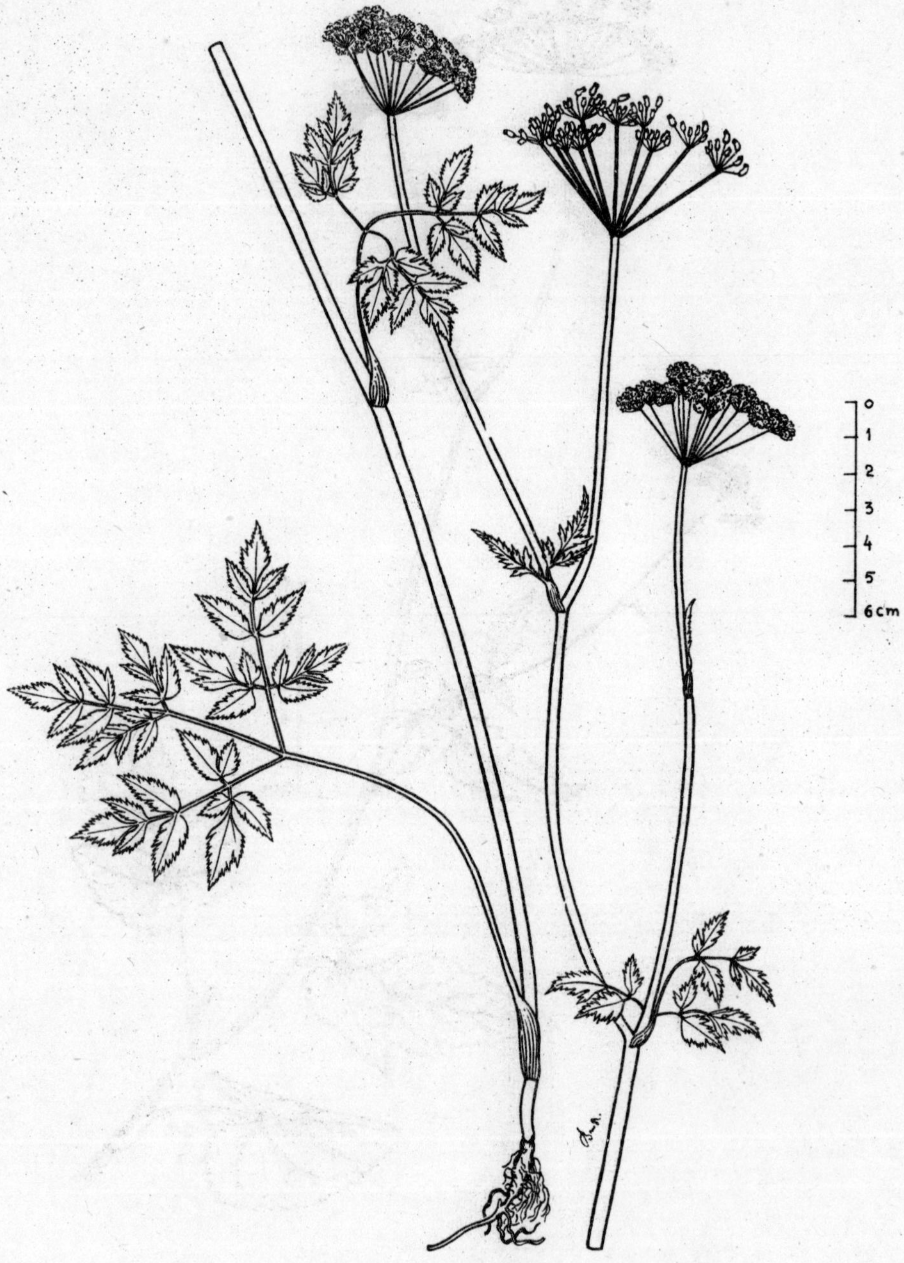

Fig. 86. *Aegopodium alpestre* Ledeb.

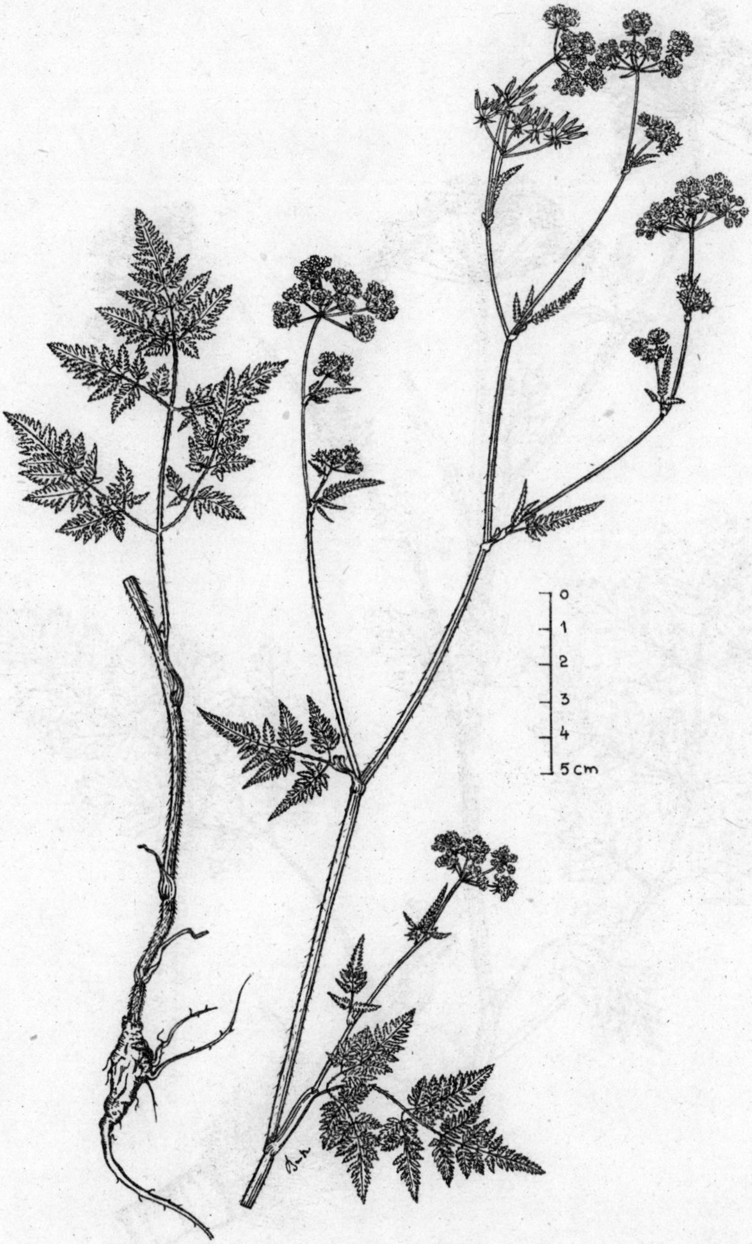

Fig. 87. *Chaerophyllum villosum* Wall. ex DC.

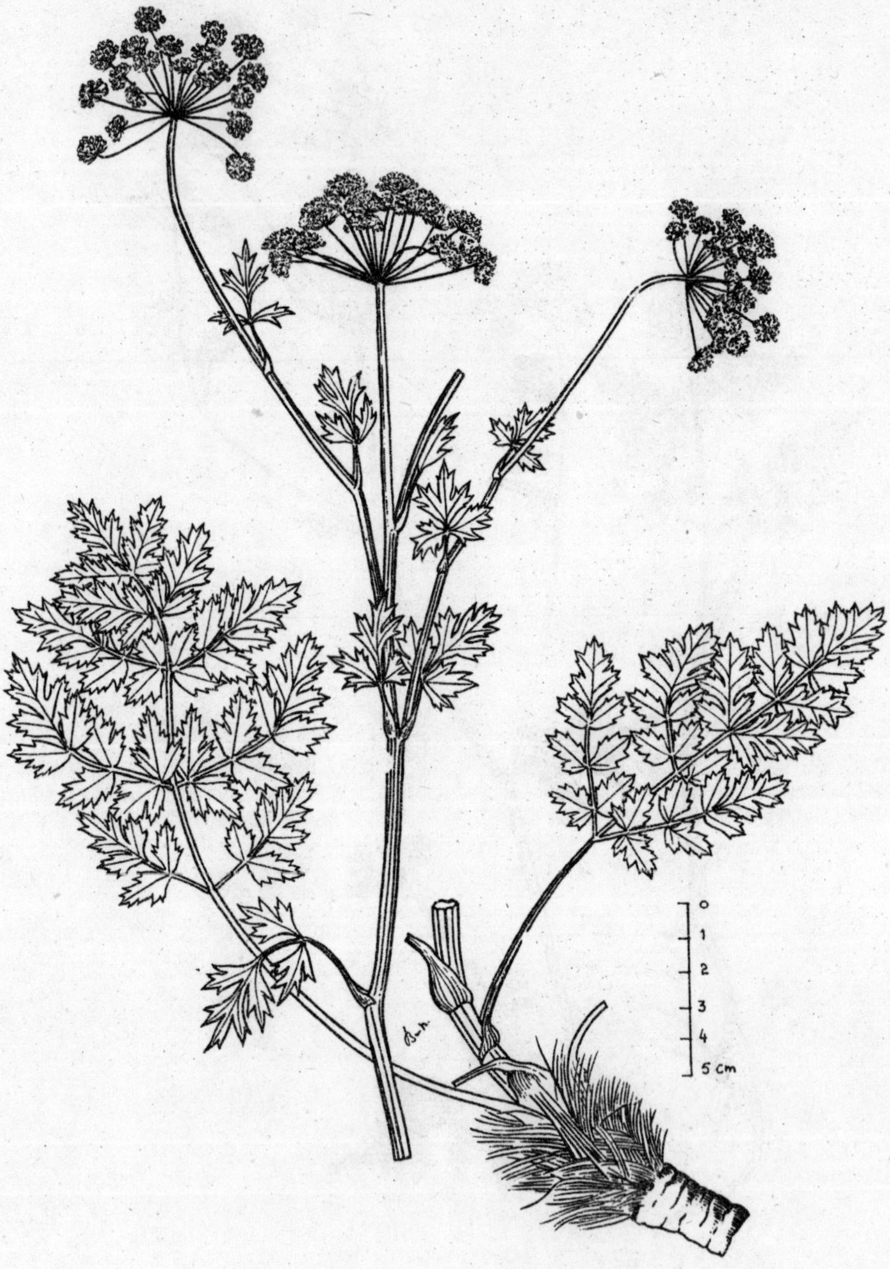

Fig. 88. *Seseli libanotis* Koch

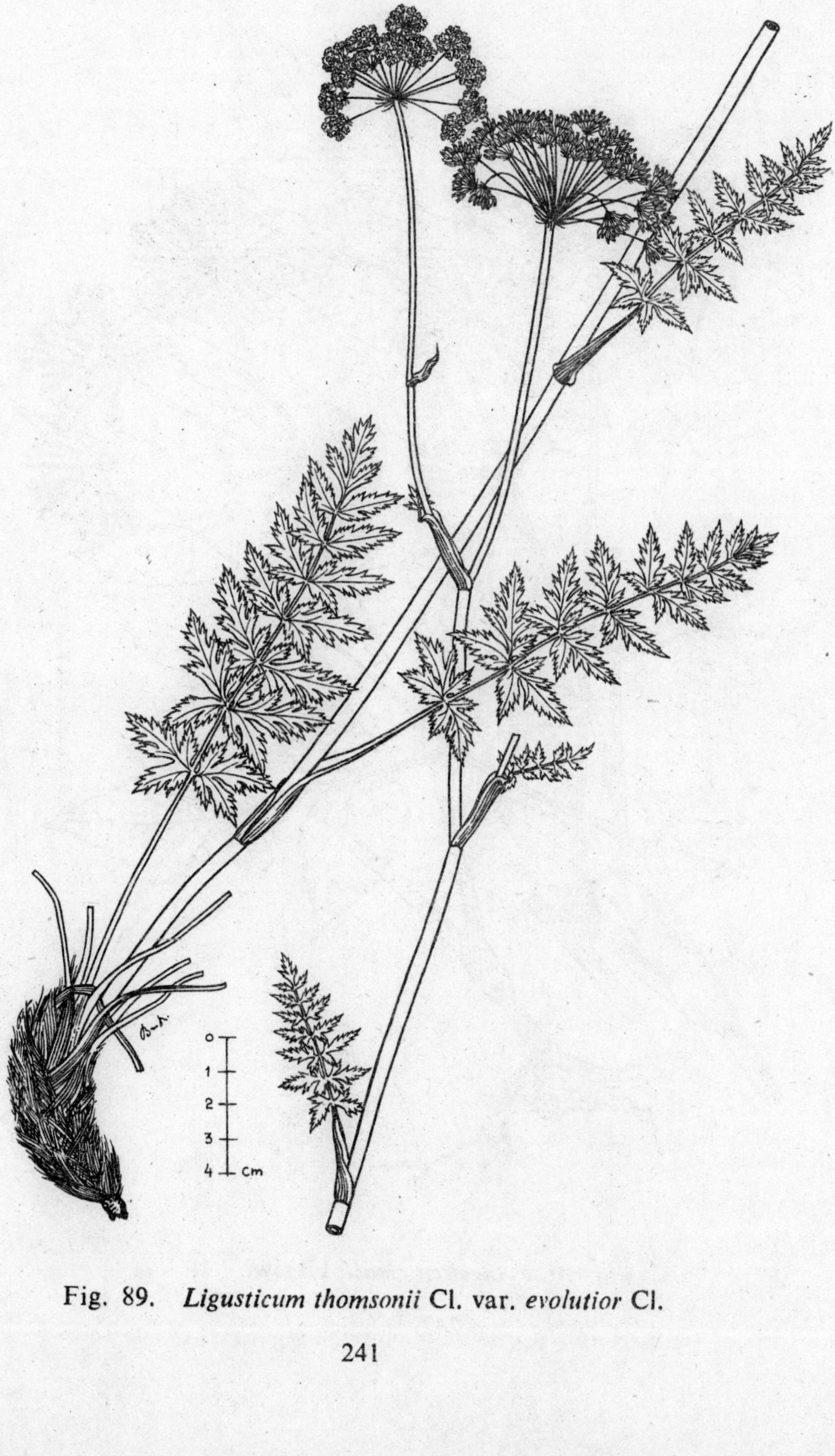

Fig. 89. *Ligusticum thomsonii* Cl. var. *evolutior* Cl.

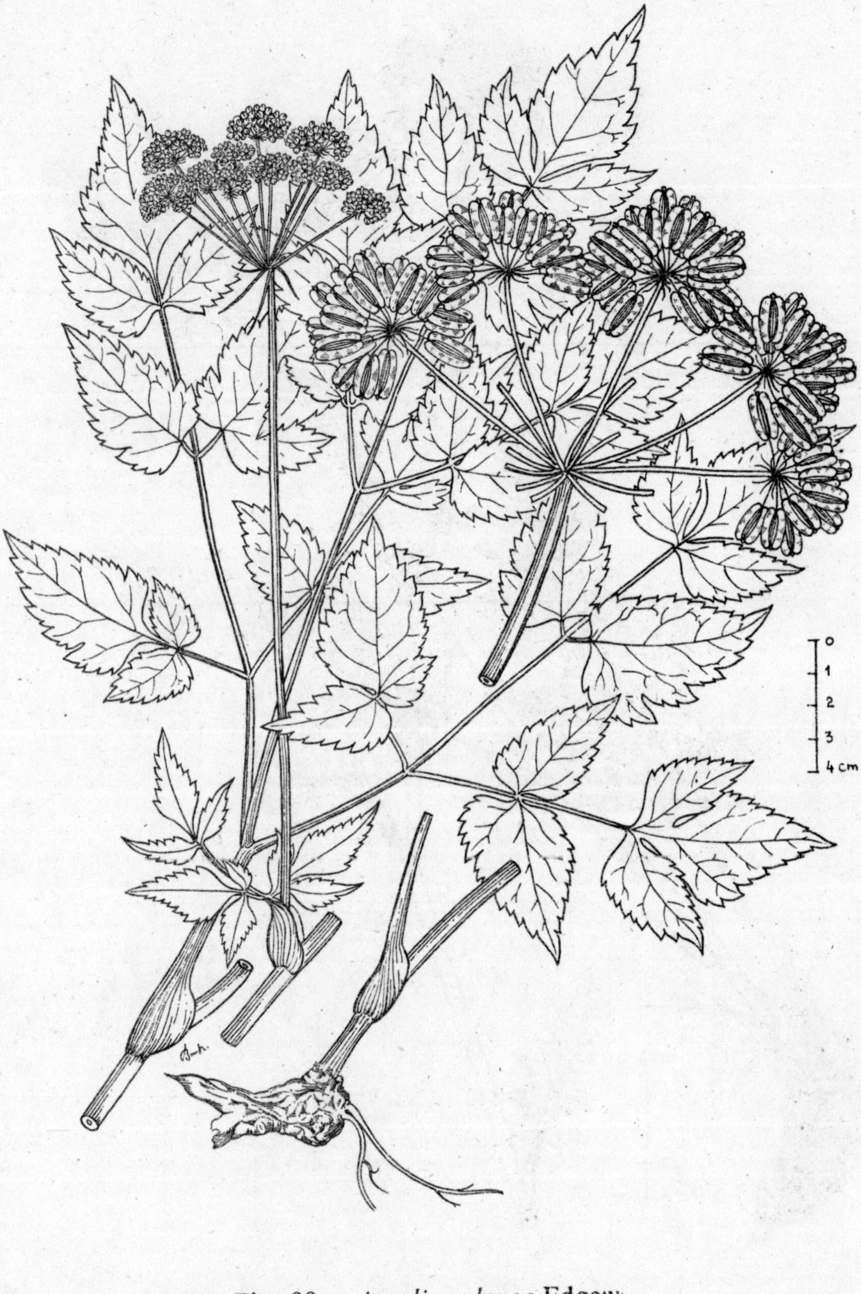

Fig. 90. *Angelica glauca* Edgew.

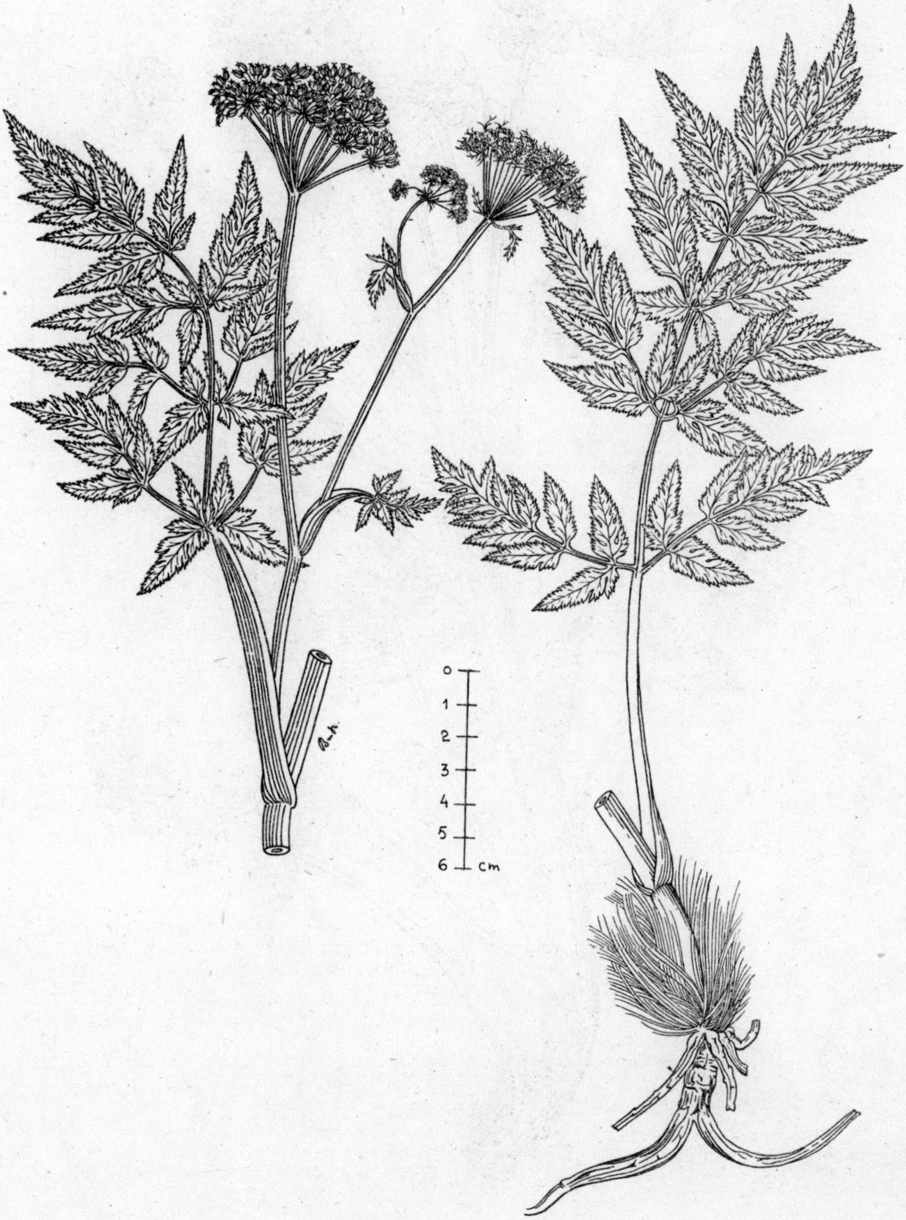

Fig. 91. *Selinum vaginatum* Cl.

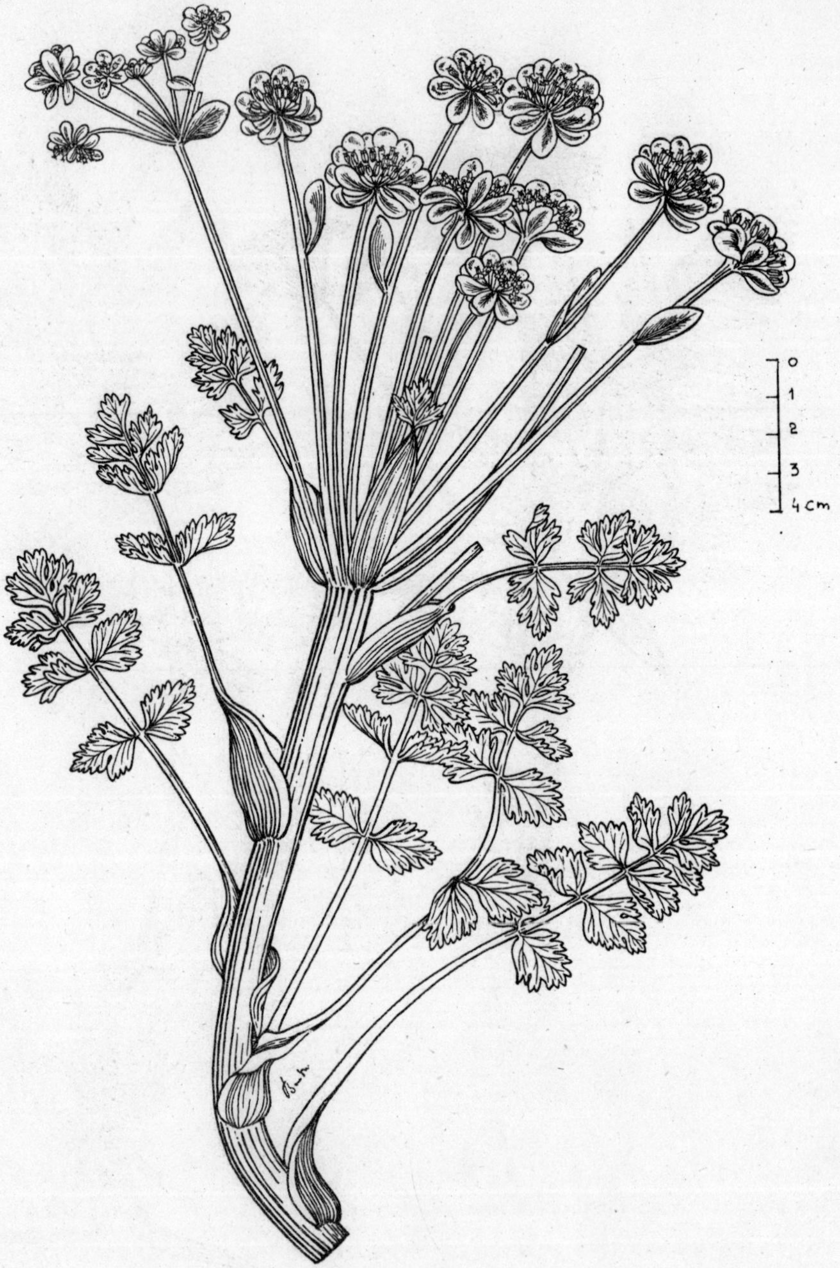

Fig. 92. *Pleurospermum candollei* Cl.

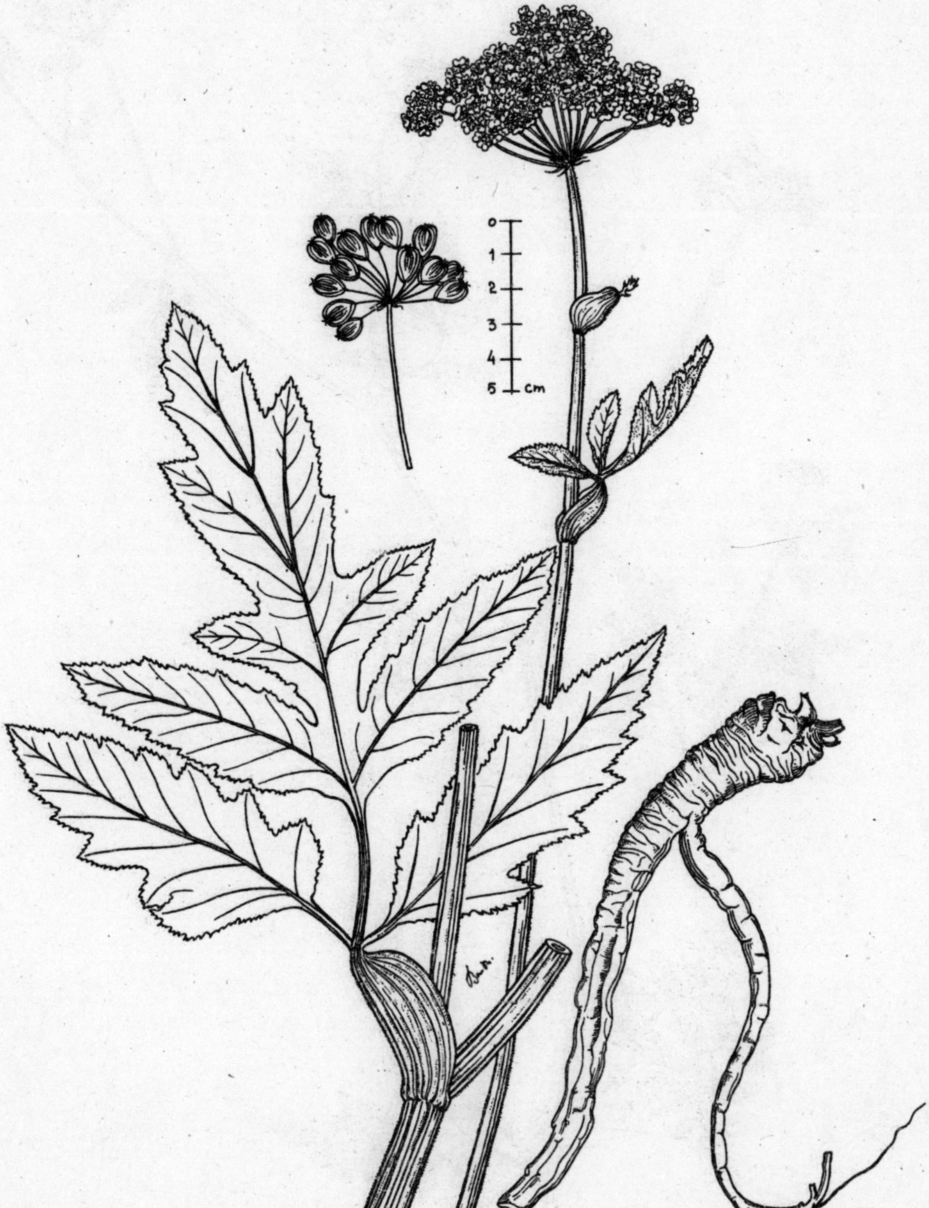

Fig. 93. *Heracleum candicans* Wall. ex DC.

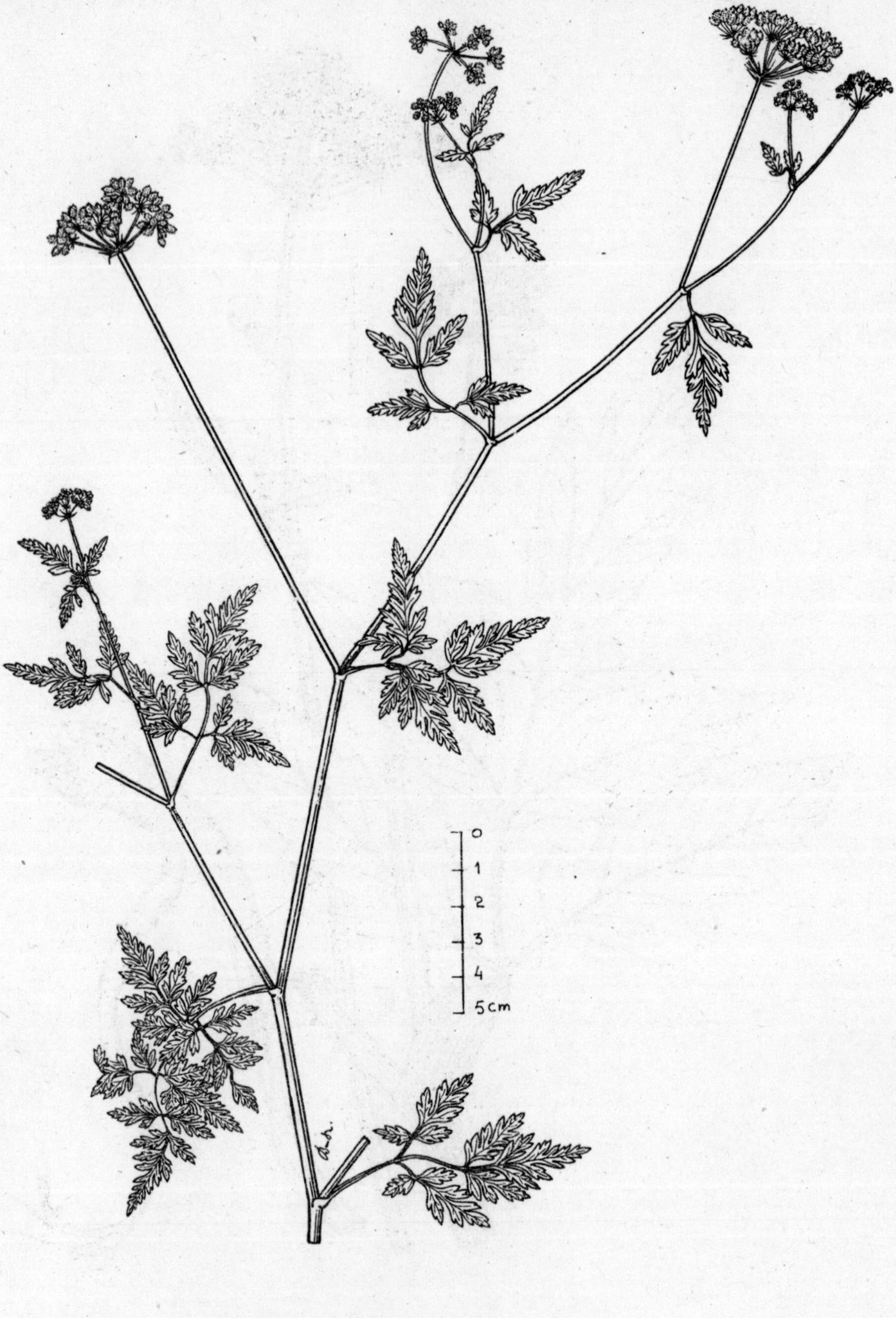

Fig. 94. *Torilis japonica* DC.

Fig. 95. *Aralia cachemirica* Decne.

Fls. May-June; *Frts.* June-July.

Aru (2420 m) *B.M. Sharma* 14381; Nangneri Nallah (2800 m) *B.M. Sharma* 14327 A & 14327 B; Aru (±2400 m) *P.S. Jamwal* 15619.

The fuits 'Caraway seeds' are carminative and used as spice. Carvone is the main constituent of their essential oil.

5. **BUNIUM** Linn.

B. persicum (Boiss.) B. Fedtsch. Restit. Turkest. 612, 1915; Wolff in Engler, Pflanzenr. 43 : 198, 1910; Raizada, Indian For. 84 (8) : 475, 1958. *Carum persicum* Boiss. in Ann. Sci. Nat. ser. 3, 138, 1844. *C. bulbocastanum* auct. non. Koch., Nov. Act. Acad. Caes. Leop. 121, 1825; Cl. in Hk.f., l.c. 681.

Perennial, glabrous, 45 cm high, branched herb from a globoid, 1 cm wide, tuberous root. Leaves finely 2-3 pinnatisect, sessile upwards; ultimate segments linear to filiform. Umbel rays unequal; bracts up to 5, like the bracteoles linear or wanting. Flowers 2.5-3 mm across, white; calyx-teeth absent.

Occ. : Collected once only from a roadside slope below Aru.

Fls. June

Below Aru (2360 m) *B.M. Sharma* 14603.

Fruits used as the condiment '*Kala Zeera*'—a product of great commercial importance.

6. **SIUM** Linn.

S. latijugum Cl. in Hk.f., FBI. 2 : 683, 1879. **Fig. 84**

Glabrous, up to 1.2 m tall, much branched, fibrous-rooted perennial in marshes or partially submerged in shallow waters; stem strongly sulcate. Leaves pinnate; pinnae ovate to lanceolate, 2-6×0.5-3 cm, serrate. Umbels terminal or lateral, 5-15 rayed; bracteoles 5-8, linear. Flowers small, white; calyx-teeth obsolete. Fruit ovoid-oblong, 5×2 mm; furrows 1-vittate.

Occ. : Abundant by water courses and other aquatic habitats in Mundlan, Aru and near Liddarwat.

Fls. July-Sept.; *Frts.* Aug.-Sept.

Aru (2412 m) *B.M. Sharma* 14083 & 14666; Lidder Valley near Aroo (9000′) *Duthie* 13478; Lidderwat (9000′) *M.V. Laurie* 18755.

7. **PIMPINELLA** Linn.

P. diversifolia DC., Prodr. 4 : 122, 1830; Cl. in Hk.f., l.c. 703. **Fig. 85**

Pubescent, 0.5-1.2 m tall perennial herb. Radical and lower most cauline leaves often simple, ovate to cordate, up to 6×6 cm, like leaflets of upper leaves coarsely toothed. Middle and upper leaves pinnate; leaflets 3-5, variable, generally ovate or lanceolate, 2.5-6×1-3 cm, sometimes incised. Umbel rays pubescent; bracteoles linear, 1-2 or like the bracts absent. Flowers 2.5-3 mm across, white, fruit ovoid, 2 mm long, pubescent,

Occ. : Common throughout along the slopes between Gurkum and Aru water mill; usually on humus rich soil.

Fls. July-Sept. ; *Frts.* Aug.-Sept.

Aru (2360 m) *B.M. Sharma* 19293; near Gurkum (2260 m) *B.M. Sharma* 14690.

8. AEGOPODIUM Linn.

A. alpestre Ledeb., Fl. Alt. 1 : 354, 1829; Hiroe, Umb. Asia 1 : 74, 1958; Stewart, l.c. 509 *Pimpinella kashmirica* R.R. Stewart ex Dunn, Kew Bull. 384, 1924. **Fig. 86**

Erect, 40-70 cm tall, sparingly leafy, glabrous perennial; stem simple or with 1 to 2 branches about or above the middle. Leaves 2-3-pinnate, basal with petiole longer than the deltoid lamina, the upper most almost sessile; segments oval, 1-3 (-4) cm long, sharply serrate to deeply incised. Umbels (6-)10-16 (-12) rayed, on long terminal and lateral stalks; bracts and bracteoles wanting. Flowers small, white; calyx teeth obsolete; petals reflexed at the apex. Fruit oblong, 3-4 mm long, ridges thin; inner seed face plane.

Occ.: Frequent among Viburnums and other bushes throughout from Liddarwat to Bara Khalen; also in similar situations in Nangneri nallah.

Fls. July; *Frts.* July-Aug.

Chhota Khalen (2900 m) *B.M. Sharma* 14159; Liddarwat (3000 m) *P.S. Jamwal* 15615.

9. CHAERORHYLLUM Linn.

1. Fruits 5-9 mm long, usually 2 or 3 ripening in an umbellate; stems hairy throughout at least along half their length from the base :
2. Leaves 2-3 pinnate; segments irregularly and finely lobed and toothed **1. C. villosum**
2. Leaves 1-2-pinnate; segments linear to lanceolate, at times pinnatifid, closely and regularly toothed **2. C. acuminatum**
1. Fruit 8-12 mm long, usually 1-(3) ripening in an umbellate; stem glabrous except near the base **3. C. reflexum**

1. C. **villosum** Wall. ex DC., Prod. 4 : 225, 1830; Cl. in Hk.f., l.c. 690. **Fig. 87**

Erect, 0.2-1 m tall perennial, from a fleshy, fusiform root; stem closely or sparsely hispid with white hairs; hairs deflexed and close set especially towards base. Leaves hairy, variable in size and segmentation; upper leaf-sheaths inflated, white-margined. Umbels 5-10-rayed; bracteoles linear to lanceolate, white-margined; flowers small, white. Fruit cylindrical, suddenly narrowed at apex, 7-9 mm long.

Occ.: Common in forest clearings, meadows and other open places throughout from Pehalgam to Dudhsar.

Fls. June-Aug.; *Frts.* Aug.

Aru (2400 m) *B. M. Sharma* 14328; Dudhsar (4000 m) *B. M. Sharma* 19385; Lidder Valley above Kainmul (10-11000′) *Duthie* 13088; Kolahoi Valley near Lidderwat (11-12000′) *Duthie* 13088; Pisu Ghati Kashmir, on way to Amarnath (10,500′) *K.M. Vaid* DD. no 118106.

2. C. **acuminatum** Lindl. in Royle, Ill. Bot. Himal. 232, 1835; Cl. in Hk.f., l.c. 691.

Hairy perennial, up to 60 cm tall. Easily distinguishable from the preceding and next species by its linear to lanceolate or oblong, 1-6×0.3-1.2 cm, closely and regularly serrate pinnae. Umbel rays up to 12, unequal; bracts 0 or few, like bracteoles long-haired on margins and soon reflexed. Flowers white; fruit cylindrical 6-8 mm long.

Occ. : Forest clearings and other open grassy places about Aru.

Fls. : June-July; *Frts.* July-Aug.

Aru (2380 m) *B. M. Sharma* 14302 & 14623.

3. **C. reflexum** Lindl. in Royle, Ill. Bot. Himal. 232, 1835; Hk.f., l.c. 691.

Perennial herb, hardly distinguishable from *C. villosum* except through the characters of the key. The latter too are not stable because intermediates establishing a link between the two are not infrequent even in the localities shared by both.

Occ. : In habitats similar to those for *C. acuminatum*. Common.

Fls. June-July; *Frts.* July-Aug.

Below Aru (2380 m) *B.M. Sharma* 14624; Badzalkhod Nallah Lidder vy. (13000′), Sangam Vy. above Lidderwat (13000′) and above Kainmul Lidder Vy., *Duthie* 13420, 13546 & 13127 respectively.

10. ANTHRISCUS Pers.

A. nemorosa (M. Bieb.) Spreng., Pl. Umb. Prodr. 27, 1813; Cl. in Hk.f., l.c. 692. *Chaerophyllum nemorosum* M. Bieb., Fl. Taur, Cauc. 1 : 232, 1808.

Nearly glabrous, 1-2.5 m tall perennial; stem striate, fistular. Leaves large (20-40 cm), 2-pinnate; leaflets broad, hairy on the nerves beneath, pinnatifid or pinnatisect into entire or toothed segments. Umbels 10-16-rayed; bracts 0 or 1-2; bracteoles linear. Flowers small, white, polygamous; calyx-teeth 0. Fruit beaked, narrowly oblong, 0.8-1 cm long, hispidulous or minutely tuberculate and bristly.

Occ. : In nitrophilous habitats and among wet rocks in Gha Shangan.

Fls. July-Aug.; *Frts.* Aug.-Sept.

Gha Shangan (2920 m) *B.M. Sharma* 19238 A & B and 19261 A & B; above Kainmul, Lidder Valley (11-12000′) *Duthie* 13098; Chatponsal nallah (10-11000′) *Duthie* 13174; Lidderwat (9-10, 000′) *Duthie* 13482.

11. SESELI Linn.

S. libanotis (Linn.) Koch, Nova Acta Acad. Leop.-Carol. 12(1) : 111, 1824; Boiss., Fl. Or. 2 : 967, 1872; Stewart, l.c. 527. *Althamanta libanotis* Linn., Sp. Pl. 244, 1753. *S. sibiricum* Cl. in Hk.f., l.c. 693; For. Fl. Sgr. 185. **Fig. 88**

Aromatic, 0.3-1 m tall, lithophytic perennial with branched, angled and ridged stem from a thick rootstock clothed with fibrous remains of old leaf-bases. Leaves 1-2-pinnate, the lower ovate and long-stalked; segments rather thick, almost prickly with sharp-toothed margins. Umbels bases, rays, bracteoles and calyces minutely bristly-pubescent; bracts generally 0. Flowers 2-2.5 mm across, white. Fruit oblong, 4 mm long, prominently ridged, pubescent.

Occ. : Not common in W. Liddar Valley but abundant on roadside rocks between Laripura and Chandanwari in the E. Liddar Valley.

Fls. July-Aug.; *Frts.* Sept.

Armiun Pathri (2850 m) *B.M. Sharma* 1429 & 14635.

12. LIGUSTICUM Linn.

Fruit dorsally compressed, broadly oval to almost orbicular; dorsal and intermediate ridges narrowly winged, lateral broadly alate **1. L. marginatum**

Fruit subterete, ellipsoid; ridges narrowly winged, the lateral nearly as broad as the dorsal and intermediate **2. L. thomsonii**

1. **L. marginatum** Cl. in Hk.f., FBI. 2 : 698, 1879.

Glabrous, 0.8-1 m tall, perennial; stem simple, hollow, clothed at the base with fibrous remains of the radical leaf sheaths. Leaves pinnate, 8-12×3-5 cm excluding petiole, the uppermost reduced to mere sheaths; pinnae ovate-lanceolate, pinnatifid into irregularly and acutely toothed lobes, puberulent on the nerves beneath. Umbels pubescent at base, on long terminal and lateral peduncles; rays 12-25, minutely bristly, 1-1.5 cm long; involucral bracts solitary, few or wanting; bracteoles linear. Flowers small, white. Fruit 3-4 mm long; inner seed face plane.

Occ. : In open stony ground beside the nallahs; not common.

Fls. July-Aug.; *Frts.* Sept.

Satlanjan (3300 m) *B.M. Sharma* 15612.

2. **L. thomsonii** Cl. in Hk.f., FBI 2 : 698, 1879.

Perennial, up to 65 cm tall, almost glabrous herb, much resembling *L. marginatum* except for the characters of the key. Leaves 6-10×1.5-4.5 cm; rays 10-20; involucral bracts few or lacking, like the bracteoles white-margined. Fruit 3×1.5 mm; fruit ridges all narrowly alate.

Occ. : In habitats similar to the preceding species; not common.

Fls. July-Aug.; *Frts.* Aug.-Sept.

Between Bara Khalen and Satlanjan (2900 m) *B.M. Sharma* 19259.

2a. **L. thomsonii** var. **evolutior** Cl. in Hk.f., l.c. 698. **Fig. 89**

Plants stouter and taller than the species, up to 1.5 m high. Leaves 8-15×2-9 cm; pinnae ovate-, or oblong-lanceolate, pinnatifid or pinnatipartite, glabrous or beneath puberulous. Rays up to 30, or even more; flowers ± mm across.

Occ. : In open stony or grassy places; sometimes in association with the species but much more frequent.

Fls. July-Aug.; *Frts.* Aug.-Sept.

Bara Khalen between Liddarwat and Satlanjan (2920 m) *B.M. Sharma* 15613; Armiun Glen (2760 m) *P.S. Jamwal* 15614.

13. ANGELICA Linn.

Under surface of leaflets waxy, glaucous. Fruit 1-1.5 cm long; dorsal and intermediate ridges aproximated; in furrows next the dorsal ridge 1-vittate **1. A. glauca**

Under surface of leaflets green. Fruit 6-8 mm long; dorsal and intermediate ridges not approximated; furrows next the dorsal ridge 2-, to multi-vittate **2. A. archangelica**

1. **A. glauca** Edgew. in Trans. Linn. Soc. 20 : 53, 1846; Cl. in Hk.f., l.c. 706. **Fig. 90**

Stout, glabrous, 0.8-2 m tall perennial; stem striate, fistular; root aromatic. Leaves inflated at the sheathing bases, 2-3 pinnate, the divisions mostly 3-nate; leaflets ovate or lanceolate, 2.5-12×2-7.5 cm, sharply toothed. Umbel rays 9-20 or more; bracts and bracteoles linear. Flowers small, white; calyx-teeth 0. Fruit oblong, flat, 0.8-1.5×0.5-0.8 cm wide; dorsal and intermediate ridges obtuse the lateral broadly alate.

Occ. : On grassy and rocky slopes in Gha Shangan and surrounding hills; nowhere common.

Fls. July; *Frts.* Aug.

Gha Shangan (2900 m) *B.M. Sharma* 19226 A, B & C; Wooded slope above Gha Shangan (3000 m) *P.S. Jamwal* 15632.

2. **A. archangelica** Linn. Sp. Pl. 250, 1753 var. **himalaica** (Clarke) E. Nasir in Nasir & Ali, Fl. W. Pak. 20 : 126, 1972; For. Fl. Sgr. 183. *Archangelica officinalis* (Moench) Hoffm., Gen. Umb. 1 : 168, 1814 var. *himalaica* Cl. in Hk.f., l.c. 707. *Angelica officinalis* Moench, Meth. 232, 1794.

Robust, aromatic, 1.5-2.5 m tall perennial; stem strongly ridged and grooved, very often flushed with purple at the base. Leaves 2-3-pinnate, with much inflated sheathing bases; leaflets sessile, lanceolate to ovate, 4-16×2-8 cm, irregularly jagged and toothed, the upper decurrent on their rachises. Umbel rays 15-60, stout, 3.5-10 cm long; bracts and bracteoles linear. Flowers small, numerous, white, calyx-teeth 0. Fruit subquadrate to oblong, 6-8×3-4 mm; lateral ridges winged.

Occ. : Very common along the banks and elevated rocky beds of the streams between Liddarwat and Satlanjan.

Fls. July.; *Frts.* Aug.-Sept.

Rewil nallah (3200 m) *B. M. Sharma* 14649; above Kainmul (11-12000′) *Duthie* 13099; Satlanjan (3100 m) *P.S. Jamwal* 19384.

14. SELINUM Linn.

Leaves 2-pinnate into lanceolate to ovate; sharply inciso-serrate leaflets; bracteoles pinnately divided; vittae obsolete in fruit **1. S. vaginatum**

Leaves 3-4 pinnately fine-dissected into narrow linear segments; bracteoles simple: vittae evident in fruit **2. filicifolium**

1. **S. vaginatum** (Edgew.) Cl. in Hk.f., FBI. 2 : 700, 1879. *Cortia vaginata* Edgew. in Trans. Linn. Soc. Bot. 20 : 55, 1851. **Fig. 91**

Stout, aromatic, 0.6-1 m tall, glabrous perennial; stem hollow, striate, at base clothed by fibres of old leaf bases. Leaves ovate to oblong in outline, lower petiolate, the upper sessile on their sheaths; segments 1-2.5 cm long. Peduncle top, rays and bracteoles pubescent; bracts 0 or 1-2, pinnate, pubescent. Flowers small, white; fruit elliptic to subquadrate, 5×2 mm.

Occ. : Among rocks between Bara Khalen and Satlanjan, also along Rewil nallah.

Fls. July-Aug.; *Frts.* Aug.-Sept.

Bara Khalen (2950 m) *B. M. Sharma* 14604; Rewil nallah (3180 m) *P.S. Jamwal* 19271 A, B & C.

2. **S. filicifolium** (Edgew.) E. Nasir in Nasir & Ali, Fl. W. Pak. 20 : 119, 1972. *Orecome filicifolia* Edgew. in Trans. Linn. Soc. Bot. 20 : 55, 1846. *S. tenuifolium* Wall. ex Cl. var. *filicifolia* Cl. in Hk.f., l.c. 700.

More or less pubescent, 25-60 cm tall perennial with 1-few, erect stems from a fibrous base; root strongly aromatic. Leaves much dissected, upper sessile on their sheaths. Umbels compound; bracts 0 or up to 10, linear, white-margined, rarely leafy and pinnate; bracteoles linear, white-margined, longer than the flowering umbellules; flowers white. Fruit 4 mm long, oblong; ridges all winged, the lateral more broadly than others.

Occ. : Open rocky and stony places throughout the right bank of stream from Liddarwat to Satlanjan, also in Nangneri and Yechi Nar. Common.

Fls. July-Aug.; *Frts.* Aug.-Sept.

Nangneri (2920 m) *B. M. Sharma* 14199; Kolahoi Valley near Lidderwat (10-11000′) *Duthie* 13525 (under *S. tenuifolium* Wall.)

15. PLEUROSPERMUM Hoffm.

Leaves 1-pinnate; fruit epicarp not transversely plicate-rugose; fruit furrows 2-3 vittate ..... **1. P. candollei**

Leaves 2-pinnate; fruit epicarp lax, distinctly transversely plicate-rugose; fruit furrows 1-vittate ..... **2. P. stylosum**

1. **P. candollei** (DC.) Cl. in Hk f., FBI. 2 : 703, 1879. *Hymenolaena candollei* DC., Prodr. 4 : 425, 1830. **Fig. 92**

Aromatic, up to 30 cm high, glabrous perennial from a thick root; stem hollow. Leaves oblong, with lanceolate-ovate basal sheaths; leaflets ovate, toothed, 0.8-2 × 0.6-1.5 cm, often carrying bract-like, elliptic or obovate, white-margined sheaths about or above their middle. Bracts 2-3, leaf-, or sheath-like or 0; bracteoles blunt or mucronulate. Flowers small, white. Fruit oblong; ridges narrowly winged.

Occ. : Moist rocky slides on mount to right bank of Kolahoi torrent, also in morainal debris near snout; rare.

Fls. July-Aug.; *Frts.* Aug.-Sept.

Kolahoi glacier snout (3725 m) *B.M. Sharma* 14644; Mt. above Kolahoi (4570 m) *B. M. Sharma* 19338; Amarnath (± 13000′) *K.M. Vaid* DD. No. 117987; Yamher Pass Kashmir, *M.V. Laurie* 16793.

2. **P. stylosum** Cl. in Hk.f., FBI. 2 : 704, 1879. *Aulacospermum stylosum* (Cl.) Rech. f. et Riedl, Biol. Skr. 13 (4) : 28, 1963.

Stout, 90 cm tall, glabrous perennial, clothed at the base with withered remains of leaf-bases; root thick. Primary pinnae lanceolate to ovate, pinnati-partite or pinnatifid into sharply serrate segments. Umbels terminal and lateral, 4-30-rayed; rays scabrid. Bracts and bracteoles white-margined, entire or divided at the apex into sharply pointed lobes. Flowers white; calyx-teeth 0. Fruit ovate, 4-5 mm long; ridges hardly winged.

Occ. : Rare. Collected only once; in humus rich soil among rocks.

Fls. July; *Frts.* late Aug.-Sept.

Gagar Pathri near Satlanjan (3140 m) *B.M. Sharma* 19315, A, B & C.

16. HERACLEUM Linn.

H. candicans Wall. ex DC., Prodr. 4 : 192, 1830. Cl. in Hk.f., l.c. 714. **Fig. 93**

Grey-pubescent, up to 1.7 m tall perennial; stem robust, sulcate; root aromatic. Leaves pinnate or pinnatifid into simple or once-lobed segments, the latter lanceolate or oblong, 4-20×1.5-8 cm, blunt or pointed, irregularly dentate, beneath pale-white or ashy-grey and densely pubescent. Umbels 15-70-rayed; rays unequal, up to 7 cm long; bracteoles linear. Flowers white, outer 0.5-1 cm across; calyx-teeth minute. Fruit flattened, obovate, 0.7-1.2×0.6-0.8 cm, laterally broadly winged.

Occ. : Common, sometimes abundant, on the islands in river bed; also met with on humus covered rocks and scrubby slopes.

Fls. June-July; *Frts.* Aug-Sept.

Gurkum (2250 m) *B.M. Sharma* 15620 Bara Khalen (2940 m) *B.M. Sharma* 19387 A & B; above Kainmul, Lidder Valley (11000′) *Duthie* 13163; Amarnath (12500′) *K.M. Vaid* DD. No. 118093; Nangneri (3000 m) *P.S. Jamwal* 14109.

Roots contain coumarins of which xanthotoxin, a furanocoumarin, is used for treatment of vitiligo.

17. TORILIS Adans.

T. japonica (Houtt.) DC., Prodr. 4 : 219, 1830; Kitam. in Fauna Fl. Nep. Himal. 189, 1955 et Fl. Afgh. 291, 1960; Raizada, Indian For. 92 (5) : 299, 1968. *Caucalis japonica* Houtt., Net. Hist. 26 : 42, 1777. *Torilis anthriscus* (Linn.) Gmel., Fl. Bad. 1 : 615, 1805 (non Gaertn., nec. Bernh.) *C. anthriscus* (Linn.) Cl. in Hk.f., l.c. 718 (non Scop. 1825) **Fig. 94**

Adpressedly hispid, 0.3-1 m high, divaricately branched, semi-erect annual. Leaves 1-2 pinnate into lanceolate leaflets pinnatifid into entire or toothed segments. Umbels 2-10-rayed, on long semi-erect peduncles; bracts 1-5, like the bracteoles linear, rarely foliaceous and divided. Flowers small, white; fruit ovoid, 3-4 mm long.

Occ. : About the fields in Aru; also sporadic among scrubs throughout on Aru-Liddarwat route.

Fls. July-Aug.; *Frts.* Aug.-Sept.

Nangneri Nallah base (2760 m) *B.M. Sharma* 14088 A & B : Fields near Aru (2350 m) *P.S. Jamwal* 15621.

XXXV. ARALIACEAE

1. ARALIA Linn.

A. cachemirica Decne. in Jacq. Voy. Bot. Inde. 72, t. 81, 1844; Cl. in Hk.f., l.c. 722. **Fig. 95**

Perennial, 1.5-2.5 m tall, shrubby herb. Leaves alternate, 1-3 pinnate; pinnae 3-9-foliolate. Leaflets pale beneath, ovate, oblong-ovate or oblong-lanceolate, 5-18×2-7 cm, from an unequal base, serrate, petiolulate or sessile, hispidulous on nerves; stipules fimbriate. Flowers small, white, in umbels forming terminal, 10-30 cm long, pubescent panicles; ovary topped by a flat disc; styles 5, free. Fruit globose, black, 5-ribbed and as many stoned, 2.5-3 mm across, aromatic.

Occ. : Fairly common under shade by the torrent below Aru; in similar situations about Satlanjan.

Fls. Aug.-Sept.; *Frts.* mainly Sept.-Oct.

Aru (2375 m) *B.M. Sharma* 19251; Bara Khalen (2950 m) *P.S. Jamwal* 15658.

SELECTED BIBLIOGRAPHY

Ahmed, M. 1968. Vegetation of the mud house tops in Srinagar. *Rep. Fac. Sci.* 55. Ege Univ. Bornova.

Bamber, C.J. 1916. *Plants of the Punjab. A Descriptive Key to the Flora of the Punjab, North-West Frontier Province and Kashmir.* Lahore. (Govt. Printing).

Blatter, E. 1927-28. *Beautiful Flowers of Kashmir.* Vols. I & II., John Bale & Danielsson Ltd., London.

Coventry. B.O. 1923-30. *Wild Flowers of Kashmir*, Ser. I & III., Raithby Lawrence & Co. Ltd., London.

De Terra, H. & Peterson, T.T. 1939. *Studies on the Ice Age in India and associated human cultures.* Carnegie Institution' Washington, D.C.

Dhar, U. & Kachroo, P. 1983. *Alpine Flora of Kashmir Himalaya.* Scientific Publishers, Jodhpur.

Duthie, J.F. 1893. Report on a botanical tour in Kashmir. *Rec. bot. Surve. India.* 1(I):1-18.

————1894. Report on a botanical tour in Kashmir. *Rec. bot. Surv. India* 3(I):25-47.

Dutt, A.K. et al. 1964. Vegetation of Srinagar (Kashmir Valley) with special reference to ecological habitat. *Proc. natn. Inst. Sci. India* 30 B (3-4) : 150-65.

Grierson, A.J.C. 1964. A revision of the *Asters* of the Himalayan area. *Notes Roy. bot. Garden Edinburgh* 26:67-163.

Gupta, V.C. & Kachroo, P. 1985. Forest flora of Pir Panjal (Western Himalaya). *J. Econ. Tax. Bot.* 6 (2): 365-96.

Hooker, J.D. 1872-97. *The Flora of British India.* Vols. I-VII., L. Reeve & Co. Ltd., London.

Inayat Ullah, M. & Tikoo, B.L. 1964. A preliminary study of the forest typology of Jammu & Kashmir. *Indian For.* 90 (6): 332-47.

Javeid, G.N. 1965. Ferns and Fern allies of Srinagar. *Kashmir Science*, 2.

————1968. Flora of Srinagar. *Kashmir Science*, 5 : 59-71.

Kachroo, P. et al. 1977. *Flora of Ladakh.* Bishen Singh Mahendra Pal Singh, Dehra Dun.

Kaul, V. & Zutshi, D.P. 1966. Vegetation of Kashmir University Campus, Srinagar. *J. Indian bot. Soc.* 45:354-64.

————1967. A study of aquatic and marshland vegetation of Srinagar. *Proc. natn. Inst. Sci. India* 33:111-27.

Kazmi, S.M.A., 1970-71. A revision of the *Boraginaceae* of West Pakistan and Kashmir. *J. Arnold Arbor.* 51 : 133-84; 52 : 110-36; 666-90.

Koul, M.K. 1971. Flowering plants of Shankracharya Hill, Srinagar, Kashmir. *Bull. bot. Surv. India* 13 (3-4):236-43.

——————1975. Contribution to the Umbellifers of Kashmir. *J. Bombay nat. Hist. Soc.* 72 (3):692-715.

——————1983. Vegetation of Tsokar Lake and Puga Valley, Trans-Himalaya. *Folia Geobot. Phytotaxon.* 18:301-8.

——————& Bakshi, S.K. 1984. Studies on the Genus *Artemisia* L. in North West Himalaya with particular reference to Kashmir. *Folia Geobot. Phytotaxon.* 19:299-316, plates 15-20.

Lambert, W.J. 1933. List of trees and shrubs from Kashmir and Jammu forest circles, Jammu & Kashmir State. *Forest Bull. No. 80. Ser.*, Calcutta.

Ludlow, F. 1951. The Primulas of Kashmir. *J. Roy. Hort. Soc.* 76:191-206.

Meebold, A.K. 1909. Eine botanische Reise durch Kashmir. *Bot. Jahrb.*, 43 (Biebl.) 63-90.

Mukerji, S.K. 1940. A revision of *Labiatae* of Indian Empire. *Rec. bot. Surv. India* 14:1-228.

Nasir, E. & Ali, S.I. 1970—*Flora of Pakistan.* Deptt. of Botany, Univ. Karachi.

Pennell, F.W. 1943. The *Scrophulariaceae* of the Western Himalayas. *Acad. Nat. Sci. Phil.*, Monogr. No. 5 : 1-163.

Polunin, O. & Stainton, A. 1984. *Flowers of the Himalaya.* Oxford Univ. Press, Delhi.

Puri, G.S. 1943. The occurrence of *Woodfordia fruticosa* (Linn.) S. Kurz in the Karewa deposits of Kashmir, with remarks on changes of altitude and climate during the Pleistocene. *J. Indian bot. Soc.* 22:125-31.

——————1947. Fossil plants and the Himalayan uplift. *J. Indian bot. Soc.* (M.O.P. Iyengar Comm. Vol.) 25:167-84.

——————1948. Historical factor and its application to forest ecology. *J. Indian bot. Soc.* 27:63-68.

——————1960. *Indian Forest Ecology.* Vols. I & II., Oxford Book House, New Delhi.

Rao, T.A. 1960. A botanical tour in Kashmir State *Rec. bot. Surv. India* 18:1-67.

——————1960. A further contribution to the Flora of Jammu and Kashmir. *Bull. bot. Surv. India* 2:387-423.

Raven, P.H. 1962. The genus *Epilobium* in the Himalayan region. *Bull. Brit. Mus. (Nat. Hist.)* 2 (12):325-82.

Royle, J.F. 1833-39. *Illustrations of the Botany and other branches of the Natural History of the Himalayan Mountains and of the Flora of Cashmere.* Vols. I & II., W.N. Allen, London.

Singh, G. 1970. Botany along Sonamarg to Trunkhal. *Kashmir Science* 7 (1–2):82-88.

———— & Kachroo, P. 1976. *Forest Flora of Srinagar*. Bishen Singh Mahendra Pal Singh, Dehra Dun.

———— & Wafai, B.A. 1973. An exploration of Phanerogamic elements of Kishtwar-Duksum route, Jammu and Kashmir state. *Indian For.* 99:310-19.

Singh, Sher 1929. The effect of climate on conifers in Kashmir. *Indian For.* 55:189-203.

Soest, J.L. van 1961. New species of *Taraxacum* from the Himalayan region. *Bull Brit Mus. (Nat. Hist.), Bot.* 2 (10):261-73.

————1963. *Taraxacum* Species from India, Pakistan and neighbouring countries. *Wentia* 10:1-91.

Stewart, R.R. 1916-17. Flora of Ladakh, Western Tibet. *Bull. Torrey bot. Club* 43:571-90 & 625-50.

———— 1945. Ferns of Kashmir. *Bull. Torry bot. Club* 72:399-426.

————1951. Ferns of Pahalgam. *J. Indian bot. Soc.* 30:137-42.

————1961. The flora of Deosi plains. *J. Pak. For.* 11 (3):225-95.

————1967. The *Cyperaceae* of Kashmir—A check List. *Bull. bot. Surv. India* 9:152-62.

————1967. The Grasses of Kashmir. *Bull. bot. Surv. India* 9:114-33.

————1970. The Grasses of West Pakistan and Kashmir. *Biologia* (Lahore), 16 (1):1-46.

————1972. An Annotated Catalogue of the Vascular Plants of West Pakistan and Kashmir. *Flora of West Pakistan*. Fakhri Printing Press, Karachi.

Vishnu-Mittre. 1963. Oaks in the Kashmir valley with remarks on their history. *Grana Palynologica* 4 (2):306-12.

Wali, M.K. & Tiku, S.N. 1964. Contribution to the Flora of Kashmir Lolab valley. *Bull. bot. Surv. India* 6 (2-4):141-49.

Younghusband, Sir Francis 1909. *Kashmir Described.* London.

INDEX

Italicized words indicate synonyms.
Numbers in bold denote figures.

INTERNATIONAL JOURNALS

JOURNAL OF ECONOMIC AND TAXONOMIC BOTANY

ISSN : 0250-9768

SCOPE

The journal publish original research and review papers in the discipline of Economic Botany (which includes Ethnobotany, Medicinal plants, Poisonous plants and plants of economic importance) and Taxonomic Botany (which includes Flora, Phytogeography, New Records of plants, New species, Nomenclature changes, Monographic work, Revisionary work and other aspects of Taxonomic Botany).

JOURNAL OF PLANT ANATOMY AND MORPHOLOGY

ISSN 250-964X

SCOPE

An international Journal which provides rapid publication outlet for all types of research and review papers encompassing the descriptive and experimental studies on plants. Short communications and Book Reviews are also accepted for publication.

ALPINE FLORA OF KASHMIR HIMALAYA

U. Dhar & P. Kachroo

University of Kashmir, Srinagar

This is a substantial work with a heavy phytogeographical emphasis ......Because of its substantial concern with phytogeography, this book should not be overlooked by persons not otherwise interested in Indian Flora. *(Taxon—33(4) 1984*

1983, XX+280 pp. Rs. 200 $ 45

81 85046 08 5

WEED FLORA OF KASHMIR VALLEY

M.K. Kaul

Srinagar

The book is practically the first definite work on the subject and is specially indispensable for the Kashmir region. Author has presented an exhaustive survey of weed flora of agricultural crops in and around Kashmir which is endowed with great scenic beauty, is also known since ancient times for specialised crops i.e. Saffron and vegetables grown on floating islands.

The book contains critical taxonomic treatment of 401 weed species belonging to 251 genera distributed over 56 families of Angiosperms. It is represented by 223 native and 178 introduced species. Proper keys for identification and important phenological data has been provided. There are a number of new records for Kashmir and some for the country as a whole.

1986, 422 pp. 100 figs. Rs. 275 $ 60

81 85046 28 X

CENTRAL ASIA AND WESTERN HIMALAYA A FORGOTTEN LINK

G.M. Buth

Kashmir University, Srinagar

Contents: Central Asia and Kashmir, : *S.M. Ahmad*. The Neolithic of Kashmir; An Appraisal, *Purashottam Singh*. Neolithic Gufkral *A.K. Sharma*. Neolithic Culture in Kashmir with reference to Central Asia, *S.L. Shali*. An Enquiry into Absence of any mesolithic site in Kashmir, *Ranjan Roy & Ashok G. Ghosh*.

Early Quarternary man culture in cis-Himalayan region by *A.K. Ghosh & A.K. Sarker*. Tile a vital link *J.L. Bhan*. Reflection on Bunzahom and Semthan excavation and later mythological period in Kashmir Valley *R.S. Bisht*. Late cenozoic Palaeoclimatic of Kashmir *D.P. Agrawal*. Antiquity of rice and its introduction in Kashmir *G.M. Buth* et al., Neolithic Links *G.M. Buth* et al. References.

1986, X+76 Pages, ills. 6 Photos.

81 85046 34 4